LIMINAL LIVES

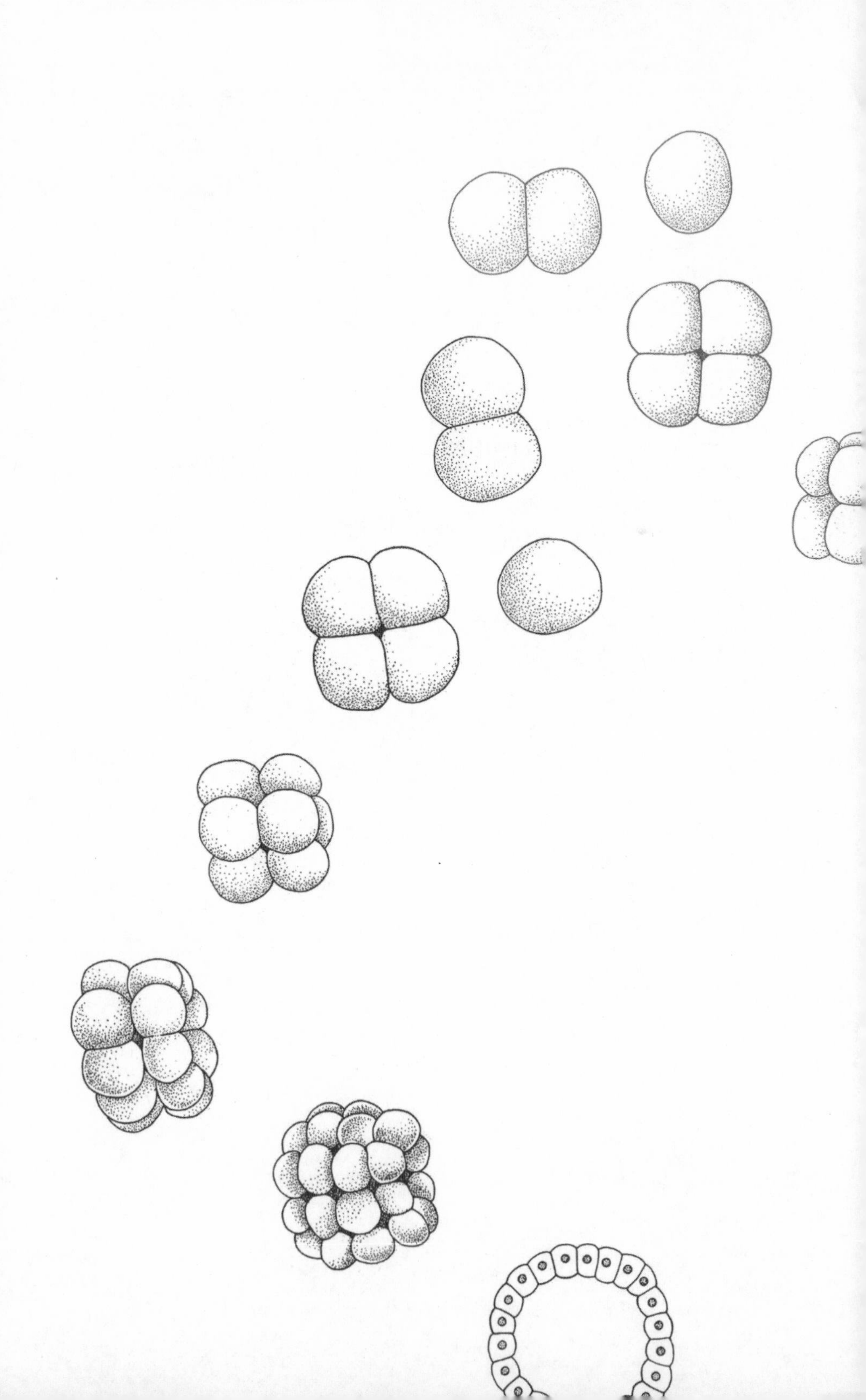

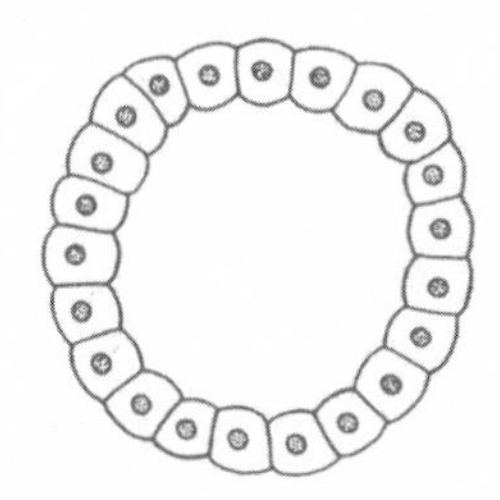

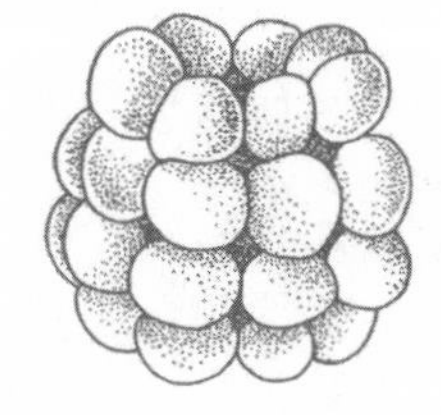

Liminal Lives

IMAGINING THE HUMAN AT

THE FRONTIERS OF BIOMEDICINE

Susan Merrill Squier

Duke University Press Durham & London

2004

Printed in the United States of America on acid-free paper ∞
Designed by C. H. Westmoreland
Typeset in Cycles with Helvetica Neue display
by Keystone Typesetting, Inc.
Republication acknowledgments appear
at the bottom of page 334.

Library of Congress Cataloging-in-Publication Data

Squier, Susan Merrill.
Liminal lives : imagining the human at the frontiers of
biomedicine / Susan Merrill Squier.
p. cm.
Includes bibliographical references and index.
ISBN 0-8223-3381-3 *(cloth : alk. paper)*
ISBN 0-8223-3366-X *(pbk. : alk. paper)*
1. Social medicine. 2. Medical ethics. 3. Medical technology.
4. Human reproductive technology. 5. Science fiction. I. Title.
[DNLM: 1. Bioethics. 2. Ethics, Medical. 3. Feminism.
4. Medicine in Literature. 5. Reproductive Techniques—ethics.
WB 60 S7731. 2004]
RA418.S76 2004
174.2'9—dc22 2004014307

I dedicate this book

with love to my parents,

CONSTANCE CHESTER SQUIER

AND JOHN DAVID SQUIER

(August 16, 1925–August 4, 2004)

All things are impermanent.

That which is born will also die; that which has met will also part;

what has been taken will be lost; what has been made will break.

Time flies past like an arrow. All is evanescent.

Is there, in this world, anything not transient?

—Eihei-ji

Contents

Illustrations

Acknowledgments

This book has taken me some time to write, and along the way I have benefited from the support, kindness, and stimulating interchange with so many people and institutions that I am sure I will forget some of them, and I beg their forgiveness. Penn State University has been a great community in which to live and to work on this book. The Rare Book Room of Pattee Library has superb holdings in science fiction, thanks to its legendary curator the late Charlie Mann, and I want to thank Sandy Stelts for her continuing responsiveness to my peculiar requests and interests. The SMTC (Science, Medicine, Technology, and Culture) group has been a long-term pleasure, and I thank Rich Doyle, and Robert Proctor and Londa Schiebinger (now at Stanford, alas). Nancy Tuana has created a great community in the Rock Ethics Institute at Penn State, with its disability studies reading group, feminist science studies reading group, and (most recently) the Agnatology Conference. Particular thanks to Nancy Tuana, Michael Bérubé, Janet Lyon, and Alexa Schriempf; and to Wenda Bauchspies, Malia Fullerton, Mark Morrison, Marianne Sommer, and Jodie Nicotra. Thanks, too, to Carolyn Sachs and the Penn State Women's Studies Program for its broad and capacious understanding of what can constitute feminist research; to the English Department at Penn State, for support of my science studies interests; and to Vincent Lankewish, Alice Sheppard, Jane Juffer, Mark Morrison, Paul Youngquist, Evan Watkins (whom we miss), Don Bialostosky, and Robert Caserio. A conversation with Stanley Weintraub when I was newly arrived at Penn State nudged my interest in the other half of the story that I began telling with *Babies in Bottles*, and I want to thank Stanley for the conversation, which alerted me to C. P. Snow's *New Lives for Old*. Thanks to the RGSO and the College of the Liberal Arts for supporting the SMTC and this project.

The Society for Literature and Science has been a core community for me since 1994, and once again I want to thank the society for the great context for interdisciplinary science studies work, and wonderful conferences. Many early versions of chapters in this book were first presented at SLS meetings, both in the United States and in Europe. Thanks to Richard Nash, Eve Keller, Hugh Crawford, Anke Ryall, Anne Balsamo, Vicki Kirby, Nina Lykke, Bruno Clarke, Linda Henderson, Stephanie Smith, Kate Hayles, Phoebe Sengers, Laura Otis, Bernice Hausman, and Jim Bono. Special thanks to Elizabeth Wilson and Cathy Waldby for sharing their work and being great long-distance interlocutors and friends. Over the years, many of these chapters have taken shape in conversation with Kathleen Woodward and the interdisciplinary communities she has created, first at the Center for Twentieth Century Studies and more recently at the Simpson Center at the University of Washington. Thanks to members of that extended community: Paul Brodwin, Stephen Katz, Alice Wexler, Anne Davis Basting, the late Glenda Laws, Philip Thurtle, Rob Mitchell, and above all Kathy and Herb. Particular thanks to Dick Blau for his photographs, and Anne Davis Basting for sharing with me the manuscript of her play *TimeSlips*. Thanks also to Kathryn Montgomery and the Medical Humanities Program at Northwestern University Medical School, where I have given a number of chapters in progress, and also to Suzanne Poirier, Tod Chambers, Ann Starr, and the late Suzanne Fleischman, members of the seminar "Case History, Narrative, and the Construction of Objectivity."

For other opportunities to present work in progress, I also want to thank Sigrid Weigel and the Einstein Forum, in Potsdam, Germany; Franziska Gygax and the Department of English and the program in Social Medicine at the University of Basel; Manuela Rossini, and participants in the Gender Talks forum in Geneva, Switzerland (especially Cathy Waldby and Anne Fausto-Sterling); Domna Pastourmatzi, Darko Suvin, and participants in the Biomedical Themes in Science Fiction conference in Thessaloniki (which I regret being unable to attend because of the attacks of 9/11); the Center for Women's Studies at the University of Oslo, Norway, and participants Nina Lykke and Pat Spallone; Anke Ryall and the Women's Studies Program at the University of Tromso, Norway; Tim Murphy and the Institute for the Humanities at the University of Illinois at Chicago; Sandy Sufian and the Medical Student Interest Group in History and Medical Humanities at the University of Illinois at Chicago Medical School; Kristi Kirschner and the Rehabilitation Institute of Chicago and Northwestern University Medical School; Kelly Ormond and the Graduate Program in Genetic Counseling at Northwestern University; Julie Uihlein and the organizers of the Paul S. Pierson Memorial Lecture on Bioethics at the Medical College of Wisconsin; Charles Garoian and Yvonne Gaudelius and the Conference on Performance Art, Technology, and the Body at Penn State University; E. Ann Kaplan, Ira Livingston, Dusa McDuff, Adrienne Munich, and the Humanities Institute at Stony Brook; Helen Longino and the organizers of the Women, Gender, and Science Conference at the University of Minnesota (and to Charis Cussins for her discussions about interspecies pregnancy); Karen Fingerman and the Gerontology Society of America; the Center for Critical Analysis of Contemporary Culture (where the opportunity to be a respondent to Kathy Woodward's work on aging was a great catalyst for my thinking); and the American Society for Bioethics and the Humanities, especially Tod Chambers, Anne Hawkins, Kathryn Montgomery, and Suzanne Poirier.

I thank Anne Hawkins, and the participants in our NEH Institute on Medicine, Literature, and Culture, for the stimulating interchanges of that intense month together, and I especially thank Susan Erikson for her responses to chapter 6, and visitors Anne Davis Basting, Terri Kapsalis, and Lisa Cartwright. I want to acknowledge the support of the Rockefeller Foundation for a research residency at the Bellagio Study and Conference Center, and in particular I thank Ann Curthoys, John Docker, Evelyn Keller, Amos Lapidoth, John Boone, and Miguel Altieri for their companionship there. Once again, I thank Lesley Hall and the Wellcome Institute for the History of Medicine (now the Wellcome Library) for being a great resource and an ideal archive in which to work, and I also thank the Library of the Philadelphia College of Physicians and Surgeons. For smart questions and good conversations, thanks to Gretchen Helfrich and other participants in the radio programs on *Odyssey* at http://www.chicagopublicradio.org/odyssey/. For bibliography help when I needed it, huge thanks to Virginia.

One of the greatest pleasures about teaching at Penn State has been working with my graduate students. I want to thank Holly Henry, Christina Jarvis, Julie Vedder, Harvey Quamen, Lisa Roney, Elizabeth Mazzolini, Megan Brown, Melissa Littlefield, and Jillian Smith for the joy of working with them and learning from them. Christina, Julie, Elizabeth, Megan, and Melissa also worked hard as my research assistants. I thank them much for their help at all stages of this book. I want to welcome my new graduate students, Marika Seigel and Shannon Walters, who have already influenced my thinking about reproduction and disability studies issues, and I thank the members of my graduate and undergraduate seminars on science studies; medicine, literature, and culture; feminist theory; and feminist perspectives on research and teaching across the disciplines.

Finally, I thank Six Rings Sangha, Dai-En Bennage and Mount Equity Zendo, Marie Secor, Syril Blechman, Jennifer Harp, Gwendolyn Thomas, and our kayaking community (especially Graham and Sandy, Elaine and Dennis, and OTT). Gowen Roper, Caitlin Squier-Roper, and Toby Squier-Roper continue to be my foundation and my delight. Virginia Squier, Lorin Nauman, Chambers Squier-Nauman, and Ian Squier-Nauman are my first phone call when I return to Centre County. Thanks, too, to Virginia for teaching me about children's literature. Reynolds Smith has been, once again, a great editor. While I happily acknowledge the help of these people and institutions—and others whom I have forgotten—the errors and failings remaining in this manuscript are my own.

Introduction

NETWORKING LIMINALITY

When claims about the epistemologically neutral
status of nature and its rigorous separation from society
are challenged by the existence of . . . ambiguous,
technologically created entities—neither alive nor dead,
both dead and alive—moralizing runs wild.
—Margaret Lock, *Twice Dead*

In March 2001, an adoption agency in California sued the United States federal government to prevent federal funding of research on embryonic stem cells. That agency, Nightlight Christian Adoptions, makes arrangements for infertile couples to "adopt" excess embryos left over from other couples' fertility treatments. As an Internet news service reported:

> [Nightlight Christian Adoptions] opposes National Institutes of Health plans to fund research using certain embryonic cells—arguing such research would cut the number of adoptable embryos and thus financially harm Nightlight and prospective parents.
>
> At issue are stem cells, the building blocks for all human tissue. Scientists say research with them could lead to revolutionary therapies for diseases from Alzheimer's to diabetes. They can be derived from aborted

> fetuses, fertility clinics' discarded embryos or adults. All types are under intense study, but embryonic stem cells generate the most scientific excitement because they appear the most flexible.
>
> Privately funded scientists have culled stem cells from embryos donated by parents—a process that does destroy the embryo—and multiplied those cells in the laboratory. The NIH plans to fund embryonic stem cell research using only lab-grown cell lines—NIH scientists can't touch additional embryos.[1]

Although its name invokes the familiar, reassuring past, when parents would leave the night light on for a worried child, this story about Nightlight Christian Adoptions informs subscribers to InfoBeat News Service of an unfamiliar, even frightening present. Over our morning coffee we can discover how the foundational categories of human life have become subject to sweeping renegotiation under the impact of contemporary biomedicine and biotechnology. This brief news story introduces us to some of the brave new beings—the *liminal lives*—populating this remarkable era. Perhaps the most prominent in this particular case is the "adoptable embryo," whose new identity challenges the accepted time frame of a human life as well as the accepted notion of civil status available to human beings, extending it back into the prebirth, and even preimplantation, period. And also very new in our conceptual universe are embryonic stem cells, which, with their scientifically exciting flexibility, challenge our fundamental sense that human life is unidirectional, proceeding ineluctably from conception to death.

In this Internet report, we encounter some of the metaphors used (usually unselfconsciously) in the representation of these liminal lives: metaphors of human construction ("building blocks for all human tissue"); of sorting the valuable from the worthless and rejected ("scientists have culled stem cells from embryos"); of being protected from unwelcome physical attention, even assault ("NIH scientists can't touch additional embryos"); of state-regulated execution ("taxpayer-funded embryo destruction"). If we look closer, we discover that these metaphors reveal how the valued "adoptable embryos" are being defined against a range of human beings who are *not* protected: criminals (who are executed), people who are physically abused, and embryos that are *culled* as the spoilage of a harvest.[2] We can find in this report a record of the conflicting forces and institutions at play in Nightlight's lawsuit: methods of financial valuation; patterns of saving or discarding; private or public structures for research; debate around the notion of donating or purchasing living material; distinctions between lab-grown cell lines and parentally produced embryos.

What do the Nightlight adoptive embryos have to do with the marginal and devalued beings that shadow their metaphoric representation? And why do I begin here, when I am interested in the ways literature and science collaborate on, and contest, a new vision of human life? Let me give provisional definitions of these terms. By literature, I mean our representations of the world in stories and language-based arts. I use the term "literature" in its unrestricted sense, as writing of any kind, although its scope has increasingly been narrowed to mean a kind of writing associated with a certain class and sensibility, restricted to imaginative works, and viewed as an important "source of cultural and economic value" (Shumway 1994). I use the term "science" as shorthand for the more cumbersome "science and technology" or Bruno Latour's term "technoscience": "all the elements tied to the scientific contents no matter how dirty, unexpected or foreign they seem" (Latour 1987, 174). I am adapting for science studies Teresa de Lauretis's modification of Foucault's notion of *technologies*: techniques and discursive strategies that are put to the service of gender production and construction. I understand both literature and science as technologies because they incorporate "institutionalized discourses, epistemologies and critical practices" to define what is knowable and to bring those objects into being (de Lauretis 1987, 2–3; Foucault 1980). Whether they are the valued recipients of civil advocacy, or the rejected products of human or agricultural production, the new entities to which we are introduced by this brief Internet news report all share what the anthropologist Victor Turner has called the liminal position. Liminality—literally, "being on a threshold"—is Turner's term for an in-between state, "betwixt-and-between the normal, day-to-day cultural and social states and processes of getting and spending, preserving law and order, and registering social status" (V. Turner 1977a, 33). We move into the realm of the liminal when we engage in cultural rituals that transcend everyday life, or to be more precise that *reframe* it in a heightened or universalized form. These rites of passage (a term coined by the folklorist Arnold Van Gennep) include weddings, births, christenings, graduations, farewell parties, wakes, and funerals. Such rituals and ceremonies help us come to grips with specific transitional moments in our lives. They generally have three stages: a separation from everyday life; a move into the margin, or *limen*, "when the subjects of ritual fall into a limbo between their past and present modes of daily existence"; and finally a return to everyday life, though at a higher level of status, consciousness, or social position (V. Turner, 1977a, 34). A space of "potency and potentiality," "experiment and play," the liminal zone escapes the fixity and regulation of clock time into a realm between what is and what may be (33–34).

The liminal lives that are the subject of this book exist in that in-between or marginal zone. Like the Nightlight "adoptable embryos," neither discarded bioproduct nor valued human being, they are participants in a rite of passage, between everyday life and a higher or different level of existence. They define a transitional civil status as well, positing an extension of legal identity into the prebirth realm of pure human possibility. Like the embryos, the embryonic stem cells are also marginal (either temporally or taxonomically) to the human being. Whether harvested from the aborted fetus, the discarded embryo, or the adult, these stem cells are both like and not like a human being. Found in the embryo or fetus, they come from a time before human birth; when they are inserted in the brain of an elderly Alzheimer's sufferer, we hope they will defer the decline to death. And though they partake of human qualities, they share with nonhuman life-forms the possibility of being harvested for a use that transcends their own life. These liminal lives test the boundaries of our vital taxonomies, whether social, ethical, biological, or economic. As medical interventions are reshaping our ways of conceiving, being born, growing, aging, and dying, liminal lives surround us—in our schools, our families, our professions, our institutions, our representations—anywhere that the expected shape or span of human life is being changed through biomedicine. If we think about the social response to in vitro fertilization, organ transplantation, and stem cell therapies, for example, we will realize that after some initial resistance (whether expressed by the press, politicians, theologians, or the law), these liminal beings are generally accepted by culture and society. As quickly as these beings are normalized, we lose awareness of them. Despite—or perhaps because of—their increasing importance to culturally dominant zones of representation and practice (science, politics, economics), they escape categorization and detection, appearing only as elements of fantasy in culturally subordinate arenas of representation and practice (literature and visual or performance art). Yet I will argue that these new beings demand our attention, because they are powerful and dangerous representatives of a transformation we are all undergoing as we become initiates in a new biomedical personhood mingling existence and nonexistence, organic and inorganic matter, life and death. They raise profound and complex ethical questions that the field of contemporary bioethics has yet to adequately address, largely because of its indifference to the epistemological power of fiction.

The disturbing undertones to the InfoBeat news story suggest that although the liminal zone can provide us with a source of creative play, possibility, and human agency, it can also generate personal, cultural, and institutional tension. As assisted reproduction, genomics, biotechnology, and other biomedical interventions become increasingly common (at least in the industrialized world), liminal lives have become symbolically privileged and troublingly unstable, even dangerous. Human beings have increasing difficulty maintaining discrete boundaries between states or realms, producing legal, psychological, social, and medical struggles explored by the popular press, government advisory groups, theologians, ethicists, and scholars. We struggle to differentiate the private fetus from the public, the "natural" fetus from the cultural, while the boundaries between fetus and gestating mother are increasingly subject to debate. The limits of a human life have been negotiated repeatedly as different definitions of death are generated. The distinction between human and animal, too, has come under pressure with the development of animal-human organ transplantation (xenotransplantation). As the impact of biotechnology grows, the variety of liminal figures to which we need to develop a response grows ever greater. Clearly we need a revision or extension of Turner's predominately cultural definition of the term.

Although he often emphasized its links to the world of art and literature, as "a play of ideas, a play of words, a play of symbols, a play of metaphors," Turner stipulated that liminality could extend to the realm of science and technology as well: "Scientific hypotheses and experiments and philosophical speculations are also forms of play, though their rules and controls are more rigorous and their relation to mundane 'indicative' reality more pointed than those of genres which proliferate in fantasy. *One might say, without too much exaggeration, that liminal phenomena are at the level of culture what variability is at the level of nature*" (V. Turner 1977a, 33; italics mine). Turner's distinction between liminality as a cultural experience and the natural process of variations addresses the cultural power of scientific hypothesis and philosophical speculation, but it fails fully to capture the imbrication of culture and biology characterizing our current condition. As Turner uses it, liminality is restricted to the cultural activities with which human beings shape and negotiate our life crises: "the movement of a man through his lifetime, from a fixed placental placement within his mother's womb to his death and ultimate fixed point of his tombstone and final containment in his grave as a dead organism—punctuated by a number of critical moments of transition which all societies ritualize and publicly mark with suitable observances to impress the significance of the individual and the group on living members of the community. These are the important times of birth, puberty, marriage, and death" (V. Turner 1967, 94). Stressing the role of cultural rituals to mark the facts of human life, Turner relies on a foundational opposition between nature and culture. Implicitly, he assumes the intransigence of biology (both gestation and burial are "fixed" in their placement) in order to focus instead on the cultural construction of meaning: the way that neophytes in a ritual are, for example, "likened to or treated as embryos, newborn infants, or sucklings by symbolic means," and the way they enact "a confusion of all the customary categories" (96–97). In the sense Turner uses the term, liminal beings are those taking part in cultural rituals that represent or construct them in certain ways, in order to cope with or control the movement over a certain fixed biological threshold. Thus, as Turner understands it, while the liminal is shifting, *life* is still stable. It is here, I argue, that contemporary biomedicine necessitates a significant revision of Turner's thesis, one that acknowledges the shifting, interconnected, and emergent quality of human life.

Turner's culturally based definition of liminality omits a crucial aspect of modern life, as is apparent if we consider it in relation to the Nightlight suit scenario.[3] When a lawsuit is filed on behalf of would-be adoptive parents of embryos, who could not by any stretch of the imagination exist as adoptive beings without the apparatus of biomedicine required for cryopreservation, thawing, implantation, gestation, and birth, the boundary between the material and scientific, and the symbolic and cultural, has become impossible to define. The interpenetration of realms and processes once believed to be separate means that it is increasingly difficult to tell whether a variation is the result of nature or culture. No longer stable, the boundaries of our human existence have become imprecise at best, contested at worst.

Indeed, the definition of life has been subject to important rethinking recently under the impact of biotechnological advances like the ones producing the Nightlight lawsuit. As anthropologist Paul Rabinow has observed, the distinction between *zoë*, "the simple fact of being alive and applied to all living beings per se," and *bios*, "the appropriate form given to a way of life of an individual or group," is coming under increased scrutiny (Rabinow 1999, 15). At stake is the distinction between the individual human being and the collectivity of all life-forms; between the unmediated and the mediated existence; or between what used to be called *natural* life, and the life we shape with human institutions, from the arts to the sciences. For those of us who work as scholars and cultural critics, these biomedical changes have social and political consequences that must be addressed. "That the new genomic knowledges will form assemblages with social and political networks is clear," Rabinow observes. But "precisely how changes in *bios* will interact with old and new forms of power relations is open to question, and the evolution must be observed and analyzed. *A pressing challenge is to find and/or invent means of doing so*" (15).

The first step in responding to that challenge, and finding ways of analyzing the new power relations produced by contemporary changes in the human bios, is to modify our working definition of the liminal. Turner's definition of liminality, as that set of ritual cultural practices with which human beings respond to crises against the backdrop of an unchanging human biological existence, omits the crucial fact about modern life that Rabinow's analysis identifies. Realms and processes that were once believed to be separate are now interpenetrating. Zoë is increasingly confused with bios, with the result that we are finding it harder and harder to define what life is, much less to decide whether we should attribute a variation we encounter to forces of nature or culture. (As I write, debates about the global decrease in sperm motility, like debates about global warming, demonstrate this category drift.) In short, we need to move beyond Turner's exclusively cultural framing to understand liminality not merely as a cultural state but as a *biocultural process*.

Along with our increasing encounters with liminal lives—from adoptive embryos to transplant recipients to elderly Alzheimer's patients hoping for stem cell therapy—comes an increasingly urgent desire to understand what these new beings mean to us, socially, politically, and ethically. While Turner emphasizes the yeasty potential of liminality, Paul Rabinow reminds us of its less-pleasant consequences, especially the responsibility we feel to make ultimate sense of this new experience of biological and social ambiguity. He diagnoses us as suffering from "purgatorial anxiety," a specific kind of liminality or in-betweenness characterized by "a chronic sense that the future is at stake; a leitmotif among scientists, intellectuals, and sectors of the public turning on redeeming past moral errors and avoiding future ones; an awareness of an urgent need to focus on a vast zone of ambiguity and shading in judging actions and actors' conduct; a heightened sense of tension between this-worldly activities and (somehow) transcendent stakes and values; and a pressing need to define a mode of relationship to these issues" (Rabinow 1999, 17–18). The distinction between Turner's liminality and Rabinow's purgatorial anxiety is worth pursuing, for the two concepts imply quite different assessments of the appropriate form for the life of an individual or a group, as well as two different time frames, reflecting the different religious domains from which they emerge. The liminal is an arena of possibility; the purgatorial is an arena of responsibility. The liminal has a referent in this world, while the purgatorial has an ultimate eschatological referent. Liminality challenges us to negotiate meaning right here and now, to invent it ad hoc, in culture and society, while purgatorial anxiety or purgatorial pressure (Rabinow uses both terms) gives rise to that "heterogeneous, heteronomic, heteromorphic" zone where the heuristic and the ultimate meanings jostle one another unsettlingly and productively (Rabinow 1999, 22). Rather than choosing between these two perspectives on our current in-between state, I find both of them illuminating. The latter reminds us of the responsibility we have to consider the long-term implications of our biomedical interventions, and the former reminds us of the inventive capacity of human beings to redirect, undo, or coopt even the most seemingly inevitable trajectories of development.

Of course, whether a new development registers as promising or frightening has something to do with the perspective from which we view it. To the Nightlight Christian Adoptions service, the possibility of adopting frozen embryos appeals, no doubt, because it provides another way of working against abortion. To a woman carrying an unwanted fetus, the same idea poses a threat of governmental intrusion into her biological autonomy, while to the parent of a childhood diabetic, the notion that embryos are conceptualized legally as protopersons, who can be adopted, may dash any hopes of using embryonic stem cells to generate a new therapy for a devastating genetic disorder.[4] Even more important, we must understand that the responses to such liminal lives are not only arrayed as a conversation between different positions (i.e., mutually exclusive); they are also quite often both mutually contradictory and simultaneously present in us, whenever we sit down with our daily paper or log in to our e-mail news services. In the ambivalence and ambiguity of our responses to them, we also confront our own in-between state.

All these meanings enter into the way I understand the idea central to this study: liminal lives. I use the term to refer to those beings marginal to human life who hold rich potential for our ongoing biomedical negotiations with, and interventions in, the paradigmatic life crises: birth, growth, aging, and death. Then I view human beings living in the era of these biomedical interventions as liminal ourselves, as we move between the old notion that the form and trajectory of any human life have certain inherent biological limits, and the new notion that both the form and the trajectory of our lives can be reshaped at will—whether our own or another's, whether for good or ill. The human bios is changing so quickly that zoë, the simple fact of being alive, is no longer stable. Turner's notion of cultural liminality superimposed safely on a solid biological life no longer applies, with destabilizing consequences that reach from our cultural analyses to our medical practices, shifting the very ground of our being. The stories we tell about our lives—whether fiction or fact—are crucial maps to this shifting ground. One of the few places where liminality isn't normalized into invisibility, these stories both register the impact of, and help us to navigate, the massive cultural and material transition produced by biomedicine. Drawing on the works of Bruno Latour, Paul Rabinow, and Michael Thompson (among others), I have come to understand that these liminal lives function relationally. In other words, as I will go on to explore in what follows, the embryos and fetuses, the dying and the newly dead, the animals and humans whose stories I explore in the chapters to come function less as *nouns*—whether subjects of experience or objects of others' actions—than they do as *verbs*, enacting a reciprocal exchange between science and culture.

Caught between our sense of the tremendous possibility of the new biomedicine, and our purgatorial anxiety to account responsibly for its implications, we have been forced to craft new ways of making sense of these new configurations and assemblages of relations around us. We can trace the distinct effects of this representational liminality in both our cultural analysis and our fiction. In the academy, as we struggle to think through what it means for us as individuals and as a species to occupy this in-between state, we are experimenting increasingly with interdisciplinarity. We are learning that we must link our contemporary medical strategies for modifying things and people with our strategies of representation. If we do so, Rabinow suggests, we may well find a reciprocal relation between the new bodies and societies generated in this era and the new modes of representation coming into being. His suggestion that our current practices function both to instantiate (embed) and to represent (embody) these new recombinant bodies and societies is a crucial one. It reminds us that to discover the sources and significance of the new forms emerging in our era, we must engage in the same kind of boundary crossing that characterizes the new biotechnologies. We must examine practices in a range of social realms, from science, to art, to law, to literature. "We are witnessing, and engaged in, contestations over how technologies of (social and bodily) recombination are to be aligned with technologies of signification. The questions then are: What forms are emerging? What practices are embedding and embodying them? What shape are the political struggles taking? What space of ethics is present?" (Rabinow 1999, 16). Just as Rabinow recommends that we investigate the contested practices that instantiate and embody these new social and biological hybrid beings, so another scholar has returned to the performances foundational to Turner's theory of liminality—the spectacles, processions, exhibitions, and so on, that society uses to negotiate rites of passage—to argue that they fail to do justice to the experience of embodied materiality, because they privilege the cultural and symbolic. Instead, Sue Broadhurst argues, we can turn to contemporary performance art to experience liminality in a wide variety of registers: spatial, temporal, topographic, existential, cultural, social, *and* biological (1999, 17). Arguing that the "experimental performative types" are engaged in a "radical transvaluation of corporeality," she finds in their performances a rethinking of embodiment that simultaneously calls attention to its representational status and offers a nonlinguistic model for bodily experience. As Broadhurst defines them, the qualities that characterize liminal performances are attention to both the immediate body and the mediated body; a hybrid and self-reflexive aesthetic; use of digital media to challenge any notion of authorial or artistic singularity and differentiation; and reliance on a transformative notion of history that replaces the unitary, singular, and teleological with the multiple, disrupted, and ontological.

This extension of Turner's theory of liminality challenges the boundaries of experience Turner took for granted. Harnessing the technoscientific realm of new media in the service of culture, Broadhurst explains how we attain the "immediacy of the body, including corporeal readings (visual, tactile, haptic, olfactory, gustatory, kinetic, proximic and so on)," via the cultural practice of performance art (Broadhurst 1999, 27). Becoming liminal is a two-way process, in other words: not only do we turn to culture to make sense of moments of biological change, but through our cultural practices—our performances—we are able to access, and indeed to produce, a range of meaningful changes in our bodies. The focus on performance can extend beyond embodiment to articulation.

Our analysis of the contestation over, and changes in, "technologies of signification" must include a reevalution of the formal structure, aesthetic valuation, generic categories, and social position of fiction. In his polemical *We Have Never Been Modern* (1993), science studies scholar and anthropologist Bruno Latour begins the task of crafting a method. In what could be a mirror image of the InfoBeat news item with which I began, Latour's opening pages survey the strange juxtapositions encountered in the daily newspaper, where news of frozen embryos jostles with stories of burning forests, the hole in the ozone layer, vials of contaminated AIDS virus, and an epidemic in sub-Saharan Africa (Latour 1993, 1–2). Latour begins with the daily newspaper because in its debt to temporality, it embodies the essence of modernity: "All its definitions point, in one way or another, to the passage of time. The adjective 'modern' designates a new regime, an acceleration, a rupture, a revolution in time. When the word 'modern,' 'modernization,' or 'modernity' appears, we are defining, by contrast, an archaic and stable past" (10).

The columns of the daily paper embody a phenomenon central to Latour's analysis: the "proliferation of hybrids" (131). What are hybrids? The category is a huge one, and while it includes the liminal lives that are my particular subject, it also exceeds them, taking in any promiscuous intermixture of nature and culture, from tissue cultures to the ozone hole, from regulations for standardizing high-definition television to the debate about the provision of pharmaceuticals to sub-Saharan Africa. Hybrids can be discursive (like newspaper articles "that sketch out imbroglios of science, politics, economy, law, religion, technology, fiction"), or they can be material; they can even be human. So Latour says he and his friends are "hybrids ourselves, installed lopsidedly within scientific institutions, half engineers and half philosophers" (3). Latour argues that hybrids are a product of modernity itself, and specifically of what he calls "the Modern Constitution": the epistemological division between "the scientific power charged with representing things and the political power charged with representing subjects" (29). This division of the world into different sectors, each subjected to a different sort of analysis and criticism, produces the specific kind of hybrids we find in each column of our daily paper, where "all of culture and all of nature get churned up again every day" (2).

Latour's accomplishment is the invention, and demonstration, of an *anthropologie symmetrique* for studying these hybrids. This is a method for integrating and understanding the interactions between nature, culture, and representation, or science, social science, and humanities, "these strange situations that the intellectual culture in which we live does not know how to categorize" (2). Mapped out in an impressive series of books from *Laboratory Life* (1979) through *Pandora's Hope* (1999), this method consists of a meticulous investigation of the tangled networks linking science and technology to culture.[5] I will discuss this practice of networking in more detail in the first chapter, but here I will just observe that its effect is to open to anthropological scrutiny the assemblage of operations linking culture and society to science, medicine, and technology, realms of human activity that have typically been exempted from analysis because they stress the individual discovery of natural fact rather than the collaborative production of cultural practices.

Latour's 1993 use of anthropologie symmetrique to trace the networks producing the hybrids he identifies may seem at first to provide the basis for precisely the method for which Rabinow called in 1999: one that would enable us to make sense of the relations between technologies of recombination and technologies of signification. And yet there is one way in which Latour falls short of his own potential. In his survey of the dizzying hybridity of his daily newspaper, he finds a small zone of soothing stasis: "Fortunately, the paper includes a few restful pages that deal purely with politics . . . and there is also the literary supplement in which novelists delight in the adventures of a few narcissistic egos ('I love you . . . you don't'). We would be dizzy without these soothing features. For the others are multiplying, those hybrid articles that sketch out imbroglios of science, politics, economy, law, religion, technology, fiction. If reading the daily paper is modern man's form of prayer, then it is a very strange man indeed who is doing the praying today while reading about these mixed-up affairs" (Latour 1993, 2). The irony is evident: the literary and the political realms, both of which are prominent agents in the creation of a human subject, whether as an individual or a collectivity, and which therefore should be the realms to which we look in our purgatorial anxiety to make some meaning out of the confusing recombinations that surround us in our new biomedical era, are instead dismissed as soothing backwaters. The slap at politics is understandable, since Latour's goal is to open up the political realm beyond the strictly human "modern constitution" to a more embracing notion of (human *and* nonhuman) agency in the Parliament of Things. But for me, the curious position held by literature indicates the most intriguingly incomplete part of Latour's method. Though he indicts *literature* for being the province of narcissistic egos, he seems to privilege fiction when he includes it among the various phenomena thrown together in the journalistic imbroglios that are his focus. And in an interview the same year, he contradicts himself even more strikingly, asking, "How can we invent literary style for science studies, and how can we pursue the fusion of social science and literature?" (Crawford 1993, 267). How can we explain this seeming inconsistency? As I will argue in coming chapters, it is precisely in the distinction between fiction and literature, enforced by hierarchies of aesthetic value and by the invocation of generic taxonomies, that we can see played out the contestatory and constructive relation between the new biotechnologies (technologies of recombination) and new narrative forms and strategies (technologies of signification).

Other science studies scholars have begun to lay the groundwork for a reevaluation of the position of literature, granting it a role far more assertive than Latour's ancillary one. While Mary Hesse and Gillian Beer brought science studies into the realm of literature by demonstrating how analogy, image, and symbol provide a space for indeterminacy and thus creativity, Donna Haraway cleared the path for those exploring the role of literature in science studies by taking fiction seriously in *Primate Visions* (1989). More recently, Catherine Waldby has come at the question from the other side, proposing a notion based on Michelle LeDoeff's formulation of the philosophical "imaginary": "the deployment of, and unacknowledged reliance on, culturally intelligible fantasies and mythologies within the terms of what claims to be a system of pure logic" (Waldby 2000, 137). Waldby proposes that scientific practice, too, should include its ambiguous, liminal, and symbolic realm, which she calls the "biomedical imaginary." As Waldby understands it, this is the broader context in and from which biomedical creativity emerges: "the speculative, propositional fabric of medical thought, the generally disavowed dream work performed by biomedical theory and innovation. . . . [the] speculative thought which supplements the more strictly systematic, properly scientific, thought of medicine, its deductive strategies and empirical epistemologies" (136).

While in her work on the metaphoric construction of AIDS, Waldby stressed the "phallocentric and homosocial" aspects of the biomedical imagination, in her later work on the visible human project, she modifies her position to introduce more flexibility into the biomedical imaginary. Arguing that scholars should examine the images driving scientific thought for the ways that they manage the often anxiety-charged process of moving from the realm of the unsystematic imagination to the systematic realm of scientific practice, Waldby describes the biomedical imaginary functioning as "a form of representational practice," "a kind of 'science fiction,'" through which the conflicts and tensions of science are expressed, preserved, or sometimes worked out (Waldby

1996, 27). "While medicine, like all sciences, bases its claims to technical precision on a strict referentiality, a truth derived from the givenness of the object, the biomedical imaginary describes those aspects of medical ideas which derive their impetus from the fictitious, the connotative and from desire" (Waldby 2000, 136–37).

Waldby's useful concept alerts us to the existence of multiple sites and forms for scientific creativity, which can be found not only in systematic, experimental science but also in images, fantasies, speculations, and fictions. We investigate the biomedical imaginary when we consider how medical issues are articulated and engaged with across all cultural fields, from medicine to government to popular culture and religion. Moreover, in its emphasis on the fictionality of medical thinking, its expression in images, metaphors, and "generally disavowed dream work," Waldby's notion of the biomedical imaginary provides a crucial new direction for investigating any biomedical development (Waldby 2000, 136). Focusing on something with so marginal a relation to medicine's self-description challenges the boundaries of the discipline. Inverting the hierarchy of disciplinary value, Waldby's questions productively open up new lines of inquiry and reveal not only how the discipline of medicine produces knowledge but also how it enforces ignorance.[6] To put it another way, the very fact that imagery and metaphor are thought to be sites extraneous to science suggests the investment science has in the marginality and obscurity enabled by those discursive modes. Thus we can look to imagery and metaphor for the expression of excess fantasy and desire, finding therein those sites of unresolved tension, cultural paradox, and stubborn ambiguity that are a crucial, if generally overlooked, aspect of biomedicine. What Donna Haraway has called the "traffic" between different discursive realms appears here as a reciprocal shaping effect. Just as scientific thinking appropriates the ambiguity afforded by nonscientific images and metaphors, so too scientific discourse and imagery can be appropriated by the broader culture, with results that can have broad, if unpredictable, results: "Once medical images leave the strictly regulated contexts of the scientific media, their debt to the imaginary, the speculative, to desire, the fictive, to particular cultural genres and stock narratives, becomes less readily ignored. The intertextuality of scientific images is more evident at these points of popularisation, and this intertextuality implies that the interpretation of images by different nonscientific audiences can lead off in a number of directions and is open to various orders of appropriation" (Waldby 2000, 138–39).

Whether they exist as fiction or as nonfiction, the narratives spawned by or produced in relation to biomedicine can be categorized as working objects, research reservoirs, or biological resources for the reconfiguration and extension of the human life span (Latour 1993; Daston and Galison 1992). Yet in calling them "working objects," I am deliberately stretching the usual meaning of the term, for it usually refers to "objects in [laboratory or scientific] process, which can be used experimentally to test out certain morphological and biotechnical propositions" (Waldby 1996, 99). In my adaptation of the term, I understand *narratives* to function as working objects, in experiments that take place not in the biomedical laboratory but in the biomedical imaginary: the rich intertidal zone where, as Waldby puts it, "biomedicine *makes things up*" (32). In other words, Waldby explains, biomedicine "realizes, or struggles to realize, these narratives through their embodiment. It *anatomizes* its narratives in the sense that it orders its images of bodies according to their logic, but it also anatomizes them in the sense that it reads into lived bodies in ways that are constitutive of important aspects of corporeality itself" (32). Considered as working objects, narratives exist in a reciprocal relation to the lived bodies that are their ultimate referent: both constituting and being constituted by them. Narratives thus provide an alternative to the impossible attempt to distinguish nature from culture, science from society; a site where we can productively consider their mutual imbrication and cogeneration. I will explore a number of narratives generated by the biomedical imaginary in the chapters that follow, including poems about tissue culture, short stories about organ transplantation, and a novel about artificially accelerated human growth. While none of these works (except possibly Wells's *Food of the Gods*) would make the cut for the category of "literature," all of them actively demonstrate how fiction operates as a technology of signification, generating biocultural meanings from the new technologies of recombination.

This book offers case studies in an adapted anthropologie symmetrique, working between the biomedical and the nonscientific, between cultural genres, stock narratives, and expert discourses. In each chapter, I consider a node of relations that combines science, social sciences, and the humanities. The liminal lives I focus on, inherently unstable, are the product of a volatile convergence of disciplines, discourses, practices, events, and people. To survey and assess the variety of new paradigms for life generated at these nodal points of biomedicine and culture, I trace the networks connecting different realms of discourse and practice, from the strictly regulated and systematic to the unsystematic and uncharted.

Though my interest lies in the way that biotechnology is reshaping the human body—that whole range of interventions including embryo culture, in vitro fertilization, growth hormone administration, interspecies fertilization as part of assisted reproduction, stem cell therapy, xenotransplantation, and fetal cell transplantation—I approach this topic not through biomedicine but through narrative, both fiction and nonfiction. Because my interest in these stories of biomedical interventions is both theoretical and methodological, let me go on to say a bit more about why I look at both fiction and nonfiction narratives.

I understand fiction as a crucial site of permitted articulation for the desires driving these new biotechnologies: I am using this term broadly enough to encompass all linguistic play with *what might be*, all transgressions of the (socially constructed) boundary of fact, including the imaginative play of poetry. Fiction gives us access to the biomedical imaginary: the zone in which experiments are carried out in narrative, and the psychic investments of biomedicine are articulated. Moreover, fiction is devalued, epistemologically and disciplinarily, by many scholars committed to objectivity and referentiality; fiction is thought not to possess truth, to be false. Ann Curthoys and John Docker point out that the struggle between history as science and history as narrative has since the eighteenth century been waged on the terrain of fiction. Thus even to ask, "Is history fiction?" is tantamount to implying the countering question: "Or does history tell the [scientific] truth?"[7]

Yet fiction, the zone where objective truth is not told, paradoxically becomes the site where one specific kind of truth is best articulated: the workings of the biomedical imaginary, the desires propelling biomedicine, can be expressed in fiction. Freud's brilliant essay "Negation" (1925) helps to clarify how this process works, by using a kind of conditional articulation to trick the forces of repression: "the subject-matter of a repressed image or thought can make its way into consciousness on condition that it is denied" (Freud [1925] 1959, 213–14).

As Freud explains, "Negation is a way of taking account of what is repressed; indeed, it is actually a removal of the repression, though not, of course, an acceptance of what is repressed" (214). It is precisely the denial of truth inherent in fiction that makes it such a good vehicle for our repressed impulses and desires. We protect ourselves from acknowledging them, and from having to deal with their consequences, by articulating them in fiction, the untrue zone. This book explores a range of narrative strategies that are engaged in subject production in two different ways, then: fictions, which monitor the production of acceptable subjects (of government, of nation) through the generation of a boundary zone beyond which facts cannot be found; and government documents, which define and thereby produce the factual subjects that they frame and hail.

To understand how the very fictional status of a narrative performs a special function in articulating the repressed of the biomedical imaginary, we need to begin by establishing the difference between any narrative and fiction, and between literature and science fiction. (As I will go on to argue, genre shapes the cultural role available to each.) To begin with, narrative consists of a story or the representation "in art" of an event or story, as distinct from fiction, which is something invented by the imagination, or the act of taking something possible as if it were a fact, irrespective of the question of its truth (*Webster's Ninth Collegiate Dictionary*, s.vv. "narrative," "fiction"). Historians have been dogged by the distinction between the two entities, worried that to narrate history was inevitably to fictionalize. Narrative for historians has depended for its authority on the assumption that it was transparent and nonfictionalizing. Yet this fiction/fact dichotomy is increasingly problematic for the discipline of history. As Curthoys and Docker point out, recently historians have been sensitized to the ways that the generic demands of narrative shape the histories that we tell. "The historical narrative points in two directions simultaneously," they observe, "toward the events reported in it, and the 'generic plot-structures conventionally used in our culture'" (Curthoys and Docker 1999, 3).[8] While the legacy of poststructuralist historiography has produced a healthy skepticism concerning the referentiality of historical narrative, it has not catalyzed a similar willingness to scrutinize the constructedness of the seemingly eternal categories of narrative genre.

The aesthetic and formal implications of the changes in human life wrought by biotechnology are registered by, as well as enforced through, literature. Yet genre variations are a crucial component of the articulation I am exploring. That is where the distinction between literature and science fiction comes in. As I will argue, we are now seeing a shift in the social valuation of science fiction, a shift in how we draw the line between "fiction" and "fact" that is related to the changing understanding of the human being produced by biomedicine. In short, the transformative processes of biomedicine are *enabled* somehow by the transformative narrative that is science fiction.

To understand how the specific subgenre of science fiction can function transformatively in culture at large, we first need to understand the literary category of genre more broadly. Rather than adopting either the deconstructive or sociological approach to literature—either textualizing all forms of social relations or positioning literature as the passive recipient of all social forces—I share with the cultural studies theorist Tony Bennett an understanding of literature as an active agent in social formation, "an institutional site providing a specific set of conditions for the operations of other social relations, just as those relations, in turn, provide the conditions for its own operation" (Bennett 1990, 108).[9] The crucial point here—and one that remarkably parallels the networking approach of Bruno Latour—is the *reciprocity* of relations between literature and other social formations (including science). "In this view, literature is regarded as itself directly a field of social relationships in its own right . . . which interacts with other fields in which social relationships are organized and constituted *in the same way as they interact with it and on the same level*" (108).

When we understand literature as a field of social relationships rather than a set of forms internal to the text itself, we are doing to texts what Latour encourages us to do to a scientific fact: to open the black box or to arrive before it has been sealed. Rather than accepting the autonomy and self-evident status of the text, Bennett suggests, we need to explore how the text consists of a set of social relations. This means not only do we understand its actual production in relation to that set of social relations, something that is fairly easy for us to do if we consider presses, readers, distributors, and reviewers, but we also need to understand literature's *form* and *consumption* as the product of social negotiations. Here is where Bennett's theory of genre is so helpful. He argues that our task, as analysts of literary genre, is neither to identify or categorize genres, nor to see how they passively take the impress of society, but rather to consider them as nodes of social practices. Genre theory should be concerned, he argues, "with the ways in which forms of writing which are culturally recognized as generically distinct . . . function within the 'forms of life'—the specific modes of organized sociality—of which they form a part" (Bennett 1990, 108–9). So our task in genre study is to see how different genres function as part of the set of networked relations—relations that extend all across the social field from the material to the semiotic—in which they are imbricated.

The rhetorician Anis Bawarshi shares the contextual approach to genre, suggesting that we think of genre not only as a classificatory term in literary and aesthetic discourse but as a category that has sociological and cultural meaning. Genre is both a regulatory and a constitutive category, he argues; it shapes how a preexisting social practice can be entered and engaged in, and it constitutes that social practice, giving us ways of understanding and engaging in it, giving us the conventions that make it possible for us to enact that practice (Bawarshi 2000, 340). The process by which genre works to regulate and constitute practices is complex. We generate genres as labels for specific rhetorical situations, so that we know one set of things as *fictional* and another as *factual*, or one set of representations as *realistic* and another as *gothic* or *science fictional.* And then, through the process of living with them, those labels become not only descriptive but prescriptive. As Bawarshi explains in "The Genre Function," "as individuals' rhetorical responses to recurrent situations become typified as genres, the genres in turn help structure the way these individuals conceptualize and experience these situations, predicting their notions of what constitutes appropriate and possible responses and actions. This is why genres are both functional and epistemological—they help us function within particular situations at the same time they help shape the ways we come to know these situations" (Bawarshi 2000, 340). In short, genres govern the structures within which we interact, as well as our actions within those social structures.[10]

Liminal Lives engages in a networked analysis of fiction and nonfiction narratives, exploring in each chapter one node in the reconfiguration of the life span, or one of the liminal lives that are my focus. I begin with a chapter setting out my methodology for working with the relations between literature and science, described in relation to the growth of two disciplines—literary criticism and feminist science studies—and demonstrated in action. The core of the book (chapters 2 through 6) offers case studies in an anthropologie symmetrique of the liminal lives that are my subject. These chapters trace the modern replotting of the human across the entire life span, from the cultured cell, the hybrid embryo, and the giant baby, to the "incubaby" (or engineered intrauterine fetus), the organ transplant recipient, and the "rejuvenate" (or artificially rejuvenated elderly person). These chapters explore the representation of certain literary and artistic fantasies—of interspecies reproduction, of growth engineering, of biotechnological interventions at the beginning and the end of life, and of preserving life eternally through organ transfer—in relation to trends in medicine expressed in government documents, scientific journals, and the consolidation of new scientific fields. Comparing transformations of the human being enacted in early-twenty-first-century biomedicine and literature to their antecedents in the twentieth century, the core chapters also explore how a frequently disavowed modernity is foundational to postmodern processes of replotting the human. These chapters also illuminate different literary methods—point of view, character construction, and plot (especially endings). They correspond roughly to life stages—beginning of life, growth, and end of life—because it is at those three major points of transition in the human life span that biomedical intervention is dramatically shifting both the shape of a human life and the shape of the stories we tell about it. Moreover, the core chapters trace the modern life strategies with which we have responded to the fact of our mortality. The changes in the human body brought about (and following from) biotechnology all tend toward the body's transformation from something with a limited life span to something with an indeterminate (and possibly unlimited) life span (Bauman 1990).

Chapter 7 explores the role of fiction in this cultural and material transformation of human life, by focusing on the final liminal moment, aging. There I consider how a shift in medical paradigms for dealing with aging—the movement from replacement medicine to regenerative medicine—is undercut by certain stabilities: a tendency to address the problem via the individual, and a tendency to rule out nonexpert, nonscientific interventions. Comparison of several literary representations of aging suggests that while literature may echo the individualistic approach of medicine (whether oriented toward replacement or regeneration), an interdisciplinary approach that puts literature in conversation with medicine can also provide an alternative model for addressing what have seemed to be intractable biomedical problems.

The coda returns to the liminal lives with which the book began: cultured cells. Examining a controversy over the topic of discussion at the first meeting of the National Bioethics Commission charged with developing an ethical policy on stem cell research, I explore the impact of denying epistemological authority to fiction. I argue that this instance of generic policing reveals both the narrowing effects of the disciplinary hierarchy accepted in contemporary society and the potential contribution literature can make to understanding the reshaping of the human body central to the notion of stem cell research. Fiction is the subject of contestation in the stem cell debate, I suggest, because it offers the unsettling and valuable opportunity to access the liminal realm of the uncertain, the undecided, the ambiguous, and the unknown, only by way of taking on the socially devalued position resulting from association with those categories.

The social theory of Ulrich Beck and Anthony Giddens, particularly their notions of risk society and the reflexive construction of the body in late modernity, undergirds the argument in *Liminal Lives* because it provides a useful analysis of the transformation of individual self-identity that is taking place on both a material and a cultural level. I extend their analyses of the impact of risk discourse, and of the reflexive construction of the embodied self, by arguing that literature contributes distinctive insight into that transformation. Because of its particular epistemological positioning between knowledge and unawareness, literature is able to hold open a zone of exploration that other mediations (political, social, scientific, and economic) foreclose. Literature thus offers an alternative to the expert discourse that, as Beck and Giddens demonstrate, has become socially and epistemologically dominant. Examination of the literature of liminal lives allows us to see the range of fantasies, desires, and fears that are articulated and enacted through the reshaping of the human life span by biomedicine. When introduced into the bioethical conversation about the reflexive construction of the body, literature can articulate an alternative to the dominant discourses of risk management and expert control that have fueled interest in stem cell research as an answer to the "problem" of mortality. In short, a reinvigorated bioethics will reposition fiction and literature as contributions to social knowledge, rather than cordoning them off into the realm of the textual and aesthetic, a zone with no purchase on the material conditions of the present.

1

The Uses of Literature for Feminist Science Studies

TRACING LIMINAL LIVES

> When Omega came it came with dramatic suddenness and was received with incredulity. Overnight, it seemed, the human race had lost its power to breed. The discovery [took place] in July 1994 that even the frozen sperm stored for experiment and artificial insemination had lost its potency.—P. D. James, *The Children of Men*

"Omega" is the moment when human fertility ends. P. D. James's novel *The Children of Men* opens into a world beyond Omega, a world in which men have lost the ability to produce viable sperm. No more babies are being born. In this world without children, adults of all ages seem not quite human, distorted in their emotional responses to the succeeding stages of life. Adults of childbearing age have become broody victims of "frustrated maternal desire," lavishing unrequited affection on increasingly intricate and expensive dolls ("the new born, the six-month-old baby, the year-old, the eighteen-month-old child able to stand and walk, intricately powered"), and the very old go, loveless, to their state-sponsored euthanasia, a mass murder/suicide called the Quietus. The newly adult, those children called "Omegas" who were born in that final fertile year, seem "incapable of human sympathy" (James 1993, 47, 34, 10). Possessing an eerie, remote perfection, these men and women "are a race apart, indulged, propitiated, feared, and regarded with a half-superstitious awe. In some countries, so we are told, they are ritually sacrificed in fertility rites resurrected after centuries of superficial civilization" (10).

With the Omega point as its premise, *The Children of Men* explores the impact of accelerating biomedical interventions on our changing understanding of human life. In its science fiction mélange of biomedical and sociocultural extrapolation, James's novel captures some of the ambiguous positions that characterize what I am calling liminal lives. The novel portrays a society vacillating between organized religion and primitive fertility ritual, longing for the now-elusive miracle of birth but condemned instead to static life and enforced death, lavishing on cherished mechanical dolls and coddled pets the nurturing energies denied not only to the children never born but to their unsatisfied would-be grandparents, the throwaway elderly men and women whose death is mourned by no one but managed by the state.

As a work of science fiction exploring both biological shifts in fertility and their sociocultural implications, James's novel embodies the modified notion of liminality that I have adapted from Victor Turner ([1969] 1995) and Arnold van Gennep (1908). The term "liminal" denotes a biological and social state of transition from a world in which human beings had a characteristic and predictable life course to a world in which neither the beginning of life, nor its flow, nor even its end has a foreseeable structure. As it addresses the social and spiritual implications of a drastic biological event, James's novel raises two issues that will be central to my exploration of the changing status of human beings in this study: the process by which technologies are naturalized, so that they fade into the background as effects not of culture but of nature, and the function and significance of genre in our representation of technologically driven changes in human life.[1] I will explore the novel's relations to other expressions (both literary and scientific) of the project of controlling the human life course, to assess the epistemological and social effects of disciplinary boundaries and interdisciplinary transgressions, and the gendered role of disciplinary discourses in constraining our sense of human possibilities. Before getting to those issues, however, I want to consider the novel's curious conjunction with another published work—not of literature but of science—in order to explore the important, and undervalued, role of literature in feminist science studies. Although it may seem a tactical error to link two such divergent realms of discourse and practice as literary representation and human reproduction, I hope to show that we need to consider both forms of subject production if we are to understand the potential and peril posed by liminal lives.

In the same year in which James's *Children of Men* appeared, Danish endocrinologist Elisabeth Carlsen and a team of Danish scientists published a study in the *British Medical Journal* that systematically reviewed the statistics on semen quality for the past half century and concluded that a significant decrease was occurring in male fertility.[2] Carlsen's essay was the first of a number of articles to be published in the next several years, in both popular and scientific journals, that would raise the alarm about a steep decline in both the quantity and quality of human sperm and would offer a wide range of theories for this sudden decline (Kolata 1996; Hertsgaard 1996; Raloff 1996). The hypothesis that male fertility decline is caused by environmental chemical pollution, in particular the proliferation of artificial estrogens, is still subject to scientific and popular debate (Oudshoorn 1994, 1996). Not debatable, however, is what one writer has called the "remarkable coincidence" that these two very different texts dealing with the same issue—the global threat of male sterility—should appear simultaneously in 1992 (Wright 1996, 55).

The choice of the label "coincidence," with its implicit assumption that literature and science are stable, discrete, and unlinked categories that only by accident share the same agenda—interests me. Because it reveals another kind of sterility—the failure of two potent fields, feminist literary criticism and feminist science studies, to merge in a fertile zone of inquiry and analysis—this is where I begin to explore the relation between literature and biomedicine in the production of liminal lives. In this chapter and those to come, I will trace the relations of literature and biomedical science from this seemingly accidental moment of intersection in the 1990s back to the century's early years, in order to investigate the sources, and suggest the remedy, for this crucial failure of intellectual and social fertilization.

Why has feminist literary criticism been so indifferent to the question of science? Why are feminist science studies so little marked by the methodology and epistemology of literary studies? We can begin to answer these questions by reviewing the gendered history of the modern discipline of literary studies, for it has shaped the current state of relations between literature and science. We need to consider the institutional relationship between feminist science studies and the analysis of literature and science, and we need to "find or invent" some tools that we can use to understand the particular assemblage of practices and social relations embodied in the James/Carlsen "coincidence" (Rabinow 1999, 15). By examining how and why these two publications converge in the highly charged zone of human reproduction, we can generate a richer analysis of the significance of biomedical interventions targeted at a zone characteristically identified with women, whether they are viewed as closer to nature than culture, or as sites of abjection.

TECHNOLOGIES FOR KNOWING AND CONSTITUTING SUBJECTS AND OBJECTS

Science and literature are the two preeminent technologies that the Enlightenment produced for constituting social subjects and objects of knowledge. Michel Foucault's analysis of the ways in which modern thought constitutes the human being as an object of knowledge, and of the interrelations of power and knowledge, is helpful here. "Man's mode of being as constituted in modern thought enables him to play two roles: he is at the same time at the foundation of all positivities and present, in a way that cannot even be termed privileged, in the element of empirical things" (Foucault 1973, 344). To reiterate some distinctions I set out in the introduction, I am using the term "science" as shorthand for the Latourian "technoscience," and the term "literature" in its broadest sense as writing of any kind (Latour 1987, 174; Shumway 1994; Williams 1977). I draw on Teresa de Lauretis's modification of Foucault's notion of technologies to conceptualize both science and literature as technologies that define what can be known and bring those objects of knowledge into being (de Lauretis 1987, 2–3; Foucault 1980). Since the eighteenth century, literature has helped us to know the self and, in a certain sense, has actually *produced* that self as a subject of knowledge. To give two examples, the literary subgenre known as domestic fiction played an important role in constructing woman as a gendered, socially positioned site of deep subjectivity, and the literary genre of sensation fiction produced certain bodily effects of physical excitation in its readers while catalyzing certain behaviors and social relations, among them the debates over the propriety or impropriety of such sexually tinged fictions (Armstrong 1998; Moore 1995). Still more specifically, Radclyffe Hall's *The Well of Loneliness* claimed territory for a powerful new subject position: the lesbian. Science engages in another kind of double process of bringing into being *and* constituting as "knowable," as sociologists of science and feminist science studies scholars have demonstrated. By processes of abstraction, demarcation, measurement, quantification, publication, and dissemination, the objects of scientific knowledge are brought into being and defined, and their properties are articulated. We can provisionally summarize the relationship I am describing between the two spheres. Science functions as the site of the construction of the *objectively known other*, while literature is the site of the construction of the *subjectively known self*.

Yet almost immediately we must qualify these assertions. Science also functions to construct the subject, bringing into being a range of new selves that are the subject of scientific scrutiny and understanding: the homosexual, the hospital patient, the criminal, the woman. The most robust product of scientific subject production, although most often exempt from scientific scrutiny, is the scientific knower, whose intellectual agency, social currency, and personal authority have been memorialized in a range of fictions and memoirs, as well as in the history of science, including the laboratory history that forms the subject of my second chapter (De Kruif 1926 [1954]; Watson 1969; and Keller 1983). And literature can also be understood as constructing objects, not only in the sense of the book as object of exchange value but also in terms of the way literary texts from children's stories to romance novels can act to produce children and women as the docile objects of social forces, whereas nineteenth-century English novels (*Jane Eyre*, *Kim*) shaped the contours of the subaltern mind.[3] Here, too, the work of Michel Foucault has sensitized us to the three modes or practices by which human beings are turned into subjects: *"classification practices, dividing practices, and self-subjectification practices"* (Katz 1996, 17).

In short, literature and science mediate social relations with material objects, as well as with subjects. Both disciplines frame and shape our understanding of the things of this world, whether the knowing subject or the object that is known is a domestic woman, an episode of hysteria, a chemical formula, or a sea urchin. Scholars expanding on the work of Foucault have amply investigated this crossover effect between the constructive forces of science and literature, demonstrating that both objectification and subjectification processes are carried out by disciplines and technologies in the arts and sciences, ranging from the cinema and literature to primatology, gerontology, and embryology. To take just two examples, Nancy Armstrong has shown how "literature provided techniques for making the individual a specific object of knowledge to himself, . . . on a mass basis," while Stephen Katz has explored the subjectification practices of the science of modern gerontology, understood as designating "the ways in which a person turns him- or herself into a social subject" (Armstrong 1987, 164; Katz 1996, 18–19). Although the process of "bringing into being" in science may seem more tangible because it frequently has material and conceptual results (a new chemical compound, for example), the imaginative practices and disciplines enforced through literature also have tangible social results, whether in the production of a literary market, or obscenity laws, or a craze for a new kind of clothing or domestic furnishings. And scientific practices also produce social subjects, often in uncanny echoes of literary predecessors. Thus, as we will see, the news in 1998 of a pioneering hand transplant, followed thirteen months later by news of the hand's amputation, was anticipated by the publication seventy years earlier of "The Black Hand," a science fiction tale of a rebellious transplanted hand (Altman 1998; Bowers 1931).

We can grasp the intertwined workings of these technologies of subjectification and objectification if we consider chapter 15 of Mary Shelley's *Frankenstein*, in which the monster relates how he learned to know himself through reading Plutarch's *Lives*, Goethe's *The Sorrows of Young Werther*, and Milton's *Paradise Lost*. These texts—critics have long agreed—shape the monster as *subject*: in society, in relation to nature, in relation to God, and to the opposite sex. Less remarked on is the fact that the monster also reads a fourth text, which shapes him as an experimental *object*: "the journal of the four months that preceded [his] creation," in which Victor Frankenstein "minutely described . . . every step [he] took in the progress of [his] work" and set down "the whole detail of that series of disgusting circumstances which produced it" (Shelley [1831] 1994, 92–93). In drawing together these different texts, Shelley opposes the formative influence of literature, particularly *Paradise Lost*'s powerful story of the origin of sin in the moment of sexual knowledge, to the constructive power of science, embodied by Frankenstein's lab book, in which he records in appalling detail the process of experimentation that led to the creature's monstrous "birth." The conjoined processes of literature and science are articulated in *Frankenstein* as two related myths of generation, one reproductive, and one replicative.

If the literature the monster reads creates him as a subject, the lab book records how his creator's scientific will to power/knowledge constituted him as an object. Indeed, these subject and object positions are troublingly conflatable and interfused. The monster models his upbringing on the De Lacey family's education of the beautiful Safie; human civilization goes astray with monstrous results. Shelley's point is worth reversing: not only can literature and science come together under abnormal circumstances to produce a monster that is the object of fear or derision, but they can also collaborate under normal circumstances to produce human beings who are valued, even idealized. And whether monstrous or normal, these acts of literary and scientific construction have concrete, material consequences; they produce a "set of effects . . . in bodies, behaviors, and social relations" (de Lauretis 1987, 3).

Science and literature are more like each other than they are different, not only because both operate in culture and society to produce subjects and objects but also because both fields have come into being through a crucial act of institutional self-creation: the creation of a disciplinary divide between scientific and literary knowledges and practices. This divide reflects the gendered nature of intellectual inquiry, as Schiebinger has demonstrated: "By the late eighteenth century, scientists and philosophers were championing a science stripped of all metaphysics, poetry, and rhetorical ornament. . . . Literature, which Claude Bernard called the 'older sister of science,' was to be distinct from science. It was banished from science under the disgraceful title of the 'feminine.' The equation of the poetic and the feminine ratified the exclusion of women from science, but also set limits to the kind of language (male) scientists could use" (Schiebinger 1989, 158–59).

We can date the beginning of the modern era from the gender-hierarchized act of partition and self-definition through which science and literature forged themselves as mutually excluding disciplines. According to Latour, that disciplinary and epistemological partition between the human and the nonhuman, the social and natural worlds, gave rise to our modern moment. Putting "intellectual life . . . out of kilter," it made "analytic continuity . . . impossible" (Latour 1993, 7). This science/literature divide has been maintained by literature *and* science, producing a kind of systematic ignorance, a product of the compartmentalization of experience that we can trace through the practices of science and literary criticism.

A BRIEF HISTORY OF MODERN LITERARY STUDIES' RELATION TO SCIENCE

Scientific discourse had a shaping effect on modernist literature, on the literature beyond or outside modernism, and on the new academic discipline of literary criticism that has definitively constructed our interpretation of each. "Literary studies as we now know it" began with the "establishment in 1917 of the first modern English course in Cambridge University" (Bloom 1993, 21). Central to the disciplinary consolidation of literary criticism was the new literary form known as modernism, which took shape in England, Europe, and the United States in accord with scientific principles of objectivity and precision. As it shaped itself around the study of modernist literature, the new discipline of literary criticism relied on science for both analytic and ethical guidelines, just as the modernism that the new literary criticism valued so highly also relied on "a hidden language of technology, chemistry and automatism" while expressing the desire to escape that technoscientific realm for a past of rural craft and classicism (21). This pattern of disciplinary consolidation through commitment to scientificity characterized literary studies as a whole: "Early scholars in English . . . made the study of literature a discipline out of their own commitment to science and research" (Shumway 1994, 7). This new field defined itself implicitly as "the last human science," reflecting the perspective of the scholar who shaped it: "the first important academic literary analyst, I. A. Richards" (Bloom 1993, 22–27). Ironically, the scientific aspirations were the product of an early disciplinary change of affiliation, for although Richards became known as the founder of modern literary criticism, he began his professional life as a psychologist (21).

Beginning with the shocking principle "a book is a machine," Richards set an agenda for contemporary literary criticism that had as its contradictory foundations both an allegiance to scientific principles and a flight from modern technoscience into nostalgia for a communitarian past. As Annan (1990) recalls, Richards's critical program was frequently linked with modernist scientificity. "Modernism affected the way we regarded life and hence our literature and art. It was a movement that both admired and rejected science. English philosophers became engrossed by the discovery of exact truth and many intellectuals were convinced that we could improve the conditions under which we lived by applying the methods of science to social problems" (10). At the dawn of literary modernism, then, critics distinguished forcibly between literature and science, if frequently in order to assert resemblances. From T. S. Eliot's claim in "Tradition and the Individual Talent" that "art may be said to approach the condition of science" (Eliot [1919] 1994, 30), to Northrop Frye's assertion that "literary works . . . are, for the critic, mute complexes of fact, like the data of science" (Frye [1949] 1994, 35), critics emphasized the objectivity, precision, the *scientificity* of literary criticism because they aspired to science's disciplinary prestige. As David Shumway points out, "While literature was held up by the culture and often by literary scholars themselves as an antidote to science and materialism, literary works are treated by scholars as material objects in need of explanation" (Shumway 1994, 110).

Science had contradictory and multiple positions to play in modern literary studies, however. Not only did it help to constitute the new discipline of literary criticism and to establish the modernist aesthetic values such critics prized, but it also catalyzed a reaction to modernist criticism and figured as a thematic element in the non-modernist popular and serious fiction of the period. That brings us to the second position we can distinguish—very broadly—in the modern literary critical response to science. Scholars coming after the initial wave of modernist criticism—and I am thinking here of Walter Benjamin and Raymond Williams most prominently—rejected the white-male-ruling-class canon of modernism. Instead they investigated the sociocultural and historical context of literature, a context in which—in the first four decades of that most scientific of centuries—science figured prominently. The increasing prestige of science in the wider culture in the 1920s and 1930s was ironically at odds with an increasing modesty within the scientific community as science realized the limitations on its knowledge. As Bertrand Russell observed: "It is a curious fact that, just when the man in the street has begun to believe thoroughly in science, the man in the laboratory has begun to lose his faith" (Russell 1931, 88). Yet while the work of Benjamin and Williams often addressed issues of science, in particular the technosciences of the printing press, photography, and methods of industrial production, their Marxist-socialist position gave short shrift to issues of gender. The categorical and disciplinary division between literature and science begun in the eighteenth century, continuing in the nineteenth, and culminating in the early twentieth century with the erection of literary criticism as a would-be counterdiscipline to science—equally objective and equally precise—achieved its definitive utterance in C. P. Snow's 1959 lecture on the two cultures. This lecture has been tagged by Williams as "the most notorious modern example" of our failure to notice how the categories "science" and "literature" constitute themselves by their mutual exclusion (Williams 1986, 10–11).

One strand of literature and science studies can be traced back to Snow's lecture (Snow [1959] 1993). The new field of "literature and science," emerging out of the conjunction of philosophy and history of science, has generally been concerned with investigating the representations of scientific thought in literary texts and in encounters in which science is taken as the stable ground, and literature the symbolic and discursive construct. However, more recent contributions have stressed the discursive character of both literature and science.[4] Although such a field has generated increasing interest, as marked by the emergence of scholarly societies, focused journals, and publishers' series, its practitioners have tended to cluster in Victorian studies and postmodern literary theory, and predominantly to address an audience of literary scholars. Furthermore, there has been little crossover between those working in the literary fields of Victorian and postmodern literary criticism and those working in feminist literary criticism, including Victorian literature. The essays in two influential collections illustrate the nearly bipolar distribution of work in the field of literature and science: George Levine's *One Culture: Essays in Science and Literature* (1987) and N. Katherine Hayles's *Chaos and Order: Complex Dynamics in Literature and Science* (1991). Although both collections include some essays dealing with modernist authors, the focus is split between the Victorians (Levine) and the postmoderns (Hayles). Only very recently have such scholars as Bruce Clarke, Linda Henderson, and Mark Morrison begun explicitly to address the relations between modern literature and modern technoscience.

A disjunction between these two groups of literary scholars contributes to what has until recently been the low profile of science studies within feminist literary criticism, as well as the relatively small amount of traffic between the interdisciplinary fields of literature and science and feminist science studies. During the seventy-five years since Virginia Woolf published *A Room of One's Own*, feminist literary criticism has devoted little attention to science, as literary trope or topic, epistemological category, culture, or discourse. The feminist critics who emerged in the early 1970s in the United States and Great Britain gave serious attention to the context of literature, but unlike the Marxist critics who were their contemporaries, their primary analytic category was not class but gender. In the decade that followed, as they were engaged in the various stages of feminist criticism, from the critique of the male-dominated field, through the recovery of forgotten women writers, to focusing on women writers exclusively, to questioning the gendered basis of the canon itself, those pioneering feminist critics devoted little time to scientific issues, themes, plots, and images. They tended to ignore science as a topic. Instead, rejecting as masculinist the entire project of objectivity and rationality, accepting rather than interrogating the gendered divide of literature and science, they investigated issues such as women's literary relations and the narrative impact of feminism (Moers 1976; Showalter 1977; Du Plessis 1985). Focusing on recovery and reconstruction, and profoundly influenced by Woolf's *A Room of One's Own*, 1970s feminist critics developed one strand of Woolf's argument (i.e., her focus on the forgotten female tradition and the unarticulated lives of women) only to ignore another (i.e., her analysis of how the "sciences" of medicine, psychology, and sexology, like the would-be science of literary criticism, established themselves as disciplines by the production of knowledge about women). As Woolf's narrator discovers during her research foray to the British library, a multidisciplinary convergence of male scholars contributed to constitute woman as object of knowledge: "Sex and its nature might well attract doctors and biologists; but what was surprising and difficult of explanation was the fact that sex—woman, that is to say—also attracts agreeable essayists, light-fingered novelists, young men who have taken the M.A. degree; men who have taken no degree; men who have no apparent qualification save that they are not women" (Woolf [1929] 1981, 27). It was only with the emergence of a historiographic strand of feminist criticism in the late 1970s and 1980s that feminist literary scholars began to consider how science helped to shape women's writing, and how women's writing might reshape science. Exemplary of that new perspective was Ludmilla Jordanova's *Languages of Nature* (1986).

FEMINIST SCIENCE STUDIES AND "LITERATURE AND SCIENCE"

Languages of Nature was designed as an introduction to "science and literature as a field" (Jordanova 1986, 16). Opening with the assertion that "science and literature are united in their shared location within cultural history," the volume was planned both to consolidate a new field of scholarship and to serve as an introductory text for the new courses that—the editor assumed—would grace this emerging field of study. The essays in this pathbreaking collection take the externalist position that both literature and science as fields are rooted in their specific historical contexts, and they turn to a range of textual and social sites to consider the relations between the two fields in the pre- and protodisciplinary moment of the eighteenth and nineteenth centuries. Literary texts including *Tristram Shandy*, *Les liasons dangereuses*, the nature poetry of Erasmus Darwin, and George Eliot's *Silas Marner* are put in conversation with science texts by Lavater, Lamarck, Darwin, and Michelet, and with social science texts by George Henry Lewes and Herbert Spencer.

Yet the literature/science categorization I have just made eludes the texts the essays examine, for as Jordanova points out in the introduction, "It is significant that the entire notion of a discipline is a recent one, having developed in the nineteenth century" (16). Moreover, *Languages of Nature* deliberately transgresses disciplinary boundaries. Rather than claiming disciplinary orthodoxy, it takes an "interdisciplinary [approach] to science and literature" and emphasizes its attention to "the discourses common to science and literature" (17). The contributors are a mix of historians and literary scholars, including Gillian Beer, whose *Darwin's Plots* (1983) had been published only three years earlier, and others who went on to make major scholarly contributions in the decade since the volume's publication (Beer 1983; Shuttleworth and Christie 1989; Keller, Jacobus, and Shuttleworth 1990; Jordanova 1990).

Although the aim of *Languages of Nature* was to establish a new field of study, drawing together literary critics and historians of science to consider the relationships between literature and science, it did not fully succeed in reorienting mainstream feminist science studies to include its methodology, the attention to the "cultures, contexts and even philosophical structures" that science shares with literature (Jordanova 1986, 17). Despite the publication of a valuable collection of essays—Benjamin's 1993 *A Question of Identity: Women, Science, and Literature*—ten years after the appearance of Jordanova's pathbreaking volume, the study of the intersection of literature and science had still not been explicitly thematized as part of the field of the gender critique of science, mapped by Evelyn Fox Keller and Helen Longino in their Oxford Readings in Feminism foundation text, *Feminism and Science* (1996). This generally excellent volume contained contributions by seven philosophers, four historians, two sociologists, one anthropologist, and one biologist but featured no study by a literary scholar, despite the fact that the coeditors devote at least a third of the book to what they described as the "large subject, the role of language in shaping research agendas," and concluded with a call for more work to be done in "studies of language and gender in the physical sciences" (Keller and Longino 1996, 6, 12). While several of the essays did address literary themes, among them the role of analogy in science, the construction of a scientific "romance," the implications of naming in science, and the epistemological role of language, and while the volume thematized the role of language in scientific practice, it never addressed the relations between science and literature as a substantial part of the gender critique of science. Thus if the 1986 Jordanova volume represented the early promise of the new field of literature and science, the 1996 Keller and Longino volume was a prescient representation of the paradox of literature and science today: the presence of literary methods of analysis coupled with the relative absence of literary scholars among the canonical or generally agreed-on representations of the field.

Even Londa Schiebinger's sweeping survey of "the current scholarship on gender and science in the United States," *Has Feminism Changed Science?* (1999), while acknowledging that "the literature on gender and science is scattered across the academy and often written in the dialect of a particular discipline," locates the expertise in the area of gender and science narrowly, among historians of science and philosophers of science (Schiebinger 1999, 13, 2). Although the text demonstrates that literary tools such as linguistic or cultural analysis can be useful in uncovering the operations of gender in science, and indeed the reader finds a rich trove of gendered images, narratives, and plots in the chapter entitled "Biology" in particular, the study avoids any direct consideration of the relations between literature and science as unlikely to yield any significant information. Schiebinger makes the obligatory reference to C. P. Snow's delineation of the gap between two cultures, the scientific and the literary, but significantly literature is important only for its analogous relation to the culture of women (Schiebinger 1999, 14). Ultimately, Schiebinger's study can meet its goal of providing useful analytical tools for the study of gender and science, despite the fact that it is strikingly representative in the restrictive boundary conditions it accepts for such tools, if we extrapolate from her concluding list of "tools for gender analysis" (186). While accepting the observation that "gender narratives are not *innocent literary devices used to abbreviate thought.* Analogies and metaphors construct as well as describe—they have both a hypothesis-creating and a proof-making function in science," we should also be alert to the implicit assumption that literature tends to be both innocent and mimetic (188–89). Taking seriously her observation that "fundamental concepts in any field should not be taken for granted but should be set within historical frameworks of meaning," a resistant reader of Schiebinger's study will ask what might motivate scholars of gender relations to accept and reproduce the hierarchically constituted two-culture divide itself. In short, why has the field of feminist science studies remained relatively uninterested in the question of the relations of literature and science? That this continues to be the case can be seen in a more recent construction of the field, *The Gender and Science Reader*, edited by Muriel Lederman and Ingrid Bartsch, which includes among its thirty-five selections of "key writings by leading scholars to provide a comprehensive feminist analysis of the nature and practice of science" the work of one scholar who devotes serious attention to literary analysis, Donna Haraway.[5]

Or take a simple survey of the scholarly marketplace. Although in the 1990s there were important series in literature and science at Stanford University Press, the University of Michigan Press, and the University of Oklahoma Press, by the beginning of the twenty-first century, in the wake of the Sokal hoax and in the climate of economic downturn, two of those series had folded. The remaining series at the University of Michigan, Studies in Literature and Science, had only an oblique relation to feminist science studies. That the tide is beginning to turn may be suggested by the appearance of an exciting new series at the University of Washington Press, In Vivo: The Cultural Mediations of Biomedical Science, which explicitly addresses the co-constitutive relation of literature and biological/medical science.[6] Yet despite the appearance of that new series, and the fact that individual literature scholars have begun to fill out our understanding of modern relations between literature and science, their wider impact on the broad field of feminist science studies remains fully to be registered (Clarke 1996; Otis 2000, 2001; Henry 2003). Arguably, the analytic edge within feminist science studies belongs not to literary critics but to more broadly systematizing disciplines such as psychology and philosophy, exemplified by the reevaluations of Darwin's work by psychologist Elizabeth Wilson and philosopher Elizabeth Grosz (Wilson 2001; Grosz 2001).

This is not to say that science studies has not paid attention to linguistic or textual or literary issues. Feminist science studies has been amply provisioned with good stories. To confirm that, we need only recall Evelyn Fox Keller's early, and crucial, delineation of the narrative foundational to science since the seventeenth-century scientific revolution—the act of unveiling and ravishing nature-as-a-woman.[7] Then there is her delineation of the self-serving, internalist "story of the rise of molecular biology . . . a drama between science and nature," and her critical, externalist counternarrative of "the transformation of biology from a science in which the language of mystery had a place not only legitimate but highly functional, to a science that tolerated no secrets, a science more like physics, predicated on the conviction that the mysteries of life were there to be unraveled" (Keller 1986, 70). And we can recall Sandra Harding's witty delineation of science's own self-construction through its "origins myth": "All of us grew up on a well-known story about the birth of modern science: who was responsible for the conception, why the labor necessary to bring forth this babe was so difficult, what its birth has meant to three centuries of European and American history, and why the mature personage this babe has become continues to be deserving of massive support in the face of competing demands for public resources" (Harding 1986, 202).[8] Finally, we can summon up the various wonderfully self-interested narratives of primate lives produced, as Donna Haraway has shown, by twentieth-century primatologists. Even leaving aside these analyses of narratives and origin stories as elements of scientific practice or scientific self-construction, science studies is increasingly replete with linguistic methodologies—from discourse analysis and studies of analogy and metaphor to interrogation of the function of emplotment (Hesse 1966; Gross 1990).

Yet the disciplines continue to have unequal weight in science studies, with literature registering least on the authority scales. Scholars have given far more time to the narratives or analogies or metaphors used by science than to the ways that the traffic between the realms of literature and science destabilizes our notions of science and literature. Donna Haraway's *Primate Visions* (1989) and *Simians, Cyborgs, and Women: The Reinvention of Nature* (1991) provide models of a science studies that attends to the traffic between literature and science in order to destabilize the very disciplinary divide. Of course, despite her huge influence in literary studies, we should remember that Haraway is trained not as a literary critic but as a historian of science. There, as in the broader field of science studies, scholars generally locate and analyze stories, metaphors, analogies, and plots in scientific practice(s), rather than science being informed by its engagement with, as well as representation in, literary practice. Science is still conceptualized as having an agency and a primacy that literature lacks; science can generate literary representations, but literature is not thought to motivate or generate scientific representations. In short, literature has yet to be seen as instrumental in the social construction of scientific facts.

Moreover, what might be called a failure of collegial cross-fertilization marks the fields of science studies and literary studies. Until recently in the United States, only the important work of N. Katherine Hayles departs from the general rule that science studies scholars are not found in, or do not emerge from, literature departments.[9] Hayles, the author of *The Cosmic Web: Scientific Field Models and Literary Strategies in the Twentieth Century* (1984), *Chaos Bound: Orderly Disorder in Contemporary Literature and Science* (1990), and *How We Became Posthuman* (1999), has played a crucial role in shaping how literature and science studies develop into the twenty-first century. Working between the fields of physics and mathematics, and literature and postmodern theory, she has mapped the common ground between literature and science. Her focus first was field theory, and later chaos theory or nonlinear dynamics; most recently, she has explored the relationship between the history of cybernetics and contemporary postmodern and hypertext fictions. As part of a critique of literary critical practices, Hayles makes the valuable observation that both cultural postmodernism and chaos theory share a commitment to what she calls the *denaturing process* (the process of depriving language, context, and time of their natural qualities), although it is crucial to distinguish between the uses to which the same process is put: the latter using it to increase control, the former to decrease it. Yet those productive analyses have not brought about the convergence between feminist literary criticism and feminist science studies that one might want. With the exception of the occasional review in sociological journals, until recently Hayles's work was taken up less by feminist literary critics than by scholars working in postmodern literary theory. Her influence was conspicuously absent—even from the index—of the benchmark volume edited by Keller and Longino (1996).

Yet Hayles makes a central point for feminist science studies in *How We Became Posthuman*. The Turing test, now understood as foundational to the distinction between human and machine, actually began by testing the ability to distinguish male from female. "By including gender, Turing implied that renegotiating the boundary between human and machine would involve more than transforming the question of 'who can think' into 'what can think.' . . . What the Turing test 'proves' is that the overlay between the enacted and the represented bodies is no longer a natural inevitability but a contingent production, mediated by a technology that has become so entwined with the production of identity that it can no longer meaningfully be separated from the human subject" (Hayles 1999, xiii). Hayles draws from this analysis of the gendered core of the Turing test the argument that distinctions between materiality and representation, the natural and the artificial, human and machine, are implicated in the culturally mediated process of gender production. There is the basis for a rapprochement between Hayles's analysis of the posthuman as a gendered category weaving together the realms of literature, cognitive science, nanotechnology, artificial life, et cetera, and the work of feminist science studies scholars to explore the gendered nature of scientific practices, but only if they take the time to challenge the general indifference toward literature and read the works of literary scholars such as Hayles.

This indifference to the literary by those working in science studies may be a reflection of a functionalist turn in the broader culture: the institutional and economic forces of the capitalist and late-capitalist eras that have shaped education toward increased emphasis on producing the fungible and highly instrumental knowledge required by industry, and have thus privileged technoscience over literature. The opposition between useful science and useless literature mystifies the functional relations between literature and political and economic power. As Richard Ohmann has pointed out, "the humanities—and high culture—[are used] within universities to harden class lines and teach the skills and habits of mind that will serve the industrial system" (Ohmann 1976, 334). Still, economic and geopolitical interests join the already existing epistemological hierarchy of the disciplines with the result that, in most cases, the linguistic turn in science studies has not dislodged the gendered relations between the disciplines themselves. The culturally enforced, hierarchized, and gendered relations between science and literature that render literature an insignificant, invisible, feminized part of the cultural project in relation to significant, visible, masculinized science remain uncontested. The result is that our critique of science is incomplete, for its gendered disciplinary boundaries remain unexamined.

In a sense, the incomplete critique of science is a function of the difficult position of feminism in the academy, even today. Alliances with nonfeminist scholars in social science and science fields are more readily cemented if the feminist scholar positions herself or himself as sharing the same commitment to *facts* as opposed to *fictions*. Because our analysis of the gendered disciplinary hierarchy is still incomplete, feminist critics of science see little to gain from an affiliation with the disciplinary position of literary studies. Instead there is a marked tendency for feminists in the social and natural sciences, like other relatively vulnerable groups, to protect their (relative) epistemological authority by retaining their disciplinary perspective and merely appropriating the discrete tools of literary study: metaphors, images, emplotment, and narrative.

This failure to engage across disciplinary boundaries results in precisely the kind of sterile thinking that labels as *coincidental* the simultaneous appearance of James's novel and Carlsen's article. Far from being a coincidence, the double publication should alert us to the persistence of a certain kind of Enlightenment epistemological logic: whenever we find an object, somewhere there's a subject (even if hidden) and vice versa. This logic implies its corollary: whenever we see literature (culturally scripted as the domain of subjectivity), we should expect that there's also science (the culturally accepted home of objectivity). For, to reiterate, literature and science operate together in culture and society to produce subjects and objects. Moreover, this epistemologically tidy, reciprocally constructed connected distinction between subject and object has been reconstituted politically as a family structure of decidedly unequal relations. That disciplinary family presents us, as feminist science studies scholars, with a dilemma that can (at least for polemical purposes) be imagined as a choice between two directions. Do we continue to demonstrate how science not only produces but depends on broadly literary representations, despite its continual work to deny and disguise the generative importance of those representations, and despite the fact that they continue to be disguised by a relentlessly masculinized disciplinary position organization and protocols? This work has indeed been pathbreaking, but it also perpetuates and even extends the Enlightenment epistemology that depends on a subject-object opposition, even as it seeks to redress the politically constructed inequality that recapitulates it. Or do we start to rethink the logic that assumes some primary demarcation of social production between subjects and objects, and some primary genre-bound division between literature and science?

While in the section that follows I map out a theoretical strategy for bridging the two fields, the convergence could also be brought about by a change not in *theory* but in the status of *literary genre*: by a repositioning of feminist science fiction in literary practice. Indeed, as I have said earlier, it seems to me most likely that such a convergence is occurring right now, fueled simultaneously by several different phenomena. A growing interest in science fiction, as exemplified by the popularity of the work of Octavia Butler, in particular, on the part of feminist science studies scholars not trained as literary critics, may well provide a bridge between feminist literary criticism and feminist science studies. Previously situated outside the canon, this generative and important genre is increasingly being given a central position in feminist teaching and criticism, with a powerful reshaping effect on practice in feminist science studies. And as we will see in a later chapter, there are complex social causes, and effects, for what could (erroneously) be understood as a merely local shift in literary taxonomies and hierarchies. Precisely because the meaning of genre is linked to larger discursive and material structures, changes in the nature and significance of genres can indeed play an important role in understanding our lives as individuals and societies.

A METHODOLOGICAL TOOL KIT FOR LITERATURE AND SCIENCE

We have not been bold enough in our approach to the intersection of literature and science, and the result has been a specific sterility. We have limited ourselves to using one field to gloss the other rather than using them to unsettle not only each other but also their mutually opposed relation. A genuinely reciprocal understanding of the ways literature and science collaborate and compete to construct the subjects of disciplinary knowledge can challenge the very organization of culture within which both fields find their place: the Enlightenment epistemology of subjects and objects. We need to shift the kinds of questions we ask away from "Is this discourse primarily scientific or literary?" or "Is it engaged primarily in the explorations of subjects or objects?" to "What kind of cultural work is this doing? Where? And in what ensemble of social relations?"

Bruno Latour has been a leader in such interdisciplinary thinking. In a 1993 interview, he put such a challenge on the agenda of science studies when he asked, "How can we invent literary style for science studies, and how can we pursue the fusion of social science and literature?" He went further still when he described his study of the failed Parisian light-rail system, *Aramis, or the Love of Technology*, as "a novel, a sort of novel but without fiction" (Crawford 1993, 267). Although Latour offered seven rules of method for analyzing the interdisciplinary domain of "science, technology, and society" in *Science in Action* (1987), his own commitment to finding "a few sets of concepts sturdy enough to stand the trip through all [the] many disciplines, periods and objects" suggests that we can transpose, or adapt, those rules to study *literature and science in action* (Latour 1987, 16). In what follows, I offer a list of those rewritten rules of method, followed by an explanation of their implications for the incorporation of literature into feminist science studies.

We study literature and science *in action* and not ready-made literature or science. To do so, either we arrive before the disciplines, facts/interpretations, and machines/texts are black-boxed, or we follow the controversies that reopen them. The study of literature and science requires awareness of the microprocesses that produced each as a discipline and gave rise to the linked and opposed entity "literature and science." Canon creation in literature, like the creation of scientific facts, requires practices of abstraction, taxonomization, and selective amnesia—that refusal to remember or reexamine origins that science studies scholars call "black boxing." When we work on "literature and science," we reexamine those uninterrogated microprocesses and reopen those black boxes. The result will be a richer, far less tidy sense of the meaning or boundaries of the terms "literature" and "science."

To determine the objectivity or subjectivity of a claim, the efficiency or perfection of a mechanism or interpretation, we look not for their *intrinsic* qualities but at all the transformations they undergo later in the hands of others. This transcends the time-honored citation index as a gauge of literary or scientific worth, since that mechanism is usually limited to the discipline in question and does not cover the traffic between disciplines. Rather, it prompts us to study the transformations and functions of a claim, theme, trope, or process as it moves between greatly dispersed disciplines, asking how it is differently constructed, situated, and deployed in its new disciplinary environment(s). It prompts us to realize that it is not the inherent qualities of a claim, theme, trope, or process but rather their trajectory and the way that they are put in play that shapes how we weigh their accuracy, efficacy, or truth value.

Because the settlement of a controversy is the *cause* of nature's representation by literature or science, not its consequence, we can never use this consequence, nature, to explain how and why a controversy has been settled. Rather than accepting as natural that some aspects of the material world fall under the purview of science and some under the purview of literature, we realize that this division into scientific and literary objects of knowledge is constructed as the solution of a past controversy over disciplinary realms and regimes. There is nothing inherently literary or scientific, only what disciplinarity makes so.

Because the settlement of a controversy in the field of literature and science is the cause of society's stability, we cannot use society to explain how and why a controversy has been settled. We should consider symmetrically the efforts to enroll human and nonhuman resources for literature and science. This rule is particularly important for those who come to the study of literature and science from a social constructionist perspective. It counsels us to examine the notion of social construction as critically as we do the notion of the natural, realizing that there is a material base to even the most seemingly socially constructed experiences or entities. It further counsels us to remember that the material world exercises a shaping effect on "the literary," as well as "the scientific." That feminist scholars are particularly prone to a "knee-jerk constructivism" indeed helps to explain the reluctance on the part of those in the humanities to engage seriously with the claims of science. As Wilson has pointed out, "A large part of the difficulty in generating politically engaging feminist critiques of the biological and behavioral sciences must be attributed to feminism's own naturalized antiessentialism. After all, how can a critical habit nurtured on antibiologism produce anything but the most cursory and negating critique of biology?" (Wilson 1998, 16).

We have to be as *undecided* as the various actors we follow as to what literature and science is made of; every time an inside/outside divide is built between literature and science, we should study the two sides simultaneously and make the list, no matter how long and heterogeneous, of those who do the work. This means being ready to investigate the scientific practices and popular scientific writings of an era, as well as its literary texts; it means understanding that influence can flow back and forth between literature and science, and that literary works (and workers) can influence science, as well as the reverse. Karen Barad has coined the term "agential reality" for that "reality within which we intra-act [that is] made up of material-discursive phenomena . . . not a fixed ontology that is independent of human practices, but . . . continually reconstituted through our material-discursive intra-actions" (Barad 1998, 19).

Confronted with the accusation of undisciplinarity, we look neither at what disciplinary rule has been broken, nor at what structure of society could explain the distortion, but to the angle and direction of the observer's *displacement from the discipline* and to the *length* of the network thus being built. Perhaps the loosest of my adaptations from Latour, this rule understands accusations of disciplinary transgression, like accusations of irrationality, as attempts to maintain disciplinary authority even at the cost of intellectual acuity. In place of disciplinary gatekeeping, it advocates making interdisciplinary networks and assemblages. The goal is to determine what the linkage of literature and science can bring into being, rather than to determine what the distinction between literature and science can prevent.

Before attributing any special quality to the mind or to the method of people involved in literature or science, let us examine first the many ways through which literary and scientific inscriptions are gathered, combined, tied together, and sent back. Only if there is something unexplained once the networks have been studied will we start to speak of cognitive factors particular to literature or science. This final rule works two ways. First, it encourages us to realize that a specific epistemology and methodology are not naturally inherent in either literature or science but rather are the product of disciplinary organization and training. Then, combating the notion of naturally discrete disciplinary practices, it encourages us to take an interdisciplinary perspective. This new perspective is profoundly decentering, a "Copernican revolution . . . a shift in what counts as center and what counts as periphery"

(Latour 1987, 226). To another scholar, taking the perspective of literary studies, it entails the Deleuzian process of "becoming minor." Minor writing "dismantles notions of value, genre, canon, etc. It travels, moves between centers and margins . . . [refusing] to admit either position as final or static" (Kaplan 1987, 188–89).

Adapting Latour's rules of method for studying "science in action" to the study of literature and science in action enables us to consider how each field would represent the other, to think of them as an ensemble of social relations, to watch them in action, to ask what they do as linked social practices, to note the traffic between and within them, and to gauge the agency of each and its relation to the other. This methodological program, like other rules of method (and I include scientific practices), is to some degree retrospectively constructed. When the study of literature and science is taxonomized and systematized, its instability and provisionality may be covered up. However, these revised rules of method embody the particular value of attention to the relations of literature and science for the field of feminist science studies: we can use them to gain access to the "biomedical imaginary," that crucial zone in which, as Waldby reminds us, we can find those "medical ideas which derive their impetus from the fictitious, the connotative and from desire" (Waldby 2000, 136).

TRACING THE NETWORKS: OMEGAS AND MR. ADAM

Let us return to the case of the simultaneous appearance of two texts dealing with a crisis in male fertility, one text literary and the other scientific, to see how these new rules of method for studying literature and science in action can be applied. Drawing on the first revised rule of method (to study literature and science *in action* rather than as ready-made, discrete objects), I refuse to label the event "a remarkable coincidence" (Wright 1996, 42). Rejecting the inside/outside divide such a label implies, as well as the notion that either literature or science is a stable and discrete category, I hypothesize instead that the simultaneity of publication caps two interwoven strands of parallel interest in male fertility, appearing in communities from the scientific to the literary, and converging (not necessarily for the first time) in the late twentieth century. I ask myself not only why these expressions of interest have converged (and why now) but also which assumptions I have made that led me to be surprised by that convergence.

This leads me to question the primary assumption—of disciplinary integrity—behind the label "coincidence." I consider the disciplinary structures that function to make these two publications seem—despite their shared status as textual representations—such different entities. Beyond the training and credentialing apparatus that differentiates Carlsen from James, these disciplinary structures also dictate the rules for literary and scientific publication, including their generic formulas. I note that each publication seems to have a hybrid identity in relation to its "home genre." Thus although *The Children of Men* shares the familiar traits of a novel of character development in its realistic narration, generally plausible academic setting, and familiar donnish protagonist, its premise—the global end to human fertility, or "Omega point"—echoes one of the central strategies of science fiction. "SF can be understood as a kind of thought experiment similar to thought experiments in science. The experimenter—the writer—begins with a hypothesis and sets up initial conditions. Following the inherent logics of these conditions (i.e., the plot) he derives some results, perhaps surprising ones. . . . Use of imagination is as central to the fictional thought experiment as to the scientific one, with the difference that the imagination of a writer is not controlled by scientific, methodological constraints, but by aesthetic, narrative principles" (Steinmuller 1997).

Analysis reveals that Carlsen's article is hybrid too, mixing scientific research and science reporting. I reflect on—and question—the disciplinary conventions that assert their kinds as incommensurable, framing the former as a realm of objective facts generated without reference to social context and the latter as a realm of subjective fictions generated as commentary on the larger social realm. In so doing, I share the understanding that "literature and science, whatever else they may be, are modes of discourse, neither of which is privileged except by the conventions of the cultures in which they are embedded" (Levine 1987, 3).

Moving to the second revised rule of method ("we look not for their *intrinsic* qualities but at all the transformations they undergo later in the hands of others"), I trace the network of transformations that Carlsen's article undergoes since its 1992 appearance in the *British Medical Journal*. The article catalyzed a number of responses, which appeared in the

British Medical Journal, Lancet, the *New England Journal of Medicine, Fertility and Sterility,* and other scientific publications. Although many debated Carlsen's statistical methods, two essays in particular are notable for the way they transform the Carlsen piece into a lesson not about statistical methods or about sperm quality but about disciplinary practices in science, and the disciplinary organization of knowledge.[10]

Farrow (1994) repositions the case of the Carlsen article as a cautionary tale for scientists, arguing that it displays errors arising not only from methodological irregularities but from disciplinary *regularities*: "Bromwich and colleagues argue that Carlsen et al. applied the wrong form of analysis and that an artefact explains nearly all of the putative 'fall' " (2). "There is inherent bias in how we define the problem in the first place," Farrow observes. Errors arise from the choice of inappropriate statistical methods, but they are also produced by the disciplinary organization of knowledge: "what we choose to collect and what we choose to leave out" (2). We risk errors, Farrow suggests, when we let one discipline monopolize the construction of knowledge (both the questions asked and the field studied); we risk error when we neglect a multidisciplinary approach to complex questions. "When inferences are being drawn over time we deserve more than simple analysis: we need to . . . seek corroborating evidence from a wide range of disciplines. Why not refer to the extensive data from veterinary research?" (1). Although Farrow's article privileges disciplinary boundary crossing as leading to greater knowledge, the title that the article is given by the *British Medical Journal*—"Falling Sperm Quality: Fact or Fiction?"—reinstates the limits on such extradisciplinary wanderings (perhaps unwittingly) and recontains what could otherwise be an unacceptable challenge to scientific authority. A similar process of recuperation might be the result of the uneasy reference in Bromwich et al. 1994: "Their conclusion received widespread recognition, including coverage by the media" (19). Both fiction and the popular press are implicitly constructed as beyond the boundaries of science. If Carlsen's finding of falling sperm quality is fiction, it is no longer the purview of science. Similarly, media acclaim for such a finding may prompt scientific skepticism.

Farrow's article is joined a year later by the contribution of Olsen et al. (1995), which makes an even more vigorous critique of the statistical methods, popular acceptance, and insufficient interdisciplinarity of

Carlsen's research. This article, too, emphasizes the popular press acclaim that has greeted Carlsen's study, observing that it "has crossed professional boundaries, appearing in publications of a variety of specialities, including medical, chemical engineering, and environmental, as well as the lay press" (Olsen et al. 1995, 887). It is worth noting that Olsen et al. list their own institutional affiliations (among them Dow Chemical Company and Shell Oil), and the list undercuts their supposed willingness to sign on to Carlsen's conclusion were it not for her flawed statistical methods. Once again, this alarming interdisciplinarity is implicitly contrasted with an approved interdisciplinarity in the concluding recommendation that there be "a thorough review of the veterinary medicine (theriogenology) literature" (892). And here, too, these two kinds of boundary crossing can be differentiated: the right kind occurs still within the broad category of science, the wrong kind is literary. Implicitly suggesting that Carlsen has adopted one of the cardinal tropes of science fiction, extrapolation into the future, Olsen et al. observe: "If extrapolated forward in time, an admittedly unscientific conclusion might hold that the linear model portends the collapse of traditional means of human procreation by the middle of the next century" (888).

This brief survey of the response to Carlsen's article in the scientific literature reveals its transformation at the hands of critical peers from scientific fact to something more like fiction. Carlsen's article has been tainted by the wrong sort of boundary crossing: scientific entry into the realm of the literary and the popular. Moreover, the articles suggest that the taint might have been prevented by the right sort of boundary crossing: a thorough examination of the literature in adjacent *scientific* fields. My point here is not that Carlsen's critics neglect serious concern for the article's scientific validity in order to attack its popularity but rather that their mode of critique inevitably reflects the disciplinary hierarchies of science, science studies, and feminist science studies, in which fictional or literary elements are at the bottom of the heap.[11] Yet to relegate to last attention works of the imagination is to overlook the biomedical imagination, which as Waldby has demonstrated "enables the importation of social narratives into biomedicine's technical narratives" (Waldby 1996, 16).

Drawing on the fifth revised rule of method ("study the two sides simultaneously and make the list, no matter how long and heterogeneous, of those who do the work"), I begin to build the list of those (no matter what their disciplinary position) who have written about a decline in sperm potency. To Carlsen's article, the responses to it in scientific journals, and P. D. James's novel, I add Pat Frank's *Mr. Adam.* Published in 1946, nearly a half century before *The Children of Men,* this novel concerns a global end to male fertility caused by a nuclear accident in Mississippi. Because I am working between the disciplines of literature and science rather than within the discipline of literary studies, in accordance with the sixth revised rule of method (look "to the angle and direction of the observer's displacement from the discipline, and to the length of the network thus being built"), I am free to examine this novel in terms of its displacement from the other novel. Thus I examine it not for its intrinsic qualities (what kind of a novel is it, is it well written, can it be called literature?) but for the similarities that appear, and transformations that occur, in the theme of sudden, global male sterility, as it travels between *Mr. Adam* and James's Omega point.

Although separated by nearly a half century, these novels are linked by their focus on a sudden crash in sperm potency. Both novels turn on the discovery of the last fertile human being. Frank's novel features the discovery of "Mr. Adam," the only man to retain his fertility after the nuclear accident, whereas James's novel focuses on the discovery of androgynously named Julian, the only woman to give birth after Omega. Both are social misfits. Frank's Homer Adam is a technoscientifically trained outsider: a shy, gangly civil engineer who had been working in a lead mine in Colorado at the time of the nuclear accident and was protected from radiation-induced sterility by the mine's thick walls. James's Julian, a young woman student named after the fourteenth-century English mystic Julian of Norwich (author of *Sixteen Revelations of Divine Love*), is a religious visionary whose unexplained pregnancy testifies to the power not of science but of divine providence.

Although they share the same premise, the two novels offer different causes and different remedies for that disastrous collapse in human reproduction, reflecting their different contexts. The jacket blurb explains that *Mr. Adam* "had been growing in [Frank's] mind since the first atomic bomb fell on Japan." Not surprisingly, the novel attributes the fertility crash to a nuclear accident. However, characteristic of the novel's optimistic era, it finds in technoscience both the frightening premise and the happy solution to the novel's dilemma. Artificial insemination—AI—looms large, and much of the novel's comedy arises from the efforts of government agencies to control not only that relatively primitive technology but also the individual who will serve as its resource. In the elaborations of that dilemma, Frank produces a remarkable anticipation of our current situation: "AI, a process that has been part of human culture for at least hundreds of years or possibly ever since the knowledge of paternity was realized, is transformed into a high-tech medical procedure, unimaginable without genetic considerations" (Spallone 1987, 181). In contrast, *The Children of Men* was written after the Chernobyl nuclear accident. As befits our era of cynical pessimism, the novel attributes the fertility crash not to nuclear but to chemical and environmental pollution and finds its solution not in medicine or science but in the countercultural and religious worlds.

Both novels portray the same initial moment of revelation, when the protagonists, checking hospital bookings, realize that there are no more babies to be born. In *Mr. Adam*, the journalist protagonist notices "that people have quit making reservations to have their babies in Polyclinic Hospital, as of June 22" (Frank 1946, 16). His investigative ardor stimulated, he gets on the telephone. "I called Rochester, Philadelphia, Miami, and New Orleans, and then desperately swung west to San Francisco. The situation was identical. . . . So far as I could discover, our July birth rate was going to be zero. . . . I began combing the Western Hemisphere. Things didn't change" (16). In *The Children of Men*, the situation is discovered by the remaining fertile woman herself:

> I was twenty-seven at Omega and working in the maternity department of the John Radcliffe. I was doing a stint in the ante-natal clinic at the time. I remember booking a patient for her next appointment and suddenly noticing that the page seven months ahead was blank. Not a single name. Women usually booked in by the time they'd missed their second period, some as soon as they'd missed one. Not a single name. I thought, what's happening to the men in this city? Then I rang a friend who was working at Queen Charlotte's. She said the same. She said she'd telephone someone she knew at the Rosie Maternity Hospital in Cambridge. She rang me back twenty minutes later. It was the same there. It was then I knew, I must have been one of the first to know. I was there at the end. Now I shall be there at the beginning. (James 1993, 148–49)

If we return to the second revised rule of method, we can note that along with the striking similarities, there are also telling transformations in the way this revelatory moment is portrayed in 1946 and 1992. When the protagonist discovers the impending end of human reproduction in *Mr. Adam,* he first assumes women have just decided to return to midwifery: "The truth is that people have just gotten damned sick and tired of kowtowing to those sacred, omnipotent institutions, the hospitals, and have decided to have their babies at home. . . . I might remind you that up until about a century ago all babies were born at home" (Frank 1946, 12). In contrast, when Julian discovers the blank pages in the hospital ledger, she immediately thinks, "What's happening to the men in this city?" (James 1993, 149). The difference in their responses not only reflects something about the two characters; it also reveals something about the social contexts of the fictions. Perhaps institutions like the profession of obstetrics were more widely accepted in 1992, less subject to female resistance, than they were in 1946. But despite the increasing confidence in obstetrical knowledge, anxiety about fertility, and male fertility in particular, seems closer to the surface in 1992 than in 1946. We see a disturbing shift in the balance of human power relations. Female agency that in 1946 led women to stay away from hospitals is replaced by a male failure of agency in 1992, as the empty hospitals testify to male reproductive incompetence.

Indeed, our consideration of Frank's novel has brought us back to our point of departure: the anxiety about fertility that surfaced in 1992 with the simultaneous publication of Carlsen's article and James's novel. Applying the revised rules of method to explore the domain of literature and science, we have traced the networks from Carlsen's article to James's novel and Pat Frank's *Mr. Adam*. Let us recap the chronology. Carlsen and her colleagues (1992) assessed the data from sixty-one papers published internationally between 1938 and 1990, with a subject pool of 14,947 men, and found that it revealed "a significant decrease in mean sperm count from 113 x 10(6)/ml in 1940 to 66 x 10(6)/ml in 1990" among men without a history of infertility (610). Six years after the first significant data began to appear, in 1946, Frank's *Mr. Adam* was published. Frank's novel dealt explicitly with the issue that science was only beginning to reveal: a decline in male fertility. It took nearly fifty more years for that data to be presented, and their implications to be considered, in Carlsen's scientific study. And when Carlsen's study finally appeared, it did so neck and neck with a second work of fiction concerning a global decline in sperm potency.

What do we make of this story of literary and scientific treatment of human reproduction? It is tempting to begin with the polemical assertion that literature leads science in the articulation of crucial cultural, social, and material issues. Literature operates as the unconscious of science, so the argument would go, as the site where culture articulates its fantasmatic investment in science, and where the implications of scientific findings speak first, despite the powerful forces of repression. Like dreams, literature functions to bring into disguised awareness issues that preoccupy us but are too frightening to be considered directly. According to this interpretation, Frank articulated in a work of fiction a concern with male reproductive potency that required literary mediation for expression, a concern that would not be explored, and articulated fully, by science for nearly a half century.

We would be wrong to give in to the temptation of this polemical assertion, however, for it participates in precisely the error that has given rise to the curiously sterile zone between feminist literary criticism and feminist science studies. It assumes a two-culture divide, with a discrete literary realm that either represents or comments on an equally discrete, and different, realm of science. And it perpetuates that familiar Enlightenment game of "see a subject, find an object," although one's disciplinary perspective determines whether the subject is located in the scientific or the literary knower, the object in nature or in the text. Rather, our application of the revised rules of method suggests that it is more productive to view literature and science not as binary opposites but as an ensemble of social relations. Emphasizing appropriation rather than possession, and transgression rather than policing, this method enables us to read the complex cultural negotiations taking place across these liminal zones. From that alternative perspective, we have found that what we might otherwise have dismissed as a "coincidence"—the simultaneous publication of two texts, one literary and one scientific, dealing with the issue of sperm potency decline—instead yielded a rich network of relations, weaving fiction and fact, literature and science, culture and nature, all concerned in their different ways with the possibility of a global fertility crash. From Carlsen's article to the scientists commenting on it, to James's and Frank's novels, that network has a familiar nonmodern hybridity, for it incorporates statistics and fiction; newspaper articles, scientific journals, hospital ledger books, and Bibles; nuclear power plants, mines, and cathedrals; the England of the future and the United States of the past; as well as disciplines as divergent as chemical engineering, microbiology, and veterinary medicine.

LATERAL ENCOUNTERS WITH LIMINAL LIVES

"Lateral encounter, between groups and individuals alive in the same time but in different initial conditions, allows fresh perceptions to thrive" (Beer 1996, 5). Taking this network of relations seriously, investigating its links as something more than merely coincidental, we have staged such a lateral encounter. The result is a description of the complex network linking all three objects under study—a network that reveals the scientific texts, the fictions, and the sperm to be not discrete objects but objects in relation. How can we best describe what kind of objects they are, and the nature of their relations? Feminist science studies and the social studies of science have offered a number of names for such objects existing at the intersection of the material and the semiotic. We could call them "boundary projects" or "cyborgs" (Haraway 1991, 201, 150); "working objects" (Daston and Galison 1992, 85); quasi subjects or quasi objects (Latour 1993, 55). Borrowing from Curthoys and Docker the particular historical and philosophical spin given this familiar literary term, we could even call them metaphors: "points of meaning—destabilizing, wayward, productive" (Curthoys and Docker 2005, 3). Yet Curthoys and Docker are exemplary in the way they (perhaps unintentionally) limit metaphor's wayward productivity and its significance to the linguistic side of the calculus. As they point out, metaphors function "not only in literary language but also in language as such, and thus necessarily in the language of scholarship as well." Similarly, Daston and Gallison (1992) limit the productivity of their "working objects" to the arena of language when they define them as "any manageable, communal representation of the sector of nature under investigation" (85). Even Latour (1993) limits the action of his quasi objects and quasi subjects to the semiotic realm: "They are discursive. . . . They are narrated, historical, passionate . . . unstable and hazardous, existential, and never forget Being" (89). In short, each term allots slightly different values to the balance between human and non-human, material and semiotic, or natural and cultural attributes indicated. And while each term marks the boundary-straddling nature of these objects, each one also restricts the objects' operations ultimately to the linguistic realm.

This inadvertent acceptance of the opposition between the realm of the material and that of language is worth questioning. While I am, as a literature scholar, committed to an understanding of the way that language helps to structure our sense of possibilities, I am also increasingly aware of the way that material conditions shape and reshape what we can put into words. The liminal lives that I explore in the chapters to come call into question both the possibility of a discrete biomedical materiality that has not felt the shaping influence of literature, and science fiction in particular, and the notion of a literature that is not registering the generic and aesthetic impact of the new physiologies and life spans that medicine is making possible. In short, in what follows I focus specifically on the ambiguous cultural and medical meanings that these new beings hold for us, the new kinds of life stories and lives that are both enabled and articulated in our era's liminal lives.

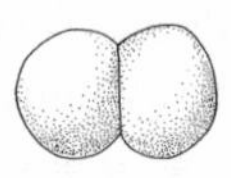

2

The Cultured Cell

LIFE AND DEATH AT STRANGEWAYS

I do not love thee, Dr. Fell
The reason why, I cannot tell
But this I know, and know full well
I do not love thee, Dr. Fell.
—Tom Brown

> The gray walls, black gowns, masks and hoods; the shining, twisted glass and pulsating colored fluids; the gleaming stainless steel, hidden steam jets, enclosed microscopes and huge witches' caldrons of the "great" laboratories of "tissue culture" have led far too many persons to consider cell culture too abstruse, recondite, and sacrosanct a field to be invaded by mere *hoi polloi*! . . . But every biology student should at some time have the dramatic experience of seeing the rhythmic beat of heart muscle, the sweep of the cilia of pulmonary epithelium, the twitching of skeletal muscle, the peristalsis of chorioallantoic or intestinal vesicles, the migration of fibroblasts, and the spread of nerve fibres. And every student can not only see these things but have the thrill of preparing them himself.[1]

"Transplant of Pig Hearts to Be Banned: Blow for Patients Awaiting Surgery" was the headline of the *London Times* I picked up at the news agent in Green Park. It was 16 January 1997, and I had come straight in from Heathrow, heading up to the hospice in Saint John's Wood where my friend was dying of end-stage lung cancer. The story took up all the top left of page 1: the British government was banning "pioneering surgery to transplant pig hearts to people after a government inquiry concluded that the procedure was too risky," reported the *Times*'s health correspondent.[2] The context for this decision was a "global organ donor crisis—at least 6,000 people are waiting for transplants in Britain and five times that number are on waiting lists in America."[3] Set into the story was a cartoon of an Argyle-sweatered man swinging from a light fixture; the twinset-clad woman in the foreground, holding a newspaper with the headline "Pig Heart Transplants Unsafe," explained to her male companion, "He opted for the monkey heart transplant instead." The government recommended the creation of an "interim authority"—a "regulatory body to control experiments," the article reported, noting that "a similar arrangement was made to control in-vitro fertilisation clinics in the 1980s before the Human Fertilisation and Embryology Act was passed in 1990." And now the in vitro fertilization (IVF) industry is in full swing in England, and perhaps xenotransplantation too is on its way to implementation, I thought.[4]

Over the next several days, I forgot the newspaper story as I sat with my friend in the hospice and as he died. But the story has come back to me now, with its tangled web of seemingly stable oppositions—birth and death, institutional power and personal choice, profit and gift, free will and bondage, first world and third world, nature and culture, human and animal—as the route in to my topic: the analysis of a biomedical technique, developed in the first three decades of the twentieth century, that has occupied a pivotal position in constructing our contemporary understanding of life and death. This technique is tissue culture.[5] As defined in *Animal Tissue into Humans*, the report of the Advisory Group on the Ethics of Xenotransplantation issued the day I arrived in London, tissue is "an organised aggregate of similar cells that perform a particular function . . . used to refer also to organs," and tissue culture is the technique of "isolating a small fragment of tissue from the body and cultivating it in a glass vessel containing suitable nutritive medium."[6] First successfully accomplished by Ross G. Harrison, who in 1907 published his discovery that amphibian nerve fiber could be made to live and grow in nutrient outside the body, the technique of tissue

"He opted for the monkey heart transplant instead." Cartoon by Pugh. Reprinted from the *Times* (London), 16 January 1997, 1.

culture was given additional impetus during World War I.[7] In the postwar period, Nobel Prize winner Alexis Carrel was one of the prominent pioneers of tissue culture. Working at the Rockefeller Institute in New York City, Carrel achieved worldwide publicity for his accomplishment in keeping a chicken heart beating in vitro for over a decade. With its extension in the development of organ culture by Honor Bridget Fell and Robert Robison in 1929, tissue culture laid the foundation for modern biomedical techniques ranging from in vitro fertilization to organ transplantation, including the animal-to-human transplantations whose press furor greeted me on my arrival in London.[8]

As it has developed, then, tissue culture exists at the nexus of two boundaries currently under renegotiation: the boundary of the species and the boundary of the human life span. Two forms of growth are typically investigated in tissue grown in vitro: organized growth and unorganized growth. Researchers have accessed them—with iconic overdetermination—by culturing the embryo and the cancer cell, potent images of life and death.[9] Since its earliest years, this Janus-faced technique has drawn on—and catalyzed—fantasies extending beyond the discrete set of cells in the laboratory culture. In lectures, scientific papers, popular science articles, and fiction, people have weighed the potential of tissue culture for the assault on aging, the prolongation of life, indeed the very fantasy of immortality. Tissue culture calls into question the definition of the individual, the boundaries of the body, the relations between species, and the authority of medical science.[10] Moreover, it also challenges the conventional or accepted structure of the human life span.[11] Cultured tissues, whether from animals or from human beings, play a crucial part in the broader twentieth-century project of reshaping the human life, in biomedicine (through interventions in conception, gestation, birth, aging, and death) and in literature and popular culture.

In the last decade, scholars have begun to hypothesize that the life course functions as a social institution, changing as other institutions have changed in response to Western processes of modernization.[12] Many now argue that the life span is "a discursive or imagined production, symbolic of a culture's beliefs about living and aging," and they understand the arc from birth to death not as merely natural but as the product of a culture's beliefs about human life and the shape of the human story (Katz 1996). However, such a constructivist position has its limits. Refuting "a view of the life course in which culture is granted the overarching power to mold nature in any form it chooses," these scholars hold instead that "human beings share with other species an embodied existence inevitably involving birth, growth, maturation and death" (Featherstone and Hepworth 1991, 375).

The biotechnology of tissue culture participates in this broader twentieth-century reconceptualization and reconstruction of the human life span. This reshaping of life is being accomplished, simultaneously, in biomedical science and in literature and popular culture, in ways I want to exemplify briefly. One kind of material and cultural reconfiguration of life is produced by biomedical practices, so that people are conceived differently, born differently, grow and live differently, and age and die differently. The new technique of cloning, or more precisely nuclear fusion, which has tissue culture as its foundation, has already provoked a rethinking of the notion of aging on the animal level that is certain to travel—even if the technology does not—to the human realm.[13] Another restructuring of human life is accomplished by fiction, science fiction, graphic fiction, popular science writing, and journalism, all part of a diverse set of inscription technologies that shape human beings to meet the needs of society. Through the vehicles of image, genre, character, and plot, literary and paraliterary texts help to set the boundary conditions for human life. Such representations help us to understand our experience as individuals, including our experience of the body. This process of identity construction shapes both the symbolic and the material realms as new kinds of life stories both catalyze and confirm new beginnings and endings.[14] A third kind of restructuring occurs at the boundary of the scientific and the literary, in the medical narratives that interweave materiality and representation, objectivity and interpretation.[15] "The journey from birth to death," Michael Holquist has observed, "serves as a biographical rhythm that entrains information into narrative so that it may be processed as meaning by men and women who are born and die."[16] Such an implicit structure lies behind even the case history and the illness narrative, so that diagnostic facts arrive already shaped invisibly by representation.[17] When the journey from birth to death is rerouted, lengthened, or curtailed, meaning too is changed. In each of these different settings (the scientific, the literary and paraliterary, and their intersection in various forms of medical writing), the practical and symbolic resources of creatures that border on the human (animals and human embryos and fetuses, as well as tissues cultured from them) are used to reshape that birth-to-death journey and thus redefine the human. This robs human beings of some old certainties and enables us to imagine new options.

If we analyze the (re)construction of the life course as carried out in symbolic process and in scientific practice, we can perhaps recover the fantasies, responses, fears, even the sense of possibilities, that may have been obscured or erased as we accepted almost without scrutiny this massive technoscientific and cultural reshaping of life. Moreover, when we transgress the disciplinary boundaries that limit knowledge, and study these acts of reconstruction in relation to each other, we can also illuminate literary and scientific practices. In what follows, I will continue this project of recovery and illumination by tracing some of the resonances of the new biotechnology of tissue culture (including its specifically literary resonances) within this larger modern and postmodern project of reshaping the course of human life. My focus is one specific site: the Strangeways Research Laboratory, in Cambridge, England.

Tissue culture came to Great Britain when Thomas Strangeways Pigg Strangeways made the technique the sole focus of the laboratory he founded.[18] Initially a research hospital investigating rheumatoid arthritis when it was established in 1905, by the early 1920s the laboratory abandoned the clinical medical component when Strangeways determined to focus on the microphysiology of disease, as revealed by this new technique he had learned from Alexis Carrel.[19] Tissue culture continued to be central to Strangeways Laboratory research throughout its founder's life. As the *Lancet* recounted in Strangeways's obituary, "By a curious irony the last lecture which he ever delivered was on Death and Immortality, and he illustrated it by a preparation bearing marvelous witness to his technical skill—a living culture of tissue made from a sausage purchased in the town."[20] With his death in 1926, the research hospital became the Strangeways Research Laboratory, funded by the Medical Research Council of Great Britain. Its directorship was assumed by twenty-eight-year-old Dr. Honor Bridget Fell, who, after completing a Ph.D. in Zoology at Edinburgh University, had studied tissue culture at Cambridge with Thomas Strangeways.[21]

The papers of the Strangeways Research Laboratory—in the 1920s and 1930s the only laboratory in the United Kingdom to concentrate exclusively on tissue culture—provide a flavor of the scientific, technical, cultural, and social issues this relatively new biotechnology raised.[22] Scientists dedicated themselves to exploring "the tissue-culture point of view," a phrase Dr. Honor Fell coined to refer to the scientific and popular mind-set bred by the micropractices of this new technique. Three different genres of documents register the impact of this new biotechnology on our understanding of the beginning and the end of life: (1) a series of lectures on tissue culture delivered by Honor Fell, (2) the press coverage of scientific experiments carried out at the Strangeways Laboratory, and (3) poems written by Strangeways researchers in response to their experimental work. I approach these documents with two questions in mind: How do they reflect the broad cultural and scientific impact of tissue culture within and beyond the laboratory? Is

Strangeways Laboratory, exterior. Reprinted by permission of Contemporary Medical Archives Centre, PP/HBF/F.15, Wellcome Institute for the History of Medicine, London.

Honor Bridget Fell.
Reprinted by permission from R. A. Peters, *History of the Strangeways Research Laboratory (formerly Cambridge Research Hospital), 1912–1962* (Cambridge: Heffer, 1962).

there a genre-linked difference in the way that impact is registered and articulated? As we begin to investigate the "tissue-culture point of view," we should hold in our minds a line from the nursery rhyme that has provided the epigraph for this essay: "I do not like thee, Dr. Fell." Although this poem originated nearly three centuries earlier, C. H. Waddington's daughter recalls reciting it as a child and assuming that it referred to her father's laboratory colleague. As her memory suggests, the point of view bred by the practices of tissue culture inspired both fascination and unease.[23]

A series of lectures Dr. Honor Fell called "Tissue Culture," which she delivered between 1936 and 1938 to vocational and postgraduate medical students, conveys the impact of the new technology on contemporary biomedical practices and cultural attitudes toward life and death. Dr. Fell is explicit about the tissue culture point of view, and her lectures set the context for the reception of the prototypical Strangeways tissue culture experiments, both within and beyond the laboratory.

The tissue culture point of view raises questions about—and catalyzes reconceptions of—the boundaries of death and life. "The tissue to be cultivated," Dr. Fell specifies dryly, "should be taken from the animal immediately after death. This, however, is not essential and tissue has been known to grow quite well in vitro when taken from the body as much as a week after death or considerably longer if the body has been kept in cold storage" (Fell 1936, 3). Perhaps to increase the drama—because this lecture is to vocational school students—Dr. Fell shifts registers as she explains the implications of this fact. She moves from the animal (for it is exclusively animal tissue that is being cultured at Strangeways) to the human: "From this we can see that when a doctor pronounces a patient 'dead' he is only using the word 'death' in a restricted sense" (3). Working to make her material accessible to a nonprofessional audience, Dr. Fell is explicit about the broader implications of tissue culture, both dramatizing and even popularizing them. "There is one very interesting and important result of this capacity for unorganized growth. . . . It makes *certain* tissues (at least) *potentially immortal*" (5).[24]

The tissue culture point of view has a conceptual and methodological scaling effect that links the human to the animal, the whole organism to the cell, and the mature to the embryonic.[25] "Scaling," or recursive symmetry, is the repetition of a pattern across different dimensions, in a graduated series.[26] Fell demonstrates this scaling event when, linking the microscopic with the human dimension, she observes, "cells in culture can be used as experimental animals. This is so true that we can even vivisect them by a kind of microsurgery."[27] The tone here is less popular and more technical, perhaps because she is addressing postgraduates rather than vocational students, and the language, which might provoke discomfort in the general public, invokes an insider audience comfortable with experimental procedures. But the invocation of vivisection also gestures to the world beyond the laboratory: Dr. Fell's inclusion of this point in a discussion of the "value of tissue culture in research" suggests that the modern trend toward miniaturization in experimental science may be, in part, a response to the late-nineteenth- and early-twentieth-century outcry against vivisection, so powerfully embodied in fiction in H. G. Wells's *Island of Dr. Moreau*, published only a decade earlier.[28]

Dr. Fell's assessment of the impact of tissue culture illustrates one sort of scaling effect: the classic scientific strategy of drawing analogies between human and animal. Indeed, because the cells that she cultures are both embryonic and mature, her scaling effect incorporates two different dimensions: the evolutionary and the (human) developmental. Strangeways cultures both embryonic and mature cells, and the cells thus cultured can stand in, Dr. Fell observes, for laboratory animals.[29] She cites this as one of the advantages of the tissue culture method: "We are . . . able to use cells as experimental animals and subject them to the influence of various experimental agents or to changes in their environment" (Fell 1937, 6).

The tissue culture point of view resituates the "body of science" from the realm of the static, graphic, and dead to that of the dynamic, photographic, and living. This shift is accomplished through visual images that resemble surveillance photography or unposed and spontaneous snapshots. The remarkable thing about tissue culture, Dr. Fell explains, is that "cells growing out of the body in this way can be watched and photographed whilst going about their ordinary business." The technique has changed the study of cell biology, she observes: "Originally histology was one of the most static of sciences: tissue culture has made it dynamic" (Fell 1936, 1).[30]

Tissue culture point of view appeals to scientists as a particularly glamorous or romantic technique, an appeal that tends to enroll more practitioners. Dr. Fell admits to her insider audience that this can be one of the drawbacks to tissue culture research: "Tissue culture often suffers from its admirers. There is something rather romantic about the idea of taking living cells out of the body and watching them living and moving in a glass vessel, like a boy watching captive tadpoles in a jar. And this has led imaginative people to express most extravagant claims about tissue culture, which our actual experience in the laboratory quite fails to justify" (Fell 1937, 5).[31] Tissue culture may be used when another method would work just as well or even better, she observes, due perhaps "to a subconscious desire to use a spectacular method in preference to a simple one" (Fell 1937, 11). Yet if Dr. Fell admits that the associations clustering around tissue culture and the subconscious desires of researchers may lead to the overuse and overestimation of this research technique, she also risks contributing to that problem in her lecture to the vocational students, when she frames the implications of the technique broadly, without the qualifications that she is careful to make in her lecture to postgraduate medical students.

The tissue culture point of view is premised on decontextualization and abstraction. Unexamined analogies from the cellular to the human, while a tempting part of the tissue culture point of view, should be resisted, Fell argues. "Results obtained from experiments and observations made on cells growing under the abnormal conditions obtaining in vitro are very possibly not applicable to cells living in their normal environment in the body, because cells in vitro may be physiologically modified and therefore not comparable with normal cells in vivo" (Fell 1937, 8). By its procedure of isolating tissues in a culture medium, the tissue culture method reduces the complex interplay of variables shaping any cell in vivo, and thus any findings should be received with caution. "One of the greatest advantages of tissue culture vis., the simplification of the environmental conditions of the cells, is at the same time one of its greatest limitations" (11). Dr. Fell's observation anticipates the position of some contemporary feminist critics of science, which has focused on the methodological limitations of laboratory science in general, in particular its strategy of contextual narrowing.[32]

If we consider how the prototypical Strangeways experiments were represented (and received) by the scientists within the laboratory and the nonscientists beyond it, we can see how the tissue culture point of view generated scientific, technical, cultural, and social issues specific to this biotechnology. Three kinds of experiments and one specific research methodology typify the laboratory's approach to studying tissue culture in the 1930s, embodying what Dr. Fell called the "tissue-culture point of view."[33]

Embryo culture, or the study of the developmental center of mammalian and avian embryos by grafting in vitro, was best exemplified at Strangeways by the work of C. H. Waddington. He provided "the first demonstration of the activity of organization centres in mammals" by grafting pieces of the primitive streak into the neural centers of rabbit and chick embryos.[34] As Waddington described his work in the *Listener* in 1936: "I found that the very young [chick] embryo would stay alive for a time if one put it on the surface of a particular sort of nutritious jelly. It starts on the jelly as a plain flat disc of cells, and it stays alive for about two days, and develops the beginning of a brain and eyes and ears and a heart which has begun to beat. It has an organiser, too. . . . In fact, probably all backboned animals have organisers in their eggs. Presumably man has. You yourself owe your brain to the working of your organiser."[35]

Waddington's explanatory passage demonstrates the scaling effect here, characteristic of the tissue culture point of view, as he draws a parallel between embryo and mature organism, animal and human. His research also embodied the double investigative focus of tissue culture research, both on the sources of life and on the causes of death. As he observed in a letter to *Nature* in 1935, "I was . . . originally responsible for the suggestion that work on embryonic organizers may throw some light on the induction of malignant growth."[36]

Organ culture, or the growth of embryonic body parts in vitro, was a major part of the laboratory's activities. Dr. Fell saw this as the center of her career.[37] Researchers at Strangeways worked on the culture of a range of organs, including ears, mammary glands, ovaries, salivary glands, the pancreas, hair, and teeth.[38] Among the experiments, one in particular garnered publicity: the culture of embryonic rats' teeth, by Miss S. Glasstone. Press coverage emphasized the prominent role played by women in the development of this new technique. As the *New York Herald* reported, "The first recorded instance of 'growing teeth' outside the body is contained in the annual report of the Medical Research Council issued here [London, 12 March 1936]. The achievement

Strangeways Laboratory, interior. Reprinted by permission of Contemporary Medical Archives Centre, PP/HBF/F.15, Wellcome Institute for the History of Medicine, London.

Staff and visitors, Cambridge Research Hospital, 1924. *Back row*, J. A. Andrews, H. B. Fell, V. C. Norfield, J. G. H. Frew; *front row*, F. G. Spear, T. S. P. Strangeways, R. Chambers, R. G. Canti. Reprinted by permission of Contemporary Medical Archives Centre, PP/HBF/F.15, Wellcome Institute for the History of Medicine, London.

Cartoon of research mission at Strangeways, from "The Begging Book," a financial appeal for building larger premises. Reprinted by permission of Contemporary Medical Archives Centre, PP/FGS/C.2, Wellcome Institute for the History of Medicine, London.

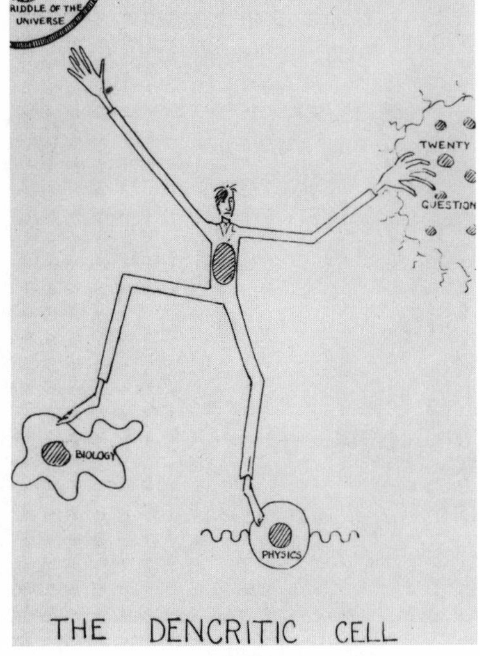

"The Dencritic Cell" (caricature of Francis Crick). Reprinted by permission of Contemporary Medical Archives Centre, PP/FGS/C.29, Wellcome Institute for the History of Medicine, London.

stands to the credit of the Strangeways Research Laboratory at Cambridge, and is regarded as an outstanding illustration of the power of the new technique of 'tissue culture.' . . . A woman, Dr. Honor B. Fell, is at the head of the small band of scientists responsible for the work, and Miss. S. Glasstone has been responsible for the laboratory's latest success."[39]

With women scientists often at the helm, the field of tissue culture seems to have been characterized by heightened gender awareness and gender play. This may explain the intriguing boundary crossing evident in a mock ceremony held at Strangeways Laboratory to mark the opening of the new x-ray department on 4 March 1937. The ceremony featured the visit of "Lady Le Bec," to present the key of the new department to Dr. Fell. Three photographs of Lady Le Bec survive in the Strangeways archive. A tall man in a white hospital gown who bears a striking resemblance to Strangeways researcher Lancelot Hogben, Lady Le Bec wears a wig, a flower-bedecked hat, and much rouge and lipstick.[40] For the male scientists, the act of growing tissue in special cultures may have seemed a humorous appropriation of a feminine, nurturant, or maternal function: they imagined themselves as patronesses or Ladies Bountiful like Lady Le Bec, or gardeners (so Arthur Hughes's image of Martinovitch culturing testicles and ovaries is shadowed, as we will see later, by the poetic question "Mary, Mary, Quite Contrary / How does your garden grow?"). Or they reimagined the task of tissue culture in more stereotypically masculine terms, as Martinovitch does when he thinks of himself as "a certain lad" sitting, Tom Sawyer–like, on the riverbank, watching the river flow. In contrast, for the women scientists, the practice of tissue culture may have seemed worthy of note precisely because of its divergence from the embodied gestational norm, as a method of "unnatural" growth.[41]

Finally, Strangeways scientists engaged in cancer research, studying unorganized cell growth and the biological effects of x-rays.[42] This research area, too, exhibits the tissue culture point of view, in its reconceptualization of life and death. While researchers working on embryo culture addressed the beginning of life, those working with wound tissue, culturing entire organs, or studying the effects of radiation on cancer cells turned their attention to the latter end of the life course, working on projects that would make significant contributions to the treatment of cancer and wartime injuries.[43]

"Lady Le Bec" and Honor Fell at the Strangeways Research Laboratory, 4 March 1937.

"Lady Le Bec" opens the X-ray Room at Strangeways, 4 March 1937.

Both photographs from the Contemporary Medical Archives Centre, PP/FGS/C18, Wellcome Institute for the History of Medicine, London. Courtesy of the Wellcome Library.

When reported in the popular press, these three representative experiments in tissue culture provided exactly the romanticized and technologically glamorous spectacle of which Honor Fell warned in her lectures. Two examples will give the flavor. In 1937, in the London *Daily Mirror*'s column "Science This Week," Harrison Hardy reported on Petar N. Martinovitch's experiments in ovary culture in vitro:

> The possibility that at some remote future time babies may be grown in bottles is an idea which cautious scientific men leave to the writer of fantasies. Though bottle babies may never be a fact, science moves slowly in their general direction. Mr. Petar N. Martinovitch, a biologist working in the Strangeways Research Laboratory, Cambridge, is growing germ-glands on watch glasses. From newborn mice and unborn rats he cuts out the sexual glands and plants them in clots of mixed fowl plasma (blood extract) and fowl embryo extract. *Female glands remained healthy after twenty-two days, male glands after seventeen days*. Mr. Martinovitch, being a serious biologist, does not talk about bottle babies, but says the results may "provide a new experimental approach to problems of sex physiology."[44]

The article reflects a familiar tension between scientific and popular discourses—the careful assessment of the scientist, and the futurological speculations of a "writer of fantasies"—while also engaging in a remarkable amount of translation and education on the methods of tissue culture.

Another article, Norah Burke's "Could You *Love* a Chemical Baby?" appearing the following year in the tabloid *Tit-Bits*, dispensed with the scientific education in favor of a futurological and fantastical presentation. Claiming that the worldwide alarm about a falling birth rate may be solved "in a few years time by the production of living babies from scientists' laboratories," this article provides a lurid survey of tissue culture. It highlights the accomplishments of a number of researchers: an unnamed British woman doctor who "has succeeded in growing teeth in a test tube," Alexis Carrel (who "has kept human hearts alive without their bodies, and bodies alive without their hearts"), Gregory Pincus (a "renowned biologist [who] has 'created' living rabbits"), and "a Russian eye-specialist [who] has kept a human eye alive for six days in an air-tight jar in an ice chest." More is to come, the article implies: "Already other parts of the human body have been grown in test-

April 16, 1936 — TIT-BITS — 3

COULD YOU LOVE a CHEMICAL BABY?

By NORAH BURKE

For That's What Science Looks Like Producing Next

WHAT! MAKE A LITTLE CHAP LIKE ME?

THERE is world alarm about the falling birth-rate. Yet the whole problem may be solved in a few years' time by the production of living babies from scientists' laboratories.

IN Russia human blood has been canned and stored for transfusions. Dr. Friederich Gottendenker, of Vienna, claims to have produced an artificial substitute for human blood. In this country a woman doctor has succeeded in growing teeth in a test-tube. Alexis Carrel has kept human hearts alive without their bodies, and bodies alive without their hearts. The diseased heart is temporarily replaced by an artificial pump, repaired, and then put back! There are people living to-day with new kidneys transplanted by Alexis Carrel. And, perhaps most amazing of all, Dr. Gregory Pincus, renowned biologist, has "created" living rabbits.

Living Without a Heart

In Russian hospitals, human blood is kept in labelled receptacles, at blood heat, marked according to its group. This bottled blood has been sent all over the Soviet Union by rail, car and 'plane. In America experiments on similar lines are being made with dogs' blood, but it is kept frozen. Suitable human blood can be obtained from a dead body, provided the blood is not affected by disease.

Canned blood, both real and artificial, will soon be available to save lives. Will it be used to create them?

Just before the 1914-18 war Alexis Carrel was ridiculed in medical papers and denounced from the pulpit because he suggested that it should be possible to take a human heart and keep it alive and beating outside the body.

But Alexis Carrel was no mere sensationalist. To begin with, he was a skilled surgeon, who had worked in Paris and at the Rockefeller Institute for Medical Research in New York. As a young medical student he employed his free time in doing exercises to make his fingers strong and supple. With two fingers he could tie a piece of surgical catgut into such a knot that other people could not untie it with both hands. With a length of gossamer silk thread, he could sew across a cigarette paper so the stitches did not show on the other side!

Carrel analysed chicken's blood, and copied it with a fluid made from chemicals. Then he took a piece of heart from an unborn chicken and preserved it in the synthetic blood. Just then war broke out, and Carrel went to France.

When he returned, five years later, the chicken heart was still alive and growing! His assistants had been obliged to cut it down, but that piece of tissue is still alive to-day. And had it been allowed to grow on would by now have become as big as the Alps!

Soon Carrel was keeping human hearts alive—the only problem now was to invent an artificial heart to insert in a body while the real one was being repaired or replaced. With the help of Charles Lindbergh, the airman, the artificial heart was invented, and has since been used several times, working successfully.

Carrel foresees, in the near future, a time when healthy hearts, kidneys, stomachs and other parts of the body will be kept alive and labelled, ready for insertion into anyone who may need them. A Russian eye-specialist has kept a human eye alive for six days in an air-tight jar in an ice-chest, at a temperature of not more than 6 degs. C. Already other parts of the human body have been grown in test-tubes.

No Need to Breathe!

Perhaps in a few years, when we have a tooth out, it will be replaced by a living synthetic tooth. Some years ago the late Dr. T. S. P. Strangeways came to the decision that he could make little real progress in his investigations of the crippling disease osteo-arthritis until more knowledge was available of the basic principles of growth. This was the beginning of an unusual speciality of the Strangeways Research Laboratory at Cambridge, where the new technique of "tissue culture" is carried out. Dr. Honor B. Fell, a woman, is at the head of the work, and another woman, Miss S. Glasstone, has been responsible for the test-tube teeth.

Minute proportions of embryonic rats' teeth were used as nuclei for synthetic teeth, which began to grow plentiful dentine. The characteristics of natural growth began to show: for instance, a deficiency of vitamin C became apparent in the same way as in normal teeth.

Tissue culture has stupendous possibilities. Already it has been shown that the healing of a test-tube wound produces normal scar-tissue, as in real life.

Another woman investigator, Miss Fischmann, has shown that bone growth left in a culture deficient in vitamin D for even so short a period as forty-eight hours is permanently affected.

It may not be long before life is possible without breathing! Test-tube human beings may be produced fully grown, by having oxygen injected direct into their blood until the lungs are ready to work.

But, with all the tissue culture in the world, with all the canned blood and preserved organs, science always has to start with a piece of living material. Is there any chance that we shall probe the ultimate secret and learn to create life? Judging by recent developments, it does seem possible.

An End to Humanity?

Dr. J. D. Bernal, Dr. I. Fankuchen, and their collaborators, of the Crystallographic Laboratory in Cambridge, have announced a tremendous discovery. They proved that a disease-producing virus affecting the tobacco plant—to conventional ideas a living thing—could be extracted from the leaves of the plant as a chemical. Have these men mapped the first great landmark in the realms where life and chemistry are one?

Then there is Dr. Gregory Pincus, renowned biologist, who has "created" living rabbits. He obtained the release of Hormone Prolan A from the pituitary gland of a doe rabbit by electrically stimulating a nerve centre in the creature's neck. Then, extracting an egg cell, he exposed it for a few minutes to a temperature of 113 degs. F. The egg, thus artificially fertilized, was transmitted to another doe rabbit, which brought it to birth.

What are the gigantic realms of discovery and of human life in the future to which these experiments may lead us? Our minds may well stand aghast and yet fascinated by these overwhelming possibilities. It may be that the human species will produce test-tube babies, or full-grown human beings, entirely chemically.

What sort of creatures will these be? All the rabbits "created" by Dr. Pincus were female. Will the test-tube human beings be unisexular, or sexless? Will the human race become like the social insects with three genders—male, female and neuter (or workers)? Will these sexless, soulless creatures of chemistry conquer the true human beings? Will our size be violently altered? Remember that chicken heart that went on growing and growing . . . .

The popular press response to the Strangeways experiments. Norah Burke, "Could You *Love* a Chemical Baby?" *Tit-Bits*, 16 April 1936, 3.

tubes."[45] Here again, the popular press emphasizes the prominence of women in the Strangeways administrative and research pool. Remarking on the "unusual specialty of the Strangeways Research Laboratory at Cambridge, where the new technique of 'tissue culture' is carried out," the article notes that "Dr. Honor B. Fell, a woman, is at the head of the work, and another woman, Miss S. Glasstone, has been responsible for the test-tube teeth."[46]

Commenting that "tissue culture has stupendous possibilities," Burke goes on to speculate: "It may not be long before life is possible without breathing! Test-tube human beings may be produced fully grown, by having oxygen injected directly into their blood until the lungs are ready to work" (3). The passage anticipates Jep Powell's short story

"The Synthetic Woman," to appear two years later in the pioneer science fiction magazine *Amazing Stories*, in which a woman is synthesized chemically and grown to full physical maturity in the laboratory, with the help of injections of "oxydyne."[47] Yet if Burke's article anticipates the science fiction genre, it also illustrates the characteristic gender tensions produced by the tissue culture point of view:

> What are the gigantic realms of discovery and of human life in the future to which these experiments may lead us? . . . It may be that the human species will produce test-tube babies, or full-grown human beings, entirely chemically.
>
> *What sort of creatures will these be? All the rabbits "created" by Dr. Pincus were female. Will the test-tube human beings be unisexular, or sexless? Will the human race become like the social insects with three genders—male, female and neuter (or workers)? Will these sexless, soulless creatures of chemistry conquer the true human beings? Will our size be violently altered? Remember that chicken heart that went on growing and growing . . .* (3)

While these newspaper stories demonstrate the broad popular interest catalyzed by the experiments in embryo and organ culture carried on at Strangeways, a different sort of publicity greeted the most significant research technique developed there: the "Canti method," as it was dubbed by the popular press. This was a method of cinema photomicrography developed by Dr. Ronald George Canti, a researcher who worked on "normal and malignant cells and their response to radiation." Canti was "the man who first turned the cinema camera in this country into an instrument of medical research," as the *Evening Standard* put it in Canti's obituary.[48] Dr. Canti, who had studied with Alexis Carrel at the Rockefeller Institute in New York, found himself frustrated—so the story goes—by the slow, clumsy techniques for culture filming used by Alessandro Fabbri, of Carrel's lab. In response, Canti devised an apparatus for photographing cell cultures in vitro, using what we would now call time-lapse photography.[49] Canti's films, with titles such as "Cultivation of the Whole Chick Embryo Femur *in Vitro*," "Beating and Fibrillation in the Chick Embryo Heart *in Vitro*," and "Investigation of the Normal and Parthenogenetic Division of the Rabbit Ovum," made in collaboration with C. H. Waddington, Honor Fell, and Gregory Pincus, captured the popular imagination.[50] The films were shown at 10 Downing Street, on television, and at scientific congresses throughout Europe.

The Canti technique played a central role in producing the tissue culture point of view, shifting science from the static and graphic to the dynamic and photographic. As a press report on the public screening of one of his films reveals, Canti's films had great rhetorical power, making scientific research appealing to the public. They may also have embodied some interesting gender dynamics, enacting and thus satisfying a fantasy of patrilineal control at the microscopic level. In June 1921, in its coverage of one of Canti's film screenings, the *World* reported:

> How often have you heard a proud mother exclaim: "I can see my boy grow."
>
> Of course she cannot see her boy grow. She sees him every day . . . and she cannot see him grow any more than she or anybody else can see a plant grow. A rank weed may spring up with wondrous speed, but no human being can see it grow. . . .
>
> If she could take one picture of her son each day and then show them in the succession in which they were taken and at the rate of sixteen a second, she could see her son's growth and so could everyone who looked at the pictures. *But she and they would see him grow much faster than he does really; so fast that if his growth were as rapid as appears on the screen he would be able to hitch his toy wagon to a star long before he was as old as Carrel's chick heart.*[51]

As this coverage indicates, Canti's films harnessed a perennial human obsession (human growth, emblematized in the growth of one's own children) to a specific set of scientific questions characteristic of tissue culture research: the growth of normal and abnormal cells. They also build on the recent desire to have not still photographs but a cinematic record of growth.

Further, the films drew on the public outcry caused by Alexis Carrel's accomplishment in the long-term culture of the chicken heart. Dr. Fell's tissue culture lectures, with which I began my discussion of the Strangeways documents, also use Carrel's experiment to capture the implications of tissue culture. However, for her, the effect is a disruption of normal size and mass, not of time: "[A] strain of heart fibroblast has been kept alive by Ebeling and Carrel for over 20 years. It has been calculated that if all the possible descendants of this patriarchal culture were alive today, the total bulk of tissue would vastly exceed the volume of the Earth" (Fell 1936, 5). Significantly, both the *World* columnist and Dr. Fell conceptualize the long-lived tissue as being male, whether a mother's beloved son, or a patriarchal culture with its descendants.[52]

The coverage in the *World* in 1936 not only linked cellular and embryonic growth to the development from child to adult but also related technological development to the wished-for human "development" that would lead to immortality. In 1938 a film collaboration between R. G. Canti and C. H. Waddington was given a public showing that catalyzed both sets of hopes: "The time may come when, by means of television, classes all over the country may be able to watch laboratory experiments more difficult and intricate than could be carried out under ordinary school conditions. In this category come the 'Artificial Immortality' tests to be conducted before the television cameras in the evening transmission . . . when Mr. C. H. Waddington, of the Strangeways Laboratory, Cambridge, will show how tissues can be kept alive indefinitely."[53]

The embryo culture to be televised could not, of course, grow fast enough for television. The new televisual technology needed the help of its more established sibling, and "a film [was] transmitted which was taken during a previous experiment at the rate of one picture a minute. When these [were] projected at the normal rate of twenty-four frames a second, the culture [could] be seen growing" ("The Seer," n.p.). In effect, viewers of the film experienced two new cultures growing parallel to each other as the film's portrait of nutrient-bathed animal tissues was transmitted by the relatively new technology of television. As press coverage revealed, this novel medium fed both the futurological curiosity of the general public and the stodgy scopophilia of the more cautious scientists. "What we are missing by having to wait for television may be gauged from the following advance description of an Alexandra Palace studio item for to-night under the title of 'Artificial Immortality' in the 'Experiments in Science' series. How a small piece of animal flesh can be kept alive indefinitely will be demonstrated in an unusual television feature. . . . Science is less interested in achieving the 'artificial immortality' than in obtaining an opportunity by these means, of watching cells while they are alive. In the living body, it is practically impossible to watch them closely. Still, why shouldn't our biologists have a night out, now and again?"[54]

If the public responded to the Strangeways experimentation with a mixture of excitement, mockery, and alarm at the ways it destabilized human identity, the life span, and human gender relations, how did the participating scientists view their activities? The official Strangeways position was one of a carefully limited statement of goals and implications: the researchers are interested only in achieving information about the behavior of certain specific cells, the problems of sex physiology, et cetera. But we see something rather different if we examine communications not from the lab to the "outside world" but between scientists within the laboratory. Three poems composed by researchers at Strangeways, whose principal interests were the three central foci of Strangeways research (embryonic development, organ culture, and processes of disorganized or cancerous growth), reveal the internal responses to the tissue culture point of view.[55]

We can begin with a poem that the embryologist C. H. Waddington composed in January 1937. This poem concerns the organizer, the element in the embryo that controls its development for the first fourteen days. As Waddington explained the concept in the same year in *The Listener*, "The actual experiments I do nearly all involve grafting little bits of the egg from one place to another. At this stage the egg has begun to develop but has hardly yet got any organs at all. It is still a round ball, and it is just beginning to fold in from the surface a sort of tube which will later turn into the gut or intestine. The region round about the edge of this tube, where it folds in from the surface, is the most important part of the egg. It is not only the place where the first organ, which is the gut, begins to appear, but it also controls the formation of all the other organs as well."[56]

Waddington's poem explores the principle of the organizer from the inside, considering, as it were, how the egg views its changed potential. Though focused on that most microscopic of protagonists, the fertilized ovum, this poem participates in a genre of visual and verbal fetal representations that move from the microcosmic to the macrocosmic, portraying the fetus as the product of social as well as scientific processes.[57] So comfortable was Waddington in making the leap from biology to politics, with its implied social constructivist viewpoint, that in a later autobiographical essay, he went even further, arguing that his particular scientific bent was shaped by an underlying metaphysic.[58] The poem begins with a celebration of the laziness of the fertilized egg:

Happy the egg, the bubble blastula,
does nothing much, but all the same,
each part as good as every other
or would be if it could,
but better its situation;
the opportunist roof, the sodden floor,
with a space between:
a capital notion, but a lazy affair.

Moving from the collectivist happiness of the "bubble blastula," the egg develops until "every separate part is tied / to particular performance but still within itself is free / To organise its own affair," and the fertilized ovum now seems to be an independent citizen possessed of autonomy and rights. In its movement from consciousness of individual freedom to functional specialization, Waddington's poem seems a miniature recap of the development of the Enlightenment individual from collectivity into class hierarchy, economic segmentation, and professionalization:

Feeling, thinking, feeding come
by limiting this freedom
of each being any,
now each for all
performs its special function.

The developing blastocyst moves from collective identity to individual functional identity and (through that) to agency. Identifying with the developing embryo, the poem affirms its change as progress.[59]

Now every separate part is tied
to particular performance
but still within itself is free
to organise its own affair

Finally, the functionally specialized embryo has attained not only economic and civic autonomy but psychic agency: "a special place among the few . . . exchanged / the possibility to be / for the act to do."

Two more poems, written on the same day in January 1938 by Strangeways researchers Petar Martinovitch, who was growing ovaries in vitro, and Arthur Hughes, who was studying vascularization in the embryonic chick, extend this identification with the objects of scientific study. Hughes instigated the exchange of poetry, writing this poem describing Martinovitch's work culturing mammalian "testicles and ovaries":

> Testicles and ovaries
> Explanted in a row
> Grown by Martinovitch
> In vitro.

The comic strategy of the Hughes poem puts the process of tissue culture in context, both institutional and personal. We seem to "pan" the laboratory, looking for evidence of growth, our vision moving from the rows of testicles and ovary explants in vitro to the other sites of development: volumes in the library; sexualized nurses at the hospital (who all wonder "should they know / About sexual maturation / In vitro"); and researchers who look forward to Martinovitch's next breakthrough ("Colleagues in the tea room / Wait for him below / To announce the new advances / In vitro"). Finally the poem rounds on itself, suggesting that tissue culture reveals something not only about the cells but about their researcher. Martinovitch is another such tricky life-form in culture, hard to get to meals, prompting his landladies to ponder "if they couldn't nourish him / In vitro." Humorously flouting the scientist's public posture of careful epistemological containment, the poem illustrates the scaling effect crucial to the tissue culture point of view, linking the cellular to the human, the laboratory experiment to its context, and sexual development to scientific and social development. Moreover, as it takes Martinovitch step-by-step out of the laboratory, it also exemplifies the tissue culture strategy of abstraction and decontextualization.

Martinovitch's poem, "A Reply," lacks the metric smoothness of Hughes's effort, but it too articulates the tissue culture point of view the Strangeways Research Laboratory produced.

1938
Jan.

Testicles and ovaries
Explanted in a row
Grown by Martinovitch
In vitro

Volumes at the libraries
See how they show
Progress in his researches
In vitro.

Copy.

Nurses at the hospital
Wonder should they know
About sexual maturation
In vitro

Colleagues in the tea room
Wait for him below
To announce the new advances
In vitro

Landladies keep pondering
Meals ready hours ago
If they couldn't nourish him
In vitro

AJWH
to P.M.

Poem by Arthur Hughes. Contemporary Medical Archives Centre, PP/FGS/C19, Wellcome Institute for the History of Medicine, London. Photograph courtesy of the Wellcome Library.

Jan 1938
(Same day)

P. Martinovitch
to A.S.W.H.

A Reply

At the Strangeways lab a certain lad
Sits on the bank (always left) of a blood vessel and follows:
Myriads of little creatures
Of similar birth, but different features
Carried by a stream of swift motion,
With powerful sweep and great commotion –
to their unknown destiny.

At the moment,
His chief concern is to show
What causes the overflow.
(Of course, you must be familiar with his previous inti-
mation [as to]
What might be the effect, if not the cause of – obliteration).

But, to make the story immortal,
The show must be filmed:
With cameras of special make
With a little twist and a little shake
With the trick A and the trick B

"A Reply," poem by Petar Martinovich.
Contemporary Medical Archives Centre, PP / FGS / C19, Wellcome
Institute for the History of Medicine, London. Photographs
courtesy of the Wellcome Library.

And behold! What do we see?
We see exactly what we have related before,
And there is no use of seeing more.

The story of circulation is an old information;
But, never before,-
This lad does not hesitate to tell,-
Has it been followed so well.

The motto of the undertaking
Of motion pictures making
Is: You must tell a tale!
So, of late, an attempt has been made
By chromosome X and his mate,
(The latter in its non-existant state),
To imitate
Micky Mouse and his gait.

If you like to see the show,-
You need not spend your dough,-
Come, and you will be told,
Precisely, what you wish to know.

At the Strangeways lab a certain lad
Sits on the bank (always left) of a blood vessel and follows;
Myriads of little creatures
Of similar birth, but different features
Carried by a stream of swift motion,
With powerful sweep and great commotion—
to their unknown destiny.

The poem reflects the double orientation toward life and death characteristic of Strangeways research. In jerky rhythms the poem identifies Arthur Hughes's research goal: through studying the circulation of the blood, to discover the "effect, if not the cause of—obliteration." The poem also indicates Hughes's method, in which the cinema played a central part; Hughes was "one of the pioneers in adapting phase-contrast cinematography to the study of cellular processes."[60] Finally, the poem's description of Hughes's work takes into account not only its scientific results but its broader popular implications. In contrast to the careful rejection of any talk of immortality when Strangeways scientists responded to public questions, the poem suggests that these scientists are hoping for professional scientific immortality: the perpetual survival of the narrative they create as scientists.

But, to make the story immortal,
The show must be filmed:
With cameras of special make
With a little twist and a little shake
With the trick A and the trick B
And behold! What do we see?
We see exactly what we have related before,
And there is no use of seeing more.

Cynicism marks Martinovitch's position here. The poem implies that the new scientific imaging that the Canti method made possible does not advance scientific knowledge ("the story of circulation is an old information"); it merely improves its communication to the public: "never before— / This lad does not hesitate to tell,— / Has it been followed so well." The ambiguity in those lines is worth pausing over. Do they emphasize Hughes's egotism, in not hesitating to tell of his success? And who is it who is described as following the narrative so well: the camera that pays such close attention to the scientific processes whose story it "tells," or the lay audience who are able to "follow" the old story of circulation better than ever before? The poem pauses, in the ambiguous gap between scientific self-aggrandizement and improved public education and communication, to consider the effect on scientific practice of a new set of rules: the disciplinary paradigms of the cinema.

The motto of the undertaking
Of motion pictures making
Is: You must tell a tale!
So, of late, an attempt has been made
By chromosome X and his mate,
(the latter in its non-existant state),
To imitate
Mickey Mouse and his gait.

Martinovitch's poetic point is worth repeating. A shift has recently occurred toward narrative in the sciences, he suggests, because of the increasing influence of the performance paradigms of cinema.[61] Not only have the objects of scientific study become subjects, but they have taken on the qualities of a celebrated and highly commodified subject: Disney's Mickey Mouse. With its allusion to *Steamboat Willie*, the famous inaugural cartoon that follows Captain Mickey in his steamboat down the Mississippi River, Martinovitch's poem grants science the power to generate a fresh kind of value: not only epistemological but commercial. This new value is premised not on originality but on (cinematic) replication.[62] We have thus shifted from the microscopic to the human, from the river of blood in an embryonic chick to the larger-than-(human)-life scale of the Mississippi.

Locus of agency has also relocated. In the course of the poem, we have moved from the lad, whose concerns and goals are the focus of the first two stanzas, to the cinematographer who "must" film the story, and whose camera twists, shakes, and tricks preoccupy the third stanza, to the viewer and producer of the film, whose skill at reception and projection is the subject of stanzas 4 and 5, and finally, in stanza 6, to the anthropomorphic chromosomes X and Y ("his" mate, curiously), who try to imitate another anthropomorphic creature: the cinema icon

Mickey Mouse. The poem closes with a summary of the effect of the turn to cinematography on the production and consumption of scientific knowledge: "If you like to see the show,— / You need not spend your dough,— / Come, and you will be told, / Precisely, what you wish to know." Again the ambiguities are multiple. Research at Strangeways was funded by nine organizations—including the Medical Research Council, the Royal Society, the British Empire Cancer Campaign, the Rockefeller Foundation, and the Wellcome Trust—and one wonders how much this prominent list of institutional sponsors colors the line "You need not spend *your* dough." And then there is the curious matter of the concluding couplet: "Come, and you will be told / Precisely, what you wish to know." What does "precisely" modify—the way of telling, or the content of what is told? Will the audience be told precisely, because this is science speaking, after all? Or does the cinema's entertainment mission drive the transfer of knowledge, so that the audience for this little film will be told "precisely, what [they] wish to know [and no more, no less]"? The passage links scientific popularization, cinematography, and what some Strangeways researchers believed to be a particularly American form of intellectual pandering.[63]

What are the implications of the tissue culture point of view, as explored in these three genres of documents? Honor B. Fell's lectures suggest that scientists are not always quite so monolithically concerned with control, management, and alienation of the embodied being—whether human or animal—as has at times been argued by the journalists who were their contemporaries or by contemporary cultural critics of science.[64] The parascientific writings on tissue culture of Dr. Fell make clear that there was a fairly high level of self-consciousness about the distortions of tissue culture, its excisions, and the links between (rather than replacement by) cells, organs, animals, and the human body. Fell's lectures also make it clear that Strangeways researchers understood the turn to tissue culture as an alternative to animal vivisection. As she put it: "Cells in culture can be used as experimental animals. . . . We can even vivisect them by a kind of microsurgery" (Fell 1937, 10). This is in vivid contrast to the "virtual test tube *in vivo*": the rabbit-ear chamber that Eliot and Elinor Clarke developed in the United States, which, rather than abstracting and decontextualizing tissue, held the rabbit captive to the viewing cylinder surgically inserted into its sensitive, translucent ear.[65] The writings about tissue culture reveal a tendency to identify with the tissue cultures as subjects rather than objects of study. In several documents, Strangeways scientists conceptualize cells as having an agency and autonomy. Fell portrays cells "going about their ordinary business" while the tissue culturist watches and photographs them; Waddington tells the story of earliest embryonic development as a condensed survey of the growth of the functionally specialized Enlightenment individual; Hughes draws parallels between the problems of feeding gonads in tissue culture and the problems of feeding the sexually and scientifically developing scientist; and Martinovitch imagines the chromosomes filmed under the microscope as performers like Mickey Mouse. As these conceptions of the cultured tissues suggest, there may have been ethical or psychological, as well as practical, reasons for preferring this method of investigation through decontextualization.[66] Finally, the poems reveal a more cynical, pragmatic, and prescient response to cinema photomicrography than the credulous embrace of the technoscientific gaze that recent scholarship has emphasized.[67]

What do we learn by comparing these different reception sites for the complex new technique of tissue culture? We have seen that distinctly different, and valuable, insights can occur when the practices of science are confronted in the discursive medium of literature. Poetry provides positions for identification that enable scientists to think against the grain of their scientific work. They can identify with the corpuscles they follow through the bloodstream, the tissues they culture, even the developing blastula. In addition, the generic demands of poetry for a concise articulation, and resolution, of tension provide the incentive for counterdisciplinary truth telling. Waddington identifies across gender to speak from the position of the ovum, articulating a nostalgia for the undemanding passive position ("to be"), if ultimately affirming the move toward agency ("to do"); Hughes gives us a scientist who—although he nurtures and disciplines the sexuality and development of other species—seems in need of both discipline and nurture; and Martinovitch admits that the Canti method doesn't so much uncover truth as enroll allies, in the Latourian sense. We learn, in short, the ambivalences inherent in the tissue culture point of view.

These ambivalences bring us back to the central ambivalence: the simultaneously productive and problematic ambiguity of Honor Fell's phrase. While in her lectures she purported to represent "the tissue-culture point of view," that was what neither she nor her researchers could do. Rather, Fell articulated the perspective of the researchers engaged in culturing tissues: the human/scientific/instrumental/agentic point of view. Despite their claims to speak for the ovum, the blastula, the cultured gonad, or the chromosome, the Strangeways researchers have no access to the point of view of the culture itself. The point of view they articulate is that of the tissue culturist. The ambiguity at the heart of Dr. Fell's phrase is both problematic, because when unacknowledged it obscures the scientists' epistemological limitations, and productive, because if acknowledged, it can encourage scientists to draw on imagination as an aid to epistemology.

As our culture increasingly explores the biotechnologies founded on tissue culture, from animal-to-human organ transplants to reproductive technologies and fetal tissue transplants, the perspective provided by examining the problems and potential embodied in the tissue culture point of view will be crucial. It can enable us to access those counteridentifications that can reveal the underside of glamorous new biotechnologies, and it can provide us with some healthy skepticism about the powers of cinema photomicrography—the Canti method—and the contemporary visualization and imaging technologies that are its legacy. Ultimately, awareness of the tissue culture point of view may give us a more complex set of responses, both emotionally and intellectually, to medical crises such as the one that brought me to London: my friend's slow dying of lung cancer. It will certainly produce a powerful respect for the basic laboratory skill of tissue culture: a competence that if used correctly makes it possible to diagnose a specific type of cancer and if not mastered—as in my friend's case—will produce misdiagnosis and inadequate treatment.[68] More than that, it may slow the rush to xenotransplantation as a feasible solution to the problem of diseased organs, in the realization that medicine first needs to deal with the epistemological, ethical, and technical issues foundational to the technique since its inception. Finally, mindfulness of the tissue culture point of view can provide a richer perspective on the issues at stake in the complex play of boundary transgressions between human and animal, prebirth and postpartum, life and death, that make up the project of reshaping the human life span.

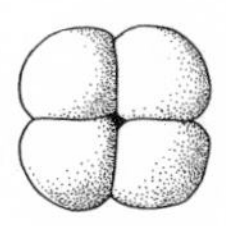

3

The Hybrid Embryo and Xenogenic Desire

Nostrils flared, ears pricked,
our son asks me if people can mate with
animals. I say it hardly
ever happens. He frowns, fur and
skin and hooves and slits and pricks and
teeth and tails whirling in his brain.
—Sharon Olds, "Bestiary"

In 1984 the British government's *Warnock Report on Human Fertilisation and Embryology*, while observing that "trans-species fertilization" was a routine part of infertility treatment, acknowledged that "the hamster tests and the possibility of other trans-species fertilisations, carried out either diagnostically or as part of a research project, have caused public concern about the prospect of developing hybrid half-human creatures."[1] Therefore the Warnock Committee proposed regulations that would prevent such an interspecies embryo being gestated or brought to term. "We recommend that where trans-species fertilisation is used as part of a recognized program for alleviating infertility or in the assessment or diagnosis of subfertility it should be subject to license and that a condition of granting any such license should be that the development of any resultant hybrid should be terminated at the two cell stage" (Warnock 1985, 71).

Ten years later, in September 1994, the United States government's NIH report of the Human Embryo Research Panel recommended a similar act of prohibition, deeming as "research considered unacceptable for federal funding" the "development of human-nonhuman . . . chimeras with or without transfer," the "cross-species fertilization except for clinical texts of the ability of sperm to penetrate eggs," and the "attempted transfer of human embryos in nonhuman animals for gestation."[2]

Something very interesting is going on here. In two government documents, handpicked committees comprising scientists and members of the educated lay community are agreeing on a position that seems contradictory: first, that interspecies fertilization exists and indeed is sanctioned as a crucial part of contemporary reproductive technology and infertility treatment, and second, that interspecies reproduction is unacceptable and unworthy of federal funding and that it should be against the law to bring interspecies hybrids to term. How do we explain this double move to approve or sanction, and then to prohibit? It seems more than merely a scientific qualification or a maneuver designed to calm public concerns. Rather, it brings to mind Freud's 1925 essay "Negation," in which he observes, "the subject-matter of a repressed image or thought can make its way into consciousness on condition that it is denied."[3] The parallel to Freud's analysis of the function of negation suggests that these government panels are doing cultural and psychic, as well as governmental and scientific, work. Invoking interspecies pregnancy in order to deny (outlaw) it, the Warnock Committee and the NIH Human Embryo Research Panel may be seen as satisfying a desire: putting into circulation the same (repressed) cultural image that they propose to legislate against.

Both the Warnock and NIH committees are concerned with the hybrid embryo, a creature resulting from interspecies reproduction, or xenogenesis. Perhaps the most unsettling of the liminal lives I will explore in this book, this is not only a new medical/scientific entity but also an image of some historical and cultural complexity. In the course of the twentieth century, the image of the hybrid embryo circulated through a number of different discourses: government documents, works of fiction, futurological tracts, natural and social history writing, popular science writing, social critique, and feminist theory. I will trace some of the embedded meanings of this liminal being, produced through xenogenesis, in order to demonstrate the intersecting disciplinary discourses and practices that (albeit unacknowledged) have given rise to the current scientific perspective on interspecies reproduction. So, for example, I will show how the Muller panel's position on interspecies reproduction is marked by some of the cultural formations within which the panel came into being, and to which it is inevitably responding. My goal is to demonstrate the value of contextualizing—in terms of both history and literature—even the most seemingly transparent scientific or medical intervention, in order to achieve the fullest understanding of its implications. Finally, by networking scientific and governmental discourses with discourses of literature and popular culture, I explore questions that are not so much scientific or governmental (though they pertain to both of those worlds) as they are imaginative. What energy fuels the son's wild speculations in Sharon Olds's poem? What practices produce, result from, and perpetuate this strange force I am calling xenogenic desire? What possibilities and new responsibilities come into being along with the hybrid embryo?

CROSS-SPECIES FERTILIZATION IN CONTEMPORARY GOVERNMENT REPORTS

The Human Embryo Research Panel, also known as the Muller panel after its chair, Stephen Muller, devoted nearly five single-spaced pages to the question of cross-species fertilization as part of human embryo research. The panel begins—as did the Warnock Committee—by acknowledging the routine acceptance of one sort of cross-species fertilization: "Fertilization of hamster eggs with human sperm is widely used in infertility clinics as a test for the fertilization competence of sperm. These eggs are used to test the competence of a particular patient's sperm to penetrate an egg. However, the fertilized eggs are not permitted to develop, nor is it likely that they would do so, due to the wide evolutionary distance between the two species. Thus the process has a clearly defined end point" (42). The reassurance obtained by affirming a sperm's competence is of course challenged if the resulting embryo were also competent, that is, if it could develop into a hybrid being: part hamster, part human. While the hybrid embryo addressed by the Muller panel provides that comfort with a reassuring end point produced by the very different morphologies of the two species, the Muller panel observes that other kinds of chimeric embryos would have no such inherent limitation on their life spans. "Because of the close evolutionary relationship between humans and some primates, for example chimpanzees, it is theoretically possible that human eggs fertilized with chimpanzee sperm might develop, at least to 14 days. Such cross-species fertilization would be unacceptable" (42).

The Muller report elaborates on why such xenogenetic—or interspecies—activities are not recommended for federal funding. "It is theoretically possible to make chimeras between human embryos and closely related primates, such as chimpanzees, but . . . the fetus would have cells derived from both species in all tissues. In other words, it might be possible for the chimeric fetus to have large parts of the brain and/or gonads derived mostly from primate cells and other parts of the body derived mostly from human cells, a situation that would, from both a medical and ethical standpoint, be totally unacceptable" (43). A later paragraph, from a section on the "development of human-nonhuman and human-human chimeras with or without transfer," is even more specific in articulating the panel's reason for "unanimously oppos[ing], on ethical and scientific grounds, the creation of heterologous, or human-nonhuman chimeras, with or without transfer": "any resulting chimera would be a mixture of both cell types in all tissues, including the brain and the gonads" (95).

In its overdetermined attention to the brain and gonads as sites that must be protected from hybridity and kept pure from any intermixture of cell types, the Muller panel's argument seems to reveal a concern that is less scientific than cultural. Language in a later section of the report, concerning the "attempted transfer of human embryos in nonhuman animals for gestation," confirms that suspicion (96). Discussing the possibility of gestating human fetuses in nonhuman animals, a possibility that received serious and extended treatment in a 1929 text I will discuss in a later section, the Muller panel "overwhelmingly concluded to prohibit such research on the basis of scientific invalidity and moral opposition" (96). Not only does the passage reveal an attention (uncharacteristic for governmental or scientific writings) to maternal-fetal interactions and the mother-infant bond (experiences often discounted when reproductive technology is being discussed), but the language is surcharged with intensity, unusual for both government and scientific discourse. As that later section reads, "There is every reason to believe that a human embryo would be immunologically rejected after transfer into another species, or, at least, that maternal-fetal placental interactions would be profoundly affected. . . . Studies of human gestation confirm the importance of maternal-fetal interactions during pregnancy. These are crucial not only for physiological development, but they also represent the beginnings of mother-child bonding and of human relationship. *The Panel finds it repugnant to experiment with such relating between a human fetus and a nonhuman gestational mother*" (96; italics mine).

As these passages reveal, the scientific image of interspecies reproduction catalyzes complex emotions. The highly charged, metaphoric language indicates the activity of what we might call the *governmental imaginary*, adapting Catherine Waldby's notion of the biomedical imaginary to designate the speculative and imaginative universe within which government, like biomedicine, functions. As Waldby points out, this realm is marked by "deployment of images and metaphoric modes of thinking as indications of excess, points at which systemic thought spills out into the speculative, the fantasmic, into desire, even while it tries to relegate such images to the status of illustration of a systematically generated idea" (Waldby 2000, 136). The two practices that this governmental panel advised be prohibited—the creation and transfer of chimeric embryos and the gestation of human fetuses in nonhuman animals—are both aspects of interspecies reproduction. As such, these practices (whether hypothetical or actual) are shadowed by the overdetermined cultural meanings they carry, meanings that form the conscious or unconscious context for any discussion of interspecies reproduction. The first act, the creation of chimeric or hybrid embryos, recalls the discourse of race from the seventeenth century through the nineteenth, while the second issue, the gestation of human fetuses in nonhuman animals, revisits a powerful site of modernist controversy over race and gender.

HYBRIDITY AND THE ANXIETY OF RACE AND SPECIES

What is the context for the Muller panel's anxiety about the creation and transfer of chimeric embryos? In particular, what might be the origins of the panel's curious obsession with the notion of gonads and brains in which both human and nonhuman cells were mixed? We can look to racial theories of the seventeenth, eighteenth, and nineteenth centuries for an answer: such theories inevitably mingled race and species because of the preoccupation with issues of origin and hierarchy, often imaged as a "chain of being" on which the species, and the races, were arranged in hierarchical order. Species discourse, like racial discourse, has long been a rich site of cultural construction. Seventeenth- and eighteenth-century naturalists toyed with the notions of apes becoming human and speculated about the intellectual and biological issue of ape-human sexual encounters, finding them a productive source of social satire and critique.[4] The European discovery of the great apes in the seventeenth and eighteenth centuries spawned numerous stories about ape-human hybrids, among them a nineteenth-century rumor that scientists had traveled from France to Africa to "experiment with breeding a male orangutan and an African woman."[5] "Many naturalists assert[ed] that women of Africa and Asia 'mixed' voluntarily or through force with male apes, and that the products of these unions had entered into both species" (Schiebinger 1993, 98). Rousseau even went so far as to suggest that a cross-breeding experiment could answer the question of whether apes were human: if the issue was fertile, his argument went, the apes' humanity would be demonstrated (98).

Interest in hybridity increased in the nineteenth century, when it became "a key issue for cultural debate," according to historian Robert Young.[6] Preoccupied by the question of monogenesis or polygenesis—whether the human race issued from a single origin or from several different "species"—mid-nineteenth-century thinkers found "hybridity" a useful term to express the anxieties produced and articulated by an evolutionary model that arranged not only species but the races of humankind hierarchically. The Victorian discourse of race expressed the enmeshed cathexis of interracial and interspecies transgression in texts portraying the Great Chain of Being, in which "predictably the

African was placed at the bottom of the human family, next to the ape, and there was some discussion as to whether the African should be categorized as belonging to the species of the ape or of the human" (Young 1995, 6–7). The definitive test for whether organisms were of the same species was whether they could successfully produce fertile offspring; if their offspring were fertile, the parents were judged to be of the same species. "The dispute over hybridity thus put the question of interracial sex at the heart of Victorian race theory," Young has observed (102). Hybridity served as a powerful site of cultural construction, connected to issues of both racial and species origins, according to Young, "because the claim that humans were one or several species (and thus equal or unequal, same or different) stood or fell over the question of hybridity, that is, intra-racial fertility" (9). In the nineteenth century, then, the term "hybridity" expressed mingled attraction and repulsion. Laden with an implicit racial as well as heterosexual ideology, the term imports into contemporary theory the unacknowledged trace of that racist past.[7] "Racial theory, which ostensibly seeks to keep races forever apart, transmutes into expressions of the clandestine, furtive forms of what can be called 'colonial desire': a covert but insistent obsession with transgressive, inter-racial sex, hybridity and miscegenation."[8]

Just as the boundary-constructing concept of race is shadowed by a desire to transgress those racial boundaries, so too the taxonomic impulse that has given us the concept of species has, as its transgressive underside, the impulse to cross species boundaries. We can modify Robert Young's formulation of "colonial desire" to theorize the existence of what we might call "xenogenic desire"—a covert but insistent obsession with transgressive, interspecies sex, hybridity, and interspecies reproduction, or xenogenesis.[9] Literature is one of the most powerful sites of the articulation of desire, precisely because—functioning like Freud's concept of negation—literature can give expression to desire while simultaneously deauthorizing it as "only fiction." The articulation, construction, and production of xenogenic desire—the fear/wish of interspecies reproduction—differs in relation to changes in the construction of the subject in modern and postmodern literature, as I will sketch by moving through a number of literary texts, first modern and then postmodern. In the modern texts, three different representations of interspecies pregnancy reflect the changing scientific understanding of the human body and subject during modernity: the surgical, the reproductive technological, and finally the genetic. With the postmodern move to the affirmation of differences and the decentered subject, xenogenic desire takes on a new, positive construction. However, here too the specific meaning of the image varies with the ideological agenda of the context, whether feminist postmodern or nonfeminist nonmodern, and with the particular notion of reproduction being deployed.

MODERN NARRATIVES OF HYBRIDITY: FROM FRANKENSTEIN'S MONSTER TO THE FIFTH CHILD

Mary Shelley's monster is an interspecies hybrid, pieced together in Victor Frankenstein's "workshop of filthy creation" out of materials stolen from "the dissecting room and the slaughter-house" (Shelley [1818] 1981, 39). He frequently functions as point of origin for the negative literary image of xenogenic desire, although the image of the hybrid and the chimera extend back to Greek and Roman mythology. In its doomed fantasy of having "a new species bless [Victor Frankenstein] as its creator and source," and in its preoccupation with the possibility that the monster might find a mate and breed "a race of devils," *Frankenstein* establishes two major themes for literary treatment of interspecies reproduction in the modern era: devolutionary anxiety linked to a hierarchized racial taxonomy, and a protomodernist notion of a tempting but dangerous scientific intervention enabling human perfectibility (39, 150).

We can trace through a range of turn-of-the-century and early-twentieth-century British literary and philosophical texts the theme of hybridity and its foundational emotions in modernity: anxiety over racial and species degeneration and an attraction to racial and species boundary crossing. In each of these texts, the notion of interspecies reproduction is linked to a fantasy of scientific control of reproduction, either to perfect the species or to annex abilities to one species that are customarily possessed by another. (Note that xenogenesis is also linked to the fantasy of achieving scientific control of death, producing the animal-human hybrid organs that are foundational to xenotransplantation and thus the goal of contemporary experimentation in cloning.)

H. G. Wells's *The Island of Dr. Moreau* (1896) is perhaps the most powerful image of a surgical xenogenesis. The vivisectionist Dr. Moreau, trained in London but now isolated on a remote Pacific island, surgically constructs Beast People, hybrids of human and animal species, to test the same human/animal boundary that preoccupied naturalists in the two centuries before. Moreau's explanation of his surgical creations combines racist and what we might call species-ist discourse. "I took a gorilla I had, and upon that, working with infinite care, and mastering difficulty after difficulty, I made my first man. All the week, night and day, I molded him. With him it was chiefly the brain that needed molding; much had to be added, much changed. I thought him a fair specimen of the negroid type when I had done him, and he lay, bandaged, bound, and motionless before me."[10]

While Wells used vivisection as his point of entry to the Victorian debate over the boundaries of race and species and the origin of humanity, writers in the early twentieth century were more likely to find their point of entry in the notion of scientifically controlled reproduction. Thus they turned to the second technique for interspecies reproduction that the Muller panel both invokes and prohibits (thereby enacting the powerful double move of negation and the gratification of desire): the idea of gestating a human fetus in a nonhuman uterus. J. B. S. Haldane's celebrated discussion of extrauterine gestation in *Daedalus, or Science and the Future* (1923) prompted Nietzschean philosopher Anthony Ludovici (1929) to respond with his own image of interspecies extrauterine gestation: a human fetus gestating in "a cow or an ass" (92). Unlike Wells's scathing portrait of the human drive to improve nature through science, Ludovici's image of human-animal hybridity expressed sexist and racist anxieties through the metaphor of human devolution:

> Science already suspects that vital fluids are not specific, and it is probable, therefore, that in the early days of extra-corporeal gestation, the fertilized human ovum will be transferred to the uterus of a cow or an ass, and left to mature as a parasite on the animal's tissues, very much as the newborn baby is now made the parasite of the cow's udder. And with this innovation, we shall probably suffer increased besotment, and intensified bovinity or asininity, according to the nature of the quadruped chosen. Thus extra-corporeal gestation, or "ectogenesis" (to use a word coined by Mr. J. B. S. Haldane for the purpose) will become a possibility, and the Feminist ideal of complete emancipation from the thraldom of sex will be realized. (92)

Lysistrata, or Woman's Future and Future Woman was philosopher and translator of Nietzsche Anthony Ludovici's contribution to the To-day and Tomorrow Series, a curious set of futurological tracts published in London in the 1920s. The volume warns that the increasing artificiality of modern life threatens masculinity and high culture in the name of a feminized, feminist mass culture. Ludovici's remarkable paranoid fantasy that feminists would seize on interspecies methods of reproduction to gain emancipation from reproductive service to the species is a rare modern prefiguration of the postmodern trend I will discuss later: the affirmative view of interspecies reproduction as positive and emancipatory, rather than negative and devolutionary.

The postmodern response to the constraints of material embodiment can be seen across the life span, from this intervention at the midlife reproductive moment to interventions at the end of life, and its motivation and outcome can frequently seem emancipatory. Thus, a postmodern model of aging, as Anne Basting has pointed out, "can offer a vision of social space in which no single image or experience of aging is deemed 'natural,' 'normal,' or perhaps most importantly, 'pathological'" (Basting 1998, 19). Just as a midlife body can turn to artificial means to enhance the options available to her to combine career and motherhood (in Ludovici's frightened fantasy), so too an aging body whose performance is enhanced by artificial means, including those that enable late-life reproduction through techniques of hybridity, may indeed exemplify this postmodern vision of aging. But as I will argue in the Coda, we need to realize that this postmodern intervention (into aging, as into other points in the life span) is but one of a number of possible culturally constructed choices for responding to the constraints of our biological existence. While we weigh the merits of the alternative choices that such a model forecloses, we should also consider what particular (and limited) conceptions of identity, biology, life span, and community provide its foundation.

Two final texts can round out the picture of modernist visions of interspecies reproduction. Five years after Ludovici's anxious fantasy of a feminist escape from childbearing, Aldous Huxley gave us the Taylorized reproductive factory in *Brave New World* (1932), where babies are mass produced in vitro, tailored to job specifications, and the term "mother" has become an obscenity. Roughly thirty years later, Roald

Dahl's macabre little short story "Royal Jelly" picks up this theme of modern science improving the human body, to tell the tale of the (aptly named) Albert Taylor, a devout reader of the *American Bee Journal*, whose interest in scientific strategies for improving the human takes a grotesque turn.[11] In an ironic nod to Gregor Samsa, Dahl's description of Albert gives us a hybrid that is not human-animal, but human-insect: a baby that turns into a bee. Yet origins are complex, for the child's father also exudes an insectlike aura, even to the loving eyes of his wife: "Looking at him now as he buzzed around in front of the bookcase with his bristly head and his hairy face and his plump pulpy body, she couldn't help thinking that somehow, in some curious way, there was a touch of the bee about this man. She had often seen women grow to look like the horses that they rode, and she had noticed that people who bred birds or bull terriers or pomeranians frequently resembled in some small but startling manner the creature of their choice. But up until now it had never occurred to her that her husband might look like a bee. It shocked her a bit" (Dahl 1962, 122).

Adapting research on rat reproductive capacities to his own fertility problem, the story's protagonist, Albert, has been taking massive doses of royal jelly—doses that, the story makes clear, have enabled him finally to impregnate his wife. When his precious newborn daughter seems to be ailing, Albert—inspired by his lifelong hobby of amateur beekeeping—decides to feed his little girl the "wonderful substance called royal jelly," which in the hive is given undiluted to those larvae "which are destined to become queens" (107). The result is a baby-becoming-bee. With characteristic Dahl mordancy, the story includes an ironic version of a set piece of such tales of hybrid birth: a scene in which the monstrous, scientifically engineered baby is seen through her horrified mother's eyes: "The woman's eyes traveled slowly downward and settled on the baby. The baby was lying naked on the table, fat and white and comatose, like some gigantic grub that was approaching the end of its larval life and would soon emerge into the world complete with mandibles and wings. 'Why don't you cover her up, Mabel,' he said. 'We don't want our little queen to catch a cold' " (130). The royal jelly has enabled the scientist-hobbyist-father–royal consort to produce his own little queen, displacing the bewildered woman who was once the mother. Dahl's story, like the stories of scientific tinkering with reproduction before it (from *Frankenstein* to *Dr. Moreau* to *Brave New World*), portrays a newborn created with scant female aid, and from which women are pointedly, uncomprehendingly, distanced.

One final modernist text of interspecies reproduction is Doris Lessing's *The Fifth Child*, a novel published in 1988 but suffused with modernist concerns. Lessing's novel expresses both the masculinist anxiety at social and biological devolution articulated earlier in Anthony Ludovici's connection between "increased besotment" and feminism, and the image of a mother's horrified response to her hybrid child, merging them in a disturbing fantasy of mothering a child who is devolved—a throwback. Lessing's heroine Harriet Lovatt finds herself pregnant with a fetus whose activity level seems monstrously high. "Phantoms and chimeras inhabited her brain. She would think, When the scientists make experiments, welding two kinds of animal together, of different sizes, then I suppose this is what the poor mother feels. She imagined pathetic botched creatures, horribly real to her, the products of a Great Dane or a borzoi with a little spaniel; a lion and a dog; a great cart horse and a little donkey; a tiger and a goat" (Lessing 1988, 41).

When her horrific pregnancy ends, Harriet gives birth to a baby whose "forehead sloped from his eyebrows to his crown," a description recalling the racist craniology of Petrus Camper, with its central notion of the "facial angle," a measurement that could be used to differentiate the ape from human beings of different races (Lessing 1988, 48–49; Schiebinger 1993, 150). Londa Schiebinger argues that this concept of the facial angle was the "central visual icon of all subsequent racism: a hierarchy of skulls passing progressively from lowliest ape and Negro to loftiest Greek" (150). Although it shares the racial subtext of nineteenth- and early-twentieth-century tales of interspecies reproduction, this story of the birth of a Neanderthal baby to a woman whose rejection of feminism makes her seem a throwback herself to an earlier time of female submissiveness and traditional values is not a cautionary tale of scientific power gone wrong. Ben, the uncanny child with the beetling brow and the bruising manner, is not the engineered product of an ape and a human; no Moreau or Frankenstein has produced him. Rather, as Lessing imagines him, he is simply a genetic accident—the unexplained reemergence of that missing link between contemporary humans and the apes that were our ancestors. "She felt she was looking, through him, at a race that reached its apex thousands and thousands of years before humanity, whatever that meant, took this stage. Did Ben's people live in caves underground while the ice age ground overhead, eating fish from dark subterranean rivers, or sneaking up into the bitter snow to snare a bear, or a bird—or even people, her (Harriet's) ancestors? Did his people rape the females of humanity's forebears? Thus making new races, which had flourished and departed, but perhaps had left their seeds in the human matrix, here and there, to appear again, as Ben had?" (130).

In this scene of the alienated mother gazing at the hybrid child, Lessing's disturbing novel recapitulates the familiar seventeenth- and eighteenth-century racist fantasy of an interspecies rape as a racial origin. Yet unlike its predecessors, *Frankenstein*, *The Island of Dr. Moreau*, and even "Royal Jelly," which moved from that fantasy to a notion of eugenic power and control, Lessing's novel provides us with virtually no control over hybridity. The act of interspecies reproduction that produces the hybrid is not a contemporary transgression (whether surgical, reproductive technological, or genetic) but rather an event in the far distant past. No act of scientific, technological, or medical abstinence can guarantee freedom from the eruption of disturbing hybrids.

FROM MODERNIST FEAR OF HYBRIDS TO POSTMODERN FASCINATION

> The monstrous fear and hope that the child will not, after all, be like the parent.—Donna Haraway, *Primate Visions*

In the arc from *Frankenstein* to *The Fifth Child*, we can see the hopes and fears articulated by the modern strand of the narrative of interspecies reproduction: the eugenic aspiration to control the development of the species through scientific intervention into reproduction, with its obsessive concern with the boundaries of race and species, and the repressed fantasy of transgressing those same racial and species boundaries through interspecies reproduction, or xenogenesis. Yet if Frankenstein's monster, Albert Taylor's little queen bee, and the Lovatts' Neanderthal child Ben all express the horrified attraction to interspecies reproduction, another novel published seven years before Lessing's modernist parable articulates a feminist vision of interspecies reproduction closer to postmodernism, for it uses xenogenesis as the platform on which to stage a critique of the Enlightenment subject of science.

Maureen Duffy's *Gor Saga* (1981) concerns the life of Gordon Bardfield, a "hominid" result of the in vitro fertilization of chimpanzee Mary by the evil scientist Forester, director of the primate section of an institute linked to the Ministry of Defense. Beginning with the technological creation of a hybrid embryo, *Gor Saga* broadens the social implications of hybridity, for Gor's physique marks him not only by race but by class. In this, Duffy may indeed reflect the demographic attributes of those who become consumers of such technologies of hybridity, whether they are medical consumers of stem cell therapies, or reproductive consumers of hybridity-enabled techniques such as IVF, or (some day) cloning. The simple fact is that class and race both function to position people as either recipients (consumers) or donors (producers) of the gametes, embryos, and technologies circulating in assisted reproduction.

Gor's origin as a by-product of IVF experimentation appropriately enough pegs him as one of the "nons," the working-class group whose cultural illiteracy is understood to place them closer to the apes. Recalling seventeenth- and eighteenth-century stories about educating an ape as a gentleman, as well as its fictional predecessor in Edgar Rice Burroughs's Tarzan stories, *Gor Saga* traces the construction of the hominid Gor as the middle-class young man Gordon. Boarded with human foster parents during his infancy, Gordon is provided with throat surgery so that he can speak, and then sent to a middle-class boarding school run on a military model. All signs are predicting a successful transition to bourgeois adult life, when puberty disrupts the experiment. "As in *Frankenstein*," Jenny Newman has observed, "it is the monster's nascent sexuality that provokes the major crisis."[12] Yet unlike Frankenstein's monster's demand for a mate, which, when unmet, unleashes murderous rage, Gor's misstep—kissing the young girl who is, unbeknownst to him, his half sister—seems less a monstrous transgression than a bit of teenage exuberance. To all, that is, except Gor's scientist-father, who decides that his little experiment is over and chases Gor with murderous intent. Gor hides himself among the "urban guerrillas" (the pun is clearly intentional) who have built a secret collectivity in the waste towns. From that hideout he investigates his origins until—upon learning the truth—he must choose between suicide and self-acceptance. The latter wins. Helping his fellow outsiders battle against the repressive forces of the army who would destroy their nascent off-grid communities in the name of a homogenized commodity culture, Gor finds himself crowned king of the day for his brave service to the new hybrid community.

Frankenstein, *Dr. Moreau*, "Royal Jelly," and *The Fifth Child* all portray the hybrid embryo, and resulting xenogenic being, as a tragedy at best, and an abomination at worst. Despite the painful testimony of Frankenstein's monster, the litany of Moreau's "beast people," and the alien buzzing or howling of the little queen and of Ben Lovatt, the hybrid beings imagined by Shelley, Wells, Dahl, and Lessing are portrayed more often as alien objects of scientific intervention than they are as speaking subjects. In contrast, Duffy's novel represents an important shift in its openness to a decentered xenogenic subject. We experience fully half the novel through Gor's eyes, and by the conclusion it is Gor's perceptions, rather than those of the government or religion, that are affirmed. Duffy's novel subjectifies the hybrid, giving us the perspective of the hominid, whose community embraces him, and rejecting the perspective of the male scientist who created him out of a perverse desire for instrumental mastery.

Although in its realistic narrative strategies *Gor Saga* is far from the self-reflexive pastiche characteristic of much postmodern fiction, in its affirmation of a hybrid subjectivity, it anticipates one of the most prominent postmodern celebrations of interspecies reproduction, Donna Haraway's *Primate Visions*. Haraway's study of the discipline of twentieth-century primatology examines the ways that as human beings we inscribe our changing self-constructions and subjectivities on the primates we study. Affirming interspecies reproduction in a casual aside, in its main argument, Haraway's study embraces hybridity as a route to increased communication and community. "*Primate Visions* does not work by prohibiting origin stories, or biological explanations of what some would insist must be exclusively cultural matters, or any other of the enabling devices among primate discourses' apparatuses of bodily production. I am not interested in policing the boundaries between nature and culture—quite the opposite, I am edified by the traffic. Indeed, I have always preferred the prospect of pregnancy with the embryo of another species" (Haraway 1989, 377).

If Haraway embraces the hybrid in the name of feminist communication, Bruno Latour does so not in the name of feminism but in the name of a retheorized modernity. In *We Have Never Been Modern* (1993), Latour argues that modernity itself has been formed through a dual process of sanction and prohibition resembling Freud's analysis of the process of negation. Latour attributes our current proliferation of hybrids to the modernist act of purification: the separation of human from nonhuman, nature from culture, and all the binary oppositions that follow from that.[13] Such acts of purification paradoxically result in the continued production of hybrids, according to Latour. If we wish to ratify, rather than to deny, the place of hybrids in contemporary life, we must renegotiate the implicit agreements for perceiving the self and the world that he dubs "the modern Constitution." Advocating that we "rethink the definition of modernity, interpret the symptom of postmodernity, and understand why we are no longer committed heart and soul to the double task of domination and emancipation," he urges us to draft another constitution, one that enfranchises more than just human beings:

> If I am right in my interpretation of the modern Constitution, if it has really allowed the development of collectives while officially forbidding what it permits in practice, how could we continue to develop quasi-objects, now that we have made their practice visible and official? By offering guarantees to replace the previous ones, are we not making impossible this double language, and thus the growth of collectives? That is precisely what we want to do. This slowing down, this moderation, this regulation, is what we expect from our morality. The fourth guarantee—perhaps the most important—is to replace the clandestine proliferation of hybrids by their regulated and commonly-agreed-upon production. It is time, perhaps, to speak of democracy again, but of a democracy extended to things themselves. (Latour 1993, 141–42)

Latour's analysis of the way that purification *produces* hybrids has a suggestive similarity to the representation of interspecies reproduction in government documents with which this chapter opened. The modern process Latour describes—the denial of hybrids and their consequent proliferation—recalls the linked fascination with, and fear of, the hybrid embryo as an emblem of interspecies reproduction, revealed in the government documents discussed earlier. Moreover, the move toward the "regulated and commonly agreed upon production" of hybrids called for by Latour may be precisely what is going on when the authors of the Warnock Report and the Muller panel's report acknowledge the production of human-animal chimeras as part of normal reproductive technology and recommend regulating their production. (A similar refusal to acknowledge the role of hybridity in the production of stem cells from donated embryos figures in the debate around that new technology, as I discuss later.)

Yet while Latour's language—like that of the Muller panel—is shadowed by the racializing discourse of the nineteenth century, his text is disturbingly blind to the gender categories that, if mapped onto his tidy diagram, would crosscut his gridded areas of purification and translation. What does Latour leave out of the picture, owing to his gender blindness? One way of understanding the project of *We Have Never Been Modern* is as a call for a more complete acknowledgment of the fullness of all being, of life itself.[14] Latour's call for wider enfranchisement, with its self-conscious rhetoric of a new "Nonmodern Constitution" and a "Parliament of Things," is an echo of the Enlightenment moment of the formation of the liberal civil state. Latour even redefines the freedom that state promised: "freedom is redefined as a capacity to sort the combinations of hybrids" (141). Yet the very discourse Latour uses invokes an era when the political subject, while newly redefined, was anything but hybrid. If it was no longer the child-subject of a patriarchal king-father, it was instead the new autonomous, homogeneous, white male individual subject of fraternal democracy. Precisely that Enlightenment echo reveals the flaw in the affirmation of interspecies reproduction put forth by both Latour and Haraway. As Carole Pateman (1988) has shown, both that government and that science are founded on the notion of the autonomous individual with the normative male body, bound in a social contract that occludes both racial and gendered others, and the sexual contract to which they are both, to different degrees, relegated.[15] A reconsideration of Haraway's treatment of the reproductive body will suggest the boundary conditions that are still in force in Latour's and Haraway's representations of xenogenesis.

BECOMING INSECT: XENOGENESIS BEYOND THE ANIMAL KINGDOM

The body as seen by the new biology is chimerical.
—Dorion Sagan, "Metametazoa"

Donna Haraway concludes her history of twentieth-century primatology with a reading of science fiction—Octavia Butler's trilogy entitled *Xenogenesis.* Concerning a species of "gene traders" called the Oankali, who must engage in interspecies reproduction to survive, Butler's novels explore the psychological and social implications of the human horror of xenogenesis, portraying humanity's fear of interspecies reproduction as a genetic flaw linked to human beings' excessive aggressiveness and hierarchical thinking. Despite her appreciation of Butler's fictional attention to "miscegenation, not reproduction of the One," and her exploration of the links between racial and species boundary crossings, Haraway judges the first volume of *Xenogenesis* to be an only partially successful narrative of an alternative to the modernist heterosexual origin story. "*Dawn* fails in its promise to tell another story, about another birth, a xenogenesis. Too much of the sacred image of the same is left intact" (Haraway 1989, 380).

Haraway's comment alerts us to another disturbing implication of our responses to the hybrid-embryo image of interspecies reproduction: we can be engaged in protecting the status quo not only when we prohibit reproduction with members of other species but also when we fail to question the *kind* of reproduction we "think with" when we imagine such transgressions. Sexual reproduction is only one, and arguably not even the predominant, kind of reproduction that is found in nature. Bacterial budding, rhizomic replication, spore production, viral infection, symbiosis, bacterial recombination—such reproductive models challenge not only our *humanness* but also (and perhaps more profoundly) our *animal-ness.* Clarice Lispector's *The Passion according to G.H.* (1978), a novel that calls into question the boundaries not only of class, race, gender, and species but of propagative methods as well, articulates the powerful disorientation or deterritorialization catalyzed by an encounter with reproduction as other than (hetero)sexual. According to Rosi Braidotti (1994), the novel was written after a failed episode of heterosexual reproduction. "Clarice Lispector acknowledges that she wrote *La Passion selon G.H.* following the experience of an abortion: consequently, the maternal is one of the horizons within which the deconstruction of Woman takes place in this story" (128). Motherhood and even womanhood being left behind in the course of G.H.'s ecstatic experience of becoming woman, the novel ends with a transfiguring moment of symbiotic connection with a cockroach. This encounter can be understood not only as a transcendence of humanness but as an act of "becoming-insect," and as such, an act of interspecies union (Braidotti 1994, 128). The result is the creation of a new sort of hybrid—"becoming-insect"—that pushes the postmodern feminist embrace of interspecies reproduction into the space beyond "the sacred image of the same," basic to sexual reproduction, that even Haraway replicates in *Primate Visions.* As Braidotti explains Deleuze and Guattari's theory of "becoming animal," it is "a question of multiplicity . . . the chain of becomings goes on: becoming-woman/child/animal/insect/vegetable/matter/molecular/imperceptible, etc., etc." (129). "The insect as a life form is a hybrid insofar as it lies at the intersection of different species: it is a winged sort of fauna, microcosmic" (127). Introducing the woman "becoming-insect," *The Passion according to G.H.* also introduces a new perspective on interspecies reproduction. G.H.'s abject, feminized experience of encounter with the cockroach, which as an insect is one of the range of abject beings that "correspond to hybrid and in-between states, and as such . . . evoke both fascination and horror, both desire and loathing," shifts the whole notion of xenogenesis from the register of heterosexual reproduction to the register of propagation "by contagion, [that] has nothing to do with filiation by heredity" (Braidotti 1994, 128; Deleuze and Guattari 1987, 241). And yet crucially, unlike Kafka's *Metamorphosis* as read by Deleuze and Guattari, this shift to a nonfiliative reproduction does *not* involve the abandonment of sexual differences, according to Braidotti: "In Clarice Lispector's story . . . the entire process of becoming, down to the crux of the encounter with the insect, is specifically sexed as female. References to sexuality, to motherhood, to body fluids, to the flow of milk, blood, and mucus are unmistakably female. At the same time, however, the structure of the successive becomings experienced by G.H. is in keeping with Deleuze's analysis of becoming as a symbiotic metamorphosis" (129).

Let me elaborate on the connections that Braidotti only implies, since she is not considering the implications of Lispector's story for the image of interspecies reproduction, between a nonheterosexual, non-animal conception of reproduction and a proliferation of differences. As Deleuze and Guattari (1987) theorize it, "filiative production or hereditary reproduction" (heterosexual reproduction, whether human or animal) is more about sameness than difference: "the only differences retained are a simple duality between sexes within the same species, and small modifications across generations" (242). Interspecies reproduction within the register of filiative production/hereditary reproduction would thus maintain the dominant sameness, adding only the difference of species to the difference of sex (male or female). In contrast, interspecies reproduction beyond that filiative register, the notion of "a propagation, a becoming that is without filiation or hereditary propagation," proliferates differences. Deleuze and Guattari illustrate this in a paragraph that is a veritable catalog of posthuman reproductive possibilities:

> We oppose epidemic to filiation, contagion to heredity, peopling by contagion to sexual reproduction, sexual production. . . . Like hybrids, which are in themselves sterile, born of a sexual union that will not reproduce itself, but which begins over again every time, gaining that much more ground. Unnatural participations or nuptials are the true Nature spanning the kingdoms of nature. . . . That is the only way Nature operates—against itself. . . . For us. . . . there are as many sexes as there are terms in symbiosis, as many differences as elements contributing to a process of contagion. (241–42)

Although they are occurring in the register of the posthuman, these new differences catalyze the same doubled pattern of "fascination and horror, both desire and loathing," suffusing the fantasies of racial and species boundary crossing with which I began (Braidotti 1994, 128). The link here seems to be the persistence of the gendered female in all hybrid and in-between states, whether or not they occur in the realm of binary gender constructions. In short, gender constructions persist even into the realm beyond gender, beyond the human. The way that the attraction/repulsion to boundary crossing is feminized recalls the epigraph from Luce Irigaray with which Braidotti begins her analysis of the feminist limitations of Deleuze's notion of the "becoming woman":

"In order to become, it is essential to have a gender or an essence (consequently a sexuate essence) as *horizon*" (111).

The act of trying to think interspecies reproduction within an alternative register of symbiotic becoming is, according to Braidotti, a "historically necessary" project for feminists, because she believes there are philosophical implications to the affirmation or refusal of the register of Oedipal sameness within which heterosexual filiative reproduction occurs (Braidotti 1994, 134). Yet because the boundary crossing such a project entails will always evoke the feminized abject, it will also always place us as feminists in "paradoxical space," the peculiar condition of simultaneously occupying "spaces that would be mutually exclusive if charted on a two-dimensional map—centre and margin, inside and outside" (Rose 1993, 140). Working between center and margin, theory and practice, we must come to terms with the appeal of a model of nonfiliative interspecies propagation that challenges the dominance of what Deleuze and Guattari rather dismissively term "the simple duality between sexes within the same species" while still retaining our attention to the material effects of that "simple duality," "the *practice* of sexual difference as a conceptual and political project" (Deleuze and Guattari 1987, 242; Braidotti 1994, 135).

WHO GETS A VOICE?

If to affirm a reproduction that is not just interspecies but nonsexual is in principle to endorse "connections, alliances, symbiosis" of multiplicity and diversity rather than uniformity, what is it in practice? (Braidotti 1994, 129). Or to put it another way, what are the practical, material implications of interspecies reproduction when we consider it as part of the more familiar realm of sexual reproduction, whether actual or potential? Returning to the government documents with which the chapter began, we can consider the implications of xenogenesis in two different contexts: interspecies reproduction within the existing program of reproductive technology, and interspecies reproduction as part of a potential program of feminist reproductive intervention. (British author and science writer Naomi Haldane Mitchison anticipated the latter in her 1975 novel of feminist reproductive technologies, *Solution Three*.)

Since the latter scenario is closest to Lispector's passionate revelation, let us begin there—by challenging its practical implications. We can ask: what would be the material consequences of a feminist enactment of interspecies reproduction? The crucial point here is that even when we are considering the feminist implications of hybrids within the realm of animal reproduction, such fantasies of disrupting the Oedipal economy of the same will have different effects depending not only on the subject position of the woman engaging in the fantasy but on the network of social relations and practices that her fantasy calls forth. To give just one example of the ways that the fantasies can differ depending on different subject positions: a white, middle-class, postchildbearing, premenopausal woman, who is socially constructed as a member of the dominant, noncriminal class, and whose political sympathies are both technophiliac and conservationist, could be drawn to interspecies reproduction because of the ways it can challenge the boundaries of class and species. That is, while surrogate mothers often imagine themselves achieving class mobility by gestating a child for a middle- or upper-class couple, interspecies reproduction (with all of its legacy of hierarchized racial and species shift) would clearly disrupt that fantasy. This fantasy depends, however, on a level of social and economic privilege (education; employment in the professional-managerial class with the associated level of individual autonomy as well as a level of biomedical technical expertise) that undercuts or recontains its socially disruptive potential. Similarly, while first world conservationists advocate the preservation of endangered species as a resource for human medical, environmental, psychological, and aesthetic needs, the act of gestating a nonhuman inverts a human-centered scenario, using the human body as a vehicle for another's species survival. In contrast, a middle-class, technophiliac, African American woman, still of childbearing age, coded as criminal in the racist West, and strategically identified with her racial group, could find the fantasy of interspecies reproduction a horrifying replay of racialized notions of race-and-species hierarchy, of devolution, and ultimately of racial extirpation, through the threatening prospect of losing an identity-relevant essentialism with the birth of a hybrid baby.[16] The additional fact that such a hybrid birth would be the product of a set of skills generated by Western scientific rationality (tissue culture, in vitro fertilization, embryo transfer) only adds to its horror.

Just as subject position, race and class location, and linked social practices shape the implications of the fantasy of a feminist creation of hybrids, so too the subject positions of the scientist/doctor and the nature and purpose of scientific/medical intervention shape the implications of xenogenesis. Returning for a moment to the Muller panel's recommendations, if the panel's regulation of hybrids represents a move to "rewrite the modern Constitution," as Latour would put it, we need to consider who is enfranchised by that new constitution, and whom it excludes. What practices and institutions serve to enforce the "modern Constitution," governing at a distance, as it were? And what are *their* implications for women, when xenogenesis is carried out in the context of existing reproductive technological practices? If the hybrid is gaining a voice as part of this increasing move to interspecies reproduction as the biomedical and legal establishment explores the potential of the new reproductive technologies, is that gain mirrored by a loss? Will there be no voice for the women who act as surrogate mothers and undertake the arduous processes of in vitro fertilization and egg and embryo harvesting?[17]

We have seen that even so charged and seemingly unambiguously negative an image as interspecies reproduction means something very different in its different discursive contexts. Contextualization—both historical and literary—is a crucial part of the project of feminist cultural studies of science. Keeping the context in mind, then, we should not embrace the hybrid or affirm xenogenic desire until and unless we are satisfied that we are not obscuring the persistence of gender hierarchies (even into the realm of the posthuman) and that we are not silencing or objectifying the reproductive experiences of women, in all of their variety, multiplicity, and diversity.

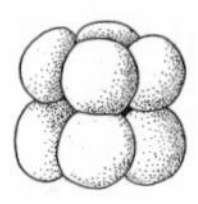

4

Giant Babies

GRAPHING GROWTH IN THE EARLY TWENTIETH CENTURY

> Growth is a fundamental and common property of nature. Its course is governed by fundamental laws common to its expression where and whenever this is manifest. The workings of these laws are to be found in the growth of crystals as in the growth of populations; in the growth of intelligence as in the growth of a root; in the growth of yeast as in that of an elephant. Growth is growth wherever found.[1]

Many medical interventions provoked vigorous debates during the twentieth century, whether over relatively normalized techniques such as tissue culture, organ transplantation, and hormone replacement therapy, or the more experimental strategies of embryo transplantation, animal-to-human gland grafting and organ transplantation, stem cell therapy, and interspecies reproduction. But one group of twentieth-century biomedical interventions have typically been slower to develop, have received much less popular attention, and have only recently begun to provoke popular debate. Those interventions focused not on the dramatic zones of birth and death but on the far more ambiguous zone of growth. The motives are clear for intervening at the two ends of the life span. For a physician to enable the conception or birth of a baby, or prevent a death, seems among the most basic of medical duties. But medical intervention in growth, that complex, multifaceted process that occupies the undefined middle of life, as the journal *Growth* described it in the introduction to its inaugural issue in 1937, has been more difficult to conceptualize, justify, and accomplish.

Partly this lag occurred because research on growth was itself slow to develop. Biometrics, the statistical analysis of variations in growth, was a relatively new technology at the turn of the century. Introduced to the public with Francis Galton's "anthropometric laboratory" at the 1884 International Health Exhibition in London, the technology was taken up in a range of applications, by Karl Pearson, professor of applied mathematics and founder of the Biometrics Laboratory at University College, London, and then by Ronald Fisher, who applied biometrics to the role of agricultural design in growth.[2] Despite this interest in the statistical aspects of growth, human, animal, and vegetable, as late as 1927 the chief medical officer of the British Board of Education felt it necessary to exhort researchers to study "the physiology of growth" ("Physiology of Growth" 1927, 269). Sir George Newman explained that this portion of the human life span had yet to receive the attention it merited: "Here is an expanding physiology, unexplored; here are the very beginnings of disease; here its cause may be searched with some hope; here it may often be prevented; here the physique and education of a nation are in rudimentary form—and whatsoever here be found by searching or proved by experiment becomes forthwith of value to the child, to the community, and to the growth of human knowledge" (269–70).[3] Despite the rich database the schools offered, Newman argued, no scholar had taken advantage of this rich trove of material until 1927, when Dr. James Kerr, the research medical officer of the London County Council, published his massive volume *The Fundamentals of School Health*.

A review in *Nature* hailed Kerr's study as a remarkable and singular effort: "It stands alone, and must for many a day stand, as an incomparable authoritative work" ("Physiology of Growth" 1927, 270). The reviewer praised the scope and thoroughness of the volume, which in its forty-three chapters covered the physical and mental health of the schoolchild, as well as issues and problems relating to school administration and the school environment. "Perhaps the most interesting," the reviewer observed, "are the early chapters dealing with problems of growth" (270). With graphs, charts, growth curves, and tables indicating the height, weight, chest circumference, cranial capacity, and a range of other growth measurements for children in England, Europe, the United States, Canada, and Australia, chapter 2 ("Growth") focuses on the ways growth can be shaped by changes in nutrition and other aspects of the manipulable environment.[4]

The medical, social, and educational communities joined the conversation on growth from about the second quarter of the twentieth century through its close. Experts in those fields offered taxonomies of normal and abnormal growth and debated the merits of intervention, part of the life strategy of controlling the entire human life span. Whether the experts were defining the range of normal growth or specifying varieties of abnormal growth, their work frequently revealed the same tensions that the *Nature* reviewer found in Kerr's study of growth: environmental models for development conflicted with those that were eugenic (or protogenetic), and descriptive generalizations about the variety of human growth processes were at war with the normalization of growth rates. In other words, the rate of growth of a human being could have different kinds of significance. It could be of importance only to that individual, reflecting the trajectory of his or her life, or it could be meaningful as an index to a general trend in the height, weight, and growth rate of the human species in general, a trend that could be labeled "normal" or "abnormal." Finally, it could have significance beyond the realm of human biology: as a comparative measure available for the study of growth rates in other species (animal, vegetable, or mineral).

Only in 1937 with the founding of the journal *Growth* did scientific researchers begin to integrate the various approaches to growth in a systematic way, asserting that "to uncover these laws and to test them; to set forth the determining chemical, physical, and genetic participants in each; to expose the processes in terms of mathematics, statistics, chemical reaction and physical attribute; to state the significance of developmental function; to review the accomplishments of the past and proffer suggestions for the future, are one and all matters of interest to the student of growth" ("Introduction to Growth" 1937, 1). In using mathematics and statistics to arrive at general statements about the significance of specifically human traits and particulars, the founders of *Growth* joined other medical and social experts in reflecting an epistemological reorientation. Since the early seventeenth century, the emergence of a new model for the production of knowledge, based on the construction and dissemination of telling and theoretically grounded facts, had specific implications for the way knowledge about human beings was framed (Foucault 1963, 1966; Latour 1993; Beck [1986] 1992; Poovey 1998). As Mary Poovey has shown, a set of rhetorical, governmental, medical, and social negotiations had converged to produce the notion of the modern fact, an entity that "could be represented either as mere data, gathered at random, or as data gathered in the light of a social or theoretical context that made them seem worth gathering" (Poovey 1998, 96). No longer did common sense and general agreement define what was thought to be a "true fact," even about human beings. Instead material particulars now had general theoretical significance. The size or growth rate of one individual was important as an index to the size and growth rate that could, through extrapolation, be projected for human beings as a collective.

Not only did social policy analysts and medical researchers stress its quantifiable and mechanistic aspects over its broad theoretical implications, but they were relatively tardy even to explore that aspect of human growth. In contrast, from the earliest years of the twentieth century, fiction (specifically science fiction) reveals a perspective on the problem of growth that is narrative and synthetic. Beginning with H. G. Wells's little-known novel *The Food of the Gods* (1904), continuing with Philip Wylie's *Gladiator* (1930) and a number of midcentury short stories, science fiction of both the United Kingdom and the United States has engaged in a thorough and rich exploration of the social and biological effects of manipulating the growth process. The kinds of interventions portrayed evolved from nutritional and hormonal enhancements to early forms of genetic engineering, but as we will see, the specific instances were varied and imaginative: food additives such as "Herakleophorbia"; under- or overstimulation of the ductless glands by the "para-Roentgen" ray; hyperstimulation of the "master glands" (the pineal body and the pituitary gland) through injections of "tethelin," a pituitary extract; the manipulation of "alkaline radicals" to accelerate the growth process; the introduction of plant or animal genetic material into the developing fetus.

Diverse as the two types of contributions were, the science fiction conversation about growth both set the agenda for, and diverged in important ways from, the approaches of biomedicine. As I will argue, the attributes of biomedicine as a field of knowledge and a set of practices, and of science fiction as a genre, shaped the different meanings attached to growth in the different venues. Fictional plots, presenting the narrative of an individual's life through time, made available experiences and meanings that were distinctly different from the statistical, aggregate, and population-based knowledges that dominate in the social sciences.[5] But in both fields, a powerful tension between the normal and the abnormal, or between the canonical and the marginal, shaped the terms and stakes in that debate with huge disciplinary and epistemological implications for the two contested categories of medicine and literature.

GROWTH IN *FOOD OF THE GODS*

H. G. Wells's *Food of the Gods* combines the two models for human life held in tension at the beginning of the twentieth century in its generic marginality and its epistemologically dominant concern for the verifiable fact. Reflecting Wells's commitment to experimentation, whether in the realm of fiction or of science, the novel is both descriptive and extrapolative, based both on random particulars and on theoretically significant specifics. Read usually (when it is read at all) as a celebration of scientific and technological power, in fact this tale of growth gone wild gives us a prophetic glimpse of the costs of the medical scientific intervention in the human life span.[6] As a novel about the dual processes by which growth and dissemination can be represented—narrative and statistics—Wells's work both enacts and invites a networked approach to literature and science.[7]

Wells's literary exploration of human growth joins related explorations in comparative anatomy, endocrinology, and sociology at the beginning of the twentieth century. Like them, the novel's methodology is spatial rather than linear, statistical rather than individual: "Our theme, which began so compactly in Mr. Bensington's study, has already spread and branched until it points this way and that, and henceforth our whole story is one of dissemination" (Wells [1904] 1965, 97). The novel both embodies and explores several different kinds of dissemination: not only the physical spread of the "Herakleophorbia," or "the food of the gods," but also the drift of expert responsibility for human welfare from the realm of medicine and the organized church to psychology, sociology, government, and finally the military. *The Food of the Gods* portrays both a scientific success, the invention of the food that accelerates growth, and a scientific and social failure (the inadequate control of the substance invented, and its dissemination to unintended populations). Thus it recalls Bruno Latour's injunction: "Every time you hear about a successful application of a science, look for the progressive extension of a network. Every time you hear about a failure of science, look for what part of which network has been punctured."[8]

We can assess the force of fiction's intervention in human growth, that of Wells's novel and the fictions that followed it, if we trace the interdisciplinary networks linking Wells's novel to Dr. Hastings Gilford's 1911 study *The Disorders of Post-natal Growth and Development*, to science fiction stories from the 1920s through the 1940s, and finally to the debate over the medical ethics of treating growth disorders. I hope to discover not only how the interest in growth continues through these different sites but also where those paths break down or, in Latour's terms, where the "network has been punctured." Though linked by the theme of intervention in human growth, these medical and narrative instances had different contexts and worked in different ways to enact a gradual reconceptualization of human growth: from an affair of the individual to an affair of the group, and from natural to constructed, whether by statistical representations or by social and economic context.

The Food of the Gods offers us the first twentieth-century fictional representation and social assessment of biomedical control as a life strategy, and for that reason it merits a close look.[9] The tale of the invention of a food that can increase the rate of growth, Wells's novel explores the massive social and cultural effects of this scientific discovery, and thus by implication of all the medical interventions into growth that the century would bring. Mr. Redwood, a professor of physiology and one of those "scientific men who are addicted to tracings and curves," has "been measuring growing things of all sorts, kittens, puppies, sunflowers, mushrooms, bean plants and (until his wife put a stop to it) his baby" (Wells [1904] 1965, 15–16). In collaboration with a chemist, Mr. Bensington, F.R.S., Redwood formulates the special food that accelerates the growth not only of plants, insects, and barnyard animals, but of the Redwood baby as well.[10] When administered by the Redwood family's pediatrician to their child, and then to other infants, the food causes them to grow at a prodigious rate, until as adults they stand nearly thirty feet tall. The novel traces the social, technological, and political effects of the invention and dissemination of "Boomfood" by following the fates of the giant children that it produced. Along the way, Wells describes the new methods of education, new technologies (motor perambulators, giant abacuses), and new standards of beauty that arise in response.

The novel offers both a critique of twentieth-century technologies and an analysis of their embedded social position. Professor Redwood was obsessed with the measurement of growth, Wells tells us, and his scientific discovery began with the introduction of new technologies for scientific calibration: "[After] his monumental work on Reaction Times (the unscientific reader is exhorted to stick to it for a little bit longer and everything will be as clear as daylight) Redwood began to turn out smoothed curves and sphygmographeries upon Growth, and it was one of his papers upon Growth that really gave Bensington his idea" (Wells [1904] 1965, 15). In a strategy that recalls Thomas Sterne's *Tristram Shandy*, another novel concerned with growth, albeit in utero, Wells provides illustrations of the growth curves that inspired Redwood and Bensington.

These images of growth curves not only inspired Redwood and Bensington's discovery of the growth-enhancing Boomfood but figure in three major technologies for representing the body that appeared in the nineteenth and twentieth centuries: the sphygmograph, "an instrument which records the movement of the pulse by means of tracings"; the discipline of statistics; and the use of visual display (graphs and charts) to represent quantitative information.[11] Although admitted to the British Association for the Advancement of Science and thus accepted as a science in 1833, in the early years of the twentieth century, the field of statistics was still a contested one. The new discipline was plagued by boundary questions that included generic hybridity ("it juxtaposed numerical, often tabular, formulations to discursive, sometimes historical or explanatory, narratives"), epistemological and ideological complexity (it mingled objective reportage with advocacy), diverse subject areas (as endorsed by the BAAS, statistics was officially divided into "economical statistics, political statistics, medical statistics, and moral and intellectual statistics"), and, most strikingly, a tension between representational parsimony and the production of ungovernable excess.[12] The new strategy of using visual displays to represent quantitative information, exemplified by Redwood's smoothed curves, was invented by the Swiss-German mathematician and scientist J. H. Lambert and the Scottish political economist William Playfair. Later, time-motion expert E. J. Marey brought the practice of graphing into his studies of animal movement, sharing with Playfair a curious fondness for graphic representations of royal genealogy.[13]

The research trajectory of Wells's Professor Redwood reflects his social and chronological positioning and parallels the developing technology of graphic representations. Just as graphing began by focusing on time (not only genealogical time but also the time-motion studies of Marey) and then turned to representing the spatial dimension, so too Redwood moves from doing reaction time studies to charting growth curves, bringing with him the concern with the temporal and spatial dimensions as well as the production of ungovernable excess (Tufte 1983, 32–40). A quintessential turn-of-the-century scientist, Redwood demonstrates in his research the nineteenth-century statistical reconceptualization of life, from an affair of individuals narrated by historians and biographers to an affair of populations represented by demographers and statisticians.[14]

The Food of the Gods illustrates—even concretizes—Mary Poovey's observation that statistical representation "both limits what it will depict and necessarily produces an uncontrollable excess" (Poovey 1993, 275). Poovey means by this that the act of paring down what counts as relevant to statistical representation (and thus to the production of a statistical fact) necessarily defines as *excessive* that which cannot be quantified. Both narrative and rhetoric thus figure as zones of excess produced by the very trend toward quantification and indexicality that excluded them as discursive modes. It is ironically fitting that the very method of research that characterized early-twentieth-century biology's approaches to growth—statistical measurement, normalization, and graphic representation—led to Boomfood, an invention producing excessive growth in plants and animals. Redwood's growth curves, which represent "growing things of all sorts," take the form of a "perfect Brock's benefit of diagrams—exactly like rocket trajectories they were, and the gist of it—so far as it had any gist—was the blood of puppies and kittens and the sap of sunflowers . . . in what he called the 'growing phase' differed as to the proportions of certain elements from their blood and sap on the days when they were not particularly growing" (Wells [1904] 1965, 16).

The diagrams producing that limited representation are a "Brock's benefit"—an archaism referring to someone "given to 'dirty tricks' "[15]—perhaps because their very abstraction inspires Redwood to play a dirty trick on the whole human race by inventing Boomfood. These new technologies for measuring and representing growth produce the desire to intervene in growth rates, and—as the novel goes on to demonstrate—the creation of a norm for growth that produces the new pathology of abnormal growth.

Although Bensington must struggle to extract the "gist" of Redwood's diagrams, he finally finds in them a link to his own research. Bensington has been researching chemicals producing one kind of excess: addiction. His focus is the alkaloids "most stimulating to the nervous system," complex organic bases whose effects range from deadly to delightful (strychnine, quinine, nicotine, morphine) (17). Redwood, for his part, has been looking into hormonal excess. Bensington suggests, "with the delightful nervous inconsecutiveness of his class," that Redwood's diagrams "might very probably be found to throw a light upon the mystery of certain of the ductless glands" (16). Redwood's compact curves and tracings stimulate Bensington to dream of ungovernable geopolitical excess on a global scale: "He dreamt he had dug a deep hole into the earth and poured in tons and tons of the Food of the Gods and the earth was swelling and swelling, and all the boundaries of the counties were bursting, and the Royal Geographical Society was all at work like one mighty guild of tailors letting out the equator"

(17–18). Bensington's dream of an all-consuming earth that splits its geographic boundaries, requiring the Royal Geographic Society to restore order, demonstrates the linkage between the discovery of Boomfood and other types of morphological, governmental, and libidinal excess. Boomfood rapidly gets out of hand, producing plants, animals, and ultimately children whose excessive rate of growth and proportions shake the foundations of civic order. The experiment gone awry tests the social, governmental, and biological networks of dissemination as an intervention targeted to agriculture insinuates its way gradually into human pediatric medicine.

There are three phases to this dissemination of the food of the gods: its invention, the spread of the food from animals and plants to humans, and finally its spread from individuals to entire populations. The *intended* recipients of Herakleophorbia, the chickens bought by Redwood and Bensington to try out their new invention, are soon followed by unintended animal and vegetable recipients: a climbing creeper that grows so quickly its twists and turns are visible, giant wasps that dive-bomb residents of neighboring towns, and finally rats as big as stallions, whose assault on the population produces a rural state of siege. The next jump is from animal to human. Although at first only Redwood was administering it, and only to his own child, Boomfood gets out of control when the Redwood child's pediatrician adapts the food as part of his medical practice. Dr. Winkles is a classic example of interventionist modern medicine. Engaged in endless specific projects of health enhancement, he lacks any larger awareness of the social, cultural, and historical context of medicine and disease. As Wells scornfully describes him, "He passed all his examinations, he had all his facts—and he had just as much knowledge—as a rotating bookshelf containing the *Times Encyclopaedia*. And he doesn't know anything *now*. . . . He is utterly void of imagination and, as a consequence, incapable of knowledge" (76–77). In his portrait of Dr. Winkles, Wells not only skewers the pared-down functionalism of the modern fact; he also satirically dramatizes two examples of what Zygmunt Bauman calls the central *life strategies* of modernity:

> alternative ways in which that primary trait of human existence—*the fact of mortality and the knowledge of it*—is dealt with and processed, so that it may turn from the condition of impossibility of meaningful life into the

> major source of life's meaning. At the end of such process death, a fact of nature, a biological phenomenon, re-emerges as a cultural artefact, and in this culturally processed form offers the primary building material for social institutions and behavioural patterns crucial to the reproduction of societies in their distinctive forms. In other words, mortality and *immortality* (as well as their imagined opposition, itself construed as a social reality through patterned thoughts and practices) become approved and practised *life strategies.*[16]

Manipulating the bodies of his small patients, Dr. Winkles fights an endless series of skirmishes to maintain the illusion of human power, control, centrality, and permanence. Using Boomfood to produce human improvement, he defers any acknowledgment of illness and death as long as possible. Dr. Winkles's stance toward Boomfood exemplifies all the attributes of the twentieth-century risk society to come: the creation of a threshold of acceptable risk, the focus on risk management, and the awareness of a limitless (if undefined) potential that resides in the combination of a potent new biomedical intervention and a set of corollary techniques for its production, management, and control (Beck 1992). So in response to organizations protesting that Boomfood will create children "six-and-thirty feet high," "Doctor Winkles, the Convincing Young Practitioner," promotes the new nutritional substance, claiming that it has insignificant (that is, acceptable) collateral risks and limitless potential: " 'These accidents,' said Winkles, when Bensington hinted at the dangers of further escapes, 'Are nothing. Nothing. The discovery is everything. Properly developed, suitably handled, sanely controlled, we have—we have something very portentous indeed in this food of ours' " (Wells [1904] 1965, 71).

Though initially the assault on morphological normalcy appears to have been rebuffed, we learn as part 2 begins that the invention of Boomfood has produced not just biological but societal anomaly. It has jumped from individuals to entire populations, and from plants and animals to human beings. As the power of boundaries gives way to the strength of networks of dissemination, the food of the gods produces another sort of excess: an excess of governmentality.[17] Royal commissions, parliamentary inquiries, and private associations and societies swing into action, energetically pursuing all sides of the question of Boomfood.[18] Different methods of resisting the use of Boomfood proliferate: societies are formed to control its administration; other societies are created to celebrate the norm it threatens to dislodge, such as the Society for the Preservation of Ancient Statures. Yet Boomfood proliferates nonetheless.

> It speedily became evident to the public mind that this time there was not simply one centre of distribution, but quite a number of centres. There was one at Ealing . . . and from that came the plague of flies and red spiders; there was one at Sunbury, productive of ferocious great eels, that could come ashore and kill sheep; and there was one in Bloomsbury that gave the world a new strain of cockroaches of a quite terrible sort. . . . Abruptly the world found itself confronted with the Hickleybrow experiences all over again, with all sorts of queer exaggerations of familiar monsters in the place of the giant hens and rats and wasps. Each centre burst out with its own characteristic local fauna and flora. (Wells [1904] 1965, 91)

Like the modern fact that they exemplify, the products of Boomfood are paradoxically both abstract and specific, general and particular. Familiar monstrosities, they all share a certain degree of morphological exaggeration while they differ by being bred true to the local fauna and flora: huge flies and spiders, ferocious eels, hideous cockroaches. Reflecting its statistical origins, the excess that is Boomfood is spatialized, keyed to specific conditions, yet still potentially graphable along a continuum. Despite universal attempts to make things safer by containing Boomfood, still the new nutrient proliferates. As an exchange between Bensington and Redwood over the impact of Boomfood illustrates, this scientific innovation gains its foothold through what might be called risk homeostasis: the process by which human imperfections (greed, self-interest, and denial) achieve a balance point with the human drive to technical control:

> "It's going to destroy the Proportions of things. It's going to dislocate—what isn't it going to dislocate?"
>
> "Whatever it dislocates," said Redwood, "my little boy must have the Food." (78)

Boomfood not only dislocates nearly all human society but also provides a potent engine for its growth. The Faustian drive for perpetual innovation, the foundation of science fiction from Shelley's *Frankenstein* to Atwood's *Oryx and Crake,* is now fueled by the new epistemological technology of statistics, precisely because it creates the possibility of *extrapolation* through linking quantifiability with futurity.

"[Anyhow], you can't help yourselves now, you've *got* to go on."

"I suppose we must," said Redwood. "Slowly—"

"No!" said Cossar, in a huge shout. "No! Make as much as you can and as soon as you can. Spread it about!"

He was inspired to a stroke of wit. He parodied one of Redwood's curves with a vast upward sweep of his arm.

"Redwood!" he said, to point the allusion, "make it SO!" (78)

With a command that both echoes Genesis and anticipates one of the late twentieth century's science fiction icons, Captain Picard of *Star Trek: The Next Generation*, Wells's novel offers us a tale of men who go "where no man has gone before." The frontier they cross is not geographic but morphological, its transport vehicle not the starship *Enterprise* but a set of medical interventions into the rate and extent of growth, their preoccupation not a new annexation of space but a new spatialization of social meanings as human growth is measured in relation to a range of contexts, from education to love to politics.

THE SCIENCE OF GROWTH

The biomedical intervention narrated in *The Food of the Gods* proceeds from the process of identifying a norm, to divergence from that norm, and finally to a process of normalization. What is initially an environmental effect (the presence of a new growth-accelerating food) becomes a eugenic one (as the new giant children set a new standard for human morphology). We see how an anomaly is recontained through recentering and standardizing a set of initially anomalous particulars: the freakish giant children gradually become the standard for all children as the dissemination of Boomfood continues and the number of giant beings increases. A similar group of issues appear in the writings of two scientists who studied human growth at roughly the same time that Wells wrote *The Food of the Gods*, the American Dr. Charles Sedgwick Minot and the British Dr. Hastings Gilford. While they share a number of traits with Wells's fictional scientists, they also reveal the necessary clinical conservatism that science fiction characters obviously need not mirror.

Minot, the James Stillman Professor of Comparative Anatomy at the Harvard Medical School, published his study *The Problem of Age,*

Growth, and Death in 1907, first serially in the *Popular Science Monthly* from June through December, and then in book form in 1908.[19] A descendant of Jonathan Edwards, Minot was trained in anatomy and embryology, first with Henry Bowditch at Harvard, then with Louis Agassiz. He finally earned a Ph.D. in science at Harvard in 1878, where, after 1883, he taught histology and embryology in the medical school. Minot's work is remarkable for its attention to growth as more than the process linking the beginning of life to the end, but as a process having importance in its own right. It was his unorthodox assertion that aging occurs most dramatically *at the beginning of life*. Thus if we want to understand aging and death, we must first understand embryological development and growth: "that which we called the condition of old age, is merely the culmination of changes which have been going on from the first stage of the germ up to the adult, the old man or woman" (Minot 1907, 358). That Minot's insight was remarkable for his time becomes evident when we consider that more than eighty years would pass before the journal *Growth* changed its name to *Growth, Development, and Aging* to reflect precisely the insight that "these are all part of the same process, so that patterns of development predict subsequent patterns of growth and aging" (*GDA* 52, no. 1 [1988]).

Rate of growth was a major preoccupation for Minot, as it was for Wells's fictional researchers Redwood and Bensington. Minot argued that to understand the mechanism of aging, we needed to study not the end of life but its beginning, because "the loss in the rate of growth is greatest in the young, least in the old, and . . . as we go back from old age towards youth, and then into the embryonic period, we find an ever-increasing power of growth, but . . . it is during the embryonic period that the loss of the power of growth is greatest" (Minot 1907, 373). Surveying the various theories currently circulating to explain aging, Professor Minot argued that we needed to investigate the mechanism of growth, including the disorders of growth central to the middle of life, if we were to understand old age: "The problem of age is . . . a biological problem in its broadest sense, and we can not study . . . the problem of age without including in it also the consideration of the problems of growth and the problems of death" (497).

Central to Minot's method was the process of statistical measurement: "With guinea-pigs I began making, years ago, a series of records, taking from day to day, later from week to week, and then, as the animals grew older, month by month, the weight of recorded individuals. There was thus obtained a body of statistics which rendered it possible to form some idea of the rapidity of growth of this species of mammal" (195). Like Redwood before him, Minot used his measurements of the guinea pigs' growth to theorize the population-based notion of a growth rate. As he explained, "Considering the rate of growth, some more definite notion must be established in our minds before we can be said to have an adequate meaning of that term. It is from the study of the statistics of the guinea-pigs . . . that we get indeed a clearer insight as to what the rate of growth really is and really means" (196). Presenting a graph of guinea pig growth rates in relation to their nutrition, Minot almost seems to be recapitulating the thought processes that led Bensington and Redwood to discover their Boomfood. Representing in tabular form the days needed to double the weight in nine different species, in relation to the protein content of the mother's milk, Minot observes, "We have . . . one of the beautiful illustrations of the teleological mechanism of the body. These various species have their characteristic rates of growth, and by an exquisite adaptation, the composition of the mother's milk has become such that it supplies the young of the species each with the proper quantum of proteid material which is needed for the rate of growth that the young offspring is capable of" (196).

If we compare Minot's thought processes in 1907 to those of Wells's character Redwood in 1904, the interwoven nature of the literary and scientific exploration of growth becomes remarkably, even humorously, evident. Working with the guinea pigs, "in the muffled and highly technical language of the really careful 'scientist,' Redwood suggested that the process of growth probably demanded the presence of a considerable quantity of some necessary substance in the blood that was only formed very slowly" (Wells [1904] 1965, 16). Although working in the pregenetic era, Minot nevertheless glimpsed a relationship between the beginning of life and its end that scientists are now rearticulating as the relationship between telomere length and rate of aging.[20] Scientific consensus is near that the length of telomeres is positively correlated to life expectancy on not only the cellular but the organismic level. As an article in *Science News* explains, "Whenever most cells divide, their telomeres shrink, which has led many scientists to view the dwindling tips as a ticking clock that reflects a cell's age. Some scientists dispute that, however, and others contest that the buildup of aged cells with shrunken telomeres explains the overall aging of an animal."[21] Questions concerning the rate of aging that emerged with the new technology of cloning were explained in one popular press account with reference to the microscopic level (telomeres) rather than the organismic level.[22]

While Minot studied normal growth at the microscopic level, tracing "the growth of animals and the progressive changes in cell structure from birth to death," British physician and researcher Dr. Hastings Gilford, F.R.C.S., studied abnormal growth at the level of the human organism.[23] Gilford's investigation of growth disorders began with the experience of an uncategorizable human being:

> Nearly twenty years ago a little shrivelled boy-man was brought into my consulting-room. It was obvious at a glance that his complaint was not of any ordinary kind. He was, indeed, a most extraordinary object; but although he was afterwards often examined with the view of finding out the nature of his disease, I could not arrive at any satisfactory conclusion. . . . Subsequently it became clear that the disease was premature old age, engrafted upon, or running side by side with, a condition of immaturity or infantilism. . . . These two morbid conditions, spontaneous and premature senile decay or "progeria," and spontaneous persistent immaturity or "ateliosis," furnished plenty of subject for cogitation. (Gilford 1911, v–vi)

As he explains in the preface to his massive work *The Disorders of Postnatal Growth and Development*, Gilford's surprising visitor raised a number of connected issues: "This opened up the larger question of the differences between growth and development, and between the growth and development which go on in the whole man, in his individual organs, and in his individual cells" (vi). The two physicians shared an understanding of the interconnectedness of the beginning and end of the life span. Minot focused on the analytic power of aggregate representations of growth rates, while Gilford's work on premature aging and abnormally extended infancy raised questions about the relation between micro- and macroprocesses of growth. Both participated together in the reconstruction of the human being also enacted by Wells's novel, for they shared its focus on the (re)constructive power of hard and soft technologies, statistics, and social context.[24]

SCIENCE FICTION RECONSIDERS GROWTH

Charles H. Rector's 1927 short story "Crystals of Growth" explored the same hormonal angle that H. G. Wells's Bensington worked twenty-three years earlier, and combined the interest in the treatment of growth disorders with an attention to the microprocesses of growth.[25] Appearing in *Amazing Stories,* Rector's story is accompanied by a caption providing the following scientific context: "What makes giants, and what makes dwarfs? Scientists of today are unanimous in the opinion that these conditions are created by the thyroid gland. If it were possible to systematically stimulate these glands, there is no question but that a race of giants could be produced. It is not at all impossible from a biological standpoint, and from experience gained in the laboratory, we know that it can be accomplished" (Rector 1927, 875). Professor Brontley, whose "theory of rapid growth" leads him to develop some crystals of growth from which a person "would derive a much greater benefit . . . than from ordinary food," acts as his own guinea pig, ingesting the substance, which produces a race of giants. He exhorts his friend the narrator to join his experiment: " 'My height,' he said, 'is now about twelve feet, a little more than double my former height. If I had wished to become still taller I could have done so by taking more of the crystals. Think what a relief these crystals will be to mankind. No more under-developed children! No more short men and short women. Tomorrow I shall show the world that I am a living proof of the existence of a superfood, the crystals of growth. Try one yourself, Jameson, and increase your size' " (876).

Yet when Brontley tries to press the superfood on his friend, the effect is anything but a relief. Struck by the "expression of mad fanaticism upon his face," Jameson evades him: "I don't need any of your growth crystals. I'm tall enough to suit me." "You little shrimp," Brontley sneers, "I'll show you whether you need them or not" (876). Brontley struggles furiously with Jameson, finally choking him senseless. When the narrator awakes, the room is empty, and Professor Brontley has disappeared. "The opinion of the majority of those who examined the laboratory was that a violent combat of some nature had taken place. But, the crystals of growth were nowhere in evidence" (877).

If Rector's scientist used hormonal methods to accelerate growth, conceptualized in terms of height, only three years later another writer imagined a form of genetic engineering that accelerated growth by producing vastly greater physical strength. Philip Wylie's little-known novel *Gladiator* (1930) shares many of the basic plot elements of Wells's *Food of the Gods*.[26] The man who would one day produce that classic of postwar misogyny *A Generation of Vipers* (1942, 1955), Wylie here turns his attention to the gendered implications of growth differentials.[27] His protagonist Abednego Danner, a chemist driven by the belief that "chemistry controls human destiny," is investigating the properties of "Alkaline radicals."[28] He is particularly convinced that "there is something that determines the quality of every muscle and nerve. Find it—transplant it—and you have the solution" (Wylie 1930, 4). After experimenting on hundreds of tadpole eggs by inserting into them "ultra-microscopic bodies . . . [that] were the 'determinants' of which he had talked," he creates a tadpole strong enough to break the plate glass of the aquarium. Danner realizes his accomplishment: "He had made a living creature abnormally strong" (5). Next he inoculates a pregnant cat with the serum and produces a kitten, Samson, so abnormally strong that a terrified Mrs. Danner guesses, "You have made all this rubbish you've been talking about strength—happen to that kitten." Voracious for meat, the kitten begins by killing large dogs and finally decimates the local cattle. At only seven weeks of age, it is fed poisoned meat by its creator, and its "dying agonies" are "Homeric" (9).[29]

Unlike Wells's Bensington and Redwood, Danner is appalled by the outcome of his experiment. Clearly strength trumps size as a threat to the status quo. "He wondered if his formulae and processes should be given to the world. But, being primarily a man of vast imagination, he foresaw hundreds of rash experiments. Suppose, he thought, that his discovery was tried on a lion, or an elephant! Such a creature would be invincible" (Wylie 1930, 9–10). Danner gives up his experimentation, until his wife tempts him with the unexpected news that she is pregnant (and thus, implicitly, a captive experimental subject). The contrast between Wells's female characters and Wylie's women is illuminating, reflecting the opposed social views of the feminist Wells and the misogynist Wylie. Bensington's aunt and Redwood's wife are both highly skeptical of the experiment in human enhancement, but Danner's wife essentially provokes it by tempting him sexually into creating the son on whom he experiments. (Before this new departure, Wylie tells us, Danner and his wife had avoided the "mechanics" of parenthood "except on those rare evenings when tranquility and the reproductive urge conspired to imbue him with courage and her with sinfulness" [11].)

"The logical step after the tadpoles and the kitten was to vaccinate the human mammal with his serum. To produce a super-child, an invulnerable man. As a scientist he was passionately intrigued by the idea. As a husband he was dubious. As a member of society he was terrified" (12). Both his own ambivalence and his awareness of the practical barriers to the experiment—the knowledge that his wife "would never allow a sticky tube of foreign animal matter to be poured into her veins"—do not dissuade Danner. Even the possibility that he could create an unnaturally strong female does not daunt him, though it does make him shudder: "He envisioned a militant reformer, an iron-bound Calvinist, remodeling the world single-handed" (12). Science exerts a force on him almost providential: "A hundred times he denied his science. A hundred and one times it begged him to be served" (12). Finally he feeds his wife opiate-laced cordial till she sleeps, and injects her with the fateful serum. The child, born on Christmas Day, seems poised between savior and Antichrist. When his wife guesses what Danner has done, his confession is tinged with defiant masculinist aspirations, originating perhaps in his henpecked timidity: "He'll be the first of a new and glorious race. A race that doesn't have to fear—because it cannot know harm. You can knock me down. You can knock me down a thousand times. . . . I have given a son whose little finger you cannot bend with a crow-bar. Oh, all these years I've listened to you and obeyed you and—yes, I've feared you a little—and God must hate me for it. Now take your son. . . . You cannot bend him to your will. He is all I might have been. All that mankind should be" (16).

While *The Food of the Gods* ends with the optimistic notion that society will accommodate the new giant human beings created by Herakleophorbia, *Gladiator* ends not with optimism but pessimism, the product in large part of its insistently individualistic focus. Wells's novel dramatizes how an intervention in growth can jump from the individual to the aggregate and in so doing produce a new norm for humanity. In contrast, Wylie's novel obsessively dramatizes an individual's ostracism on the grounds of his inhumanity. Hugo is consistently disappointed in his attempts to make good use of his extraordinary strength. Whether on the football field in college, as a Coney Island strongman, a sailor, a soldier in the French foreign legion, a steelworker, a farmhand, a political organizer, or a member of an archaeological expedition to the Yucatán, Hugo nets only suspicion, ostracism, and fear from this gift that his father had hoped would work to the benefit of all mankind.

Finally, called to his father's deathbed, the son cannot bear to admit that his father's experiment has failed. Instead he responds to his father's implicit questions with falsehoods: "Hugo knew what those questions would be. Here, on his death-bed, his father was still a scientist. His [Hugo's] soul flinched from giving its account. He saw suddenly that he could never tell his father the truth; pity, kindredship, kindness, moved him. 'I know what you wanted to ask, father. Am I still strong? . . . I am. I grew constantly stronger when I left you. In college I was strong. At sea I was strong. In the war. First I wanted to be mighty in games and I was. Then I wanted to do services. And I did, because I could' " (Wylie 1930, 155).

When his father marvels, "To be you must be splendid," Hugo responds with an ironically compassionate evasion: "The most splendid thing on earth! And I have you to thank, you and your genius to tender gratitude to. I am merely the agent. It is you that created and the whole world that benefits" (158). But if Hugo intended to put his dying father at ease, he has ironically succeeded in unsettling himself even more, when his joyful father bequeaths to Hugo the problem that has tormented him day and night: "Should there be made more men like you—and women like you?" (158). In an ironic reversal of the monster's petition to Frankenstein, Hugo must consider whether he wishes to have a friend or a mate. Moreover, he is offered the choice of merging monster and scientist, when his father informs him of the existence of six lab notebooks in which he recorded the experiment that created his superstrong son.

> That is my life-work, Hugo. It is the secret—of you. Given those books, a good laboratory worker could go through all the experiments and repeat each with the same success. I tried a little myself. I found out things—for example, the effect of the process is not inherited by the future generations. It must be done over each time. It has seemed to me that those six little books . . . are a terrible explosive. They can rip the world apart and wipe humanity from it. In malicious hands they would end life. . . . You

will not avoid this, the greatest of your responsibilities. Since the days when I made those notes—what days!—biology has made great strides. For a time I was anxious. For a time I thought that my research might be rediscovered. But it cannot be. The fact of you, at best, may remain always no more than a theory. . . . My findings were a combination of accidents almost outside the bounds of mathematical probability. It is you who must bear the light. (158)

The conclusion of Wylie's novel is a curious equivocation, tracing, with its final turn, the conflict between biological enhancement and social restriction. Hugo becomes an unpaid participant in an expedition to the Yucatán launched by Professor Daniel Hardin, a noted archaeologist. Hugo's new role is grounded in the conviction that "of all human beings alive, the scientists were the only ones who retained imagination, ideals, and a sincere interest in the larger world. It was to them he should give his allegiance, not to the statesmen, not to industry or commerce or war" (177–78). This final episode, in which Hugo accompanies Professor Hardin to the Yucatán, convinced that he will find there the traces of an earlier civilization whose people, too, had been extraordinarily strong, provides the platform for an impassioned argument for the eugenic improvement of the human race. When Hugo confides his history to Hardin, the archaeologist seizes on the vision of a new civilization it offers:

Other men like you. Not one or two. Scores, hundreds. And women. All picked with the utmost care. Eugenic offspring. Cultivated and reared in secret by a society for the purpose. Not necessarily your children, but the children of the best parents. Perfect bodies, intellectual minds, your strength. Don't you see it, Hugo? You are not the reformer of the old world. You are the beginning of the new. We begin with a thousand of you. Living by yourselves and multiplying, you produce your own arts and industries and ideas. The New Titans! Then—slowly—you dominate the world. Conquer and stamp out all these things to which you and I and all men of intelligence object. In the end—you are alone and supreme. . . . A city in the jungle—the jungle had harbored races before: not only these Mayas, but the Incas, Khmers, and others. A modern city for dwellings, and these tremendous ruins would be the blocks for the nursery. They would teach them art and architecture—and science. Engineering, medicine—their own, undiscovered medicine—the new Titans, the sons of dawn—so ran their inspired imaginations. (184–85)

With this shared vision, we can see the impact of growth jump from the individual to the aggregate. In a sense, the father's intervention in his son's life can achieve its purpose only if it spreads from one person to a whole civilization, so that what appears abnormal in the single individual is reconceptualized as the new (and vastly more powerful) norm. Yet both Hugo and the entire novel retreat from this eugenic utopian vision. Anguished and doubtful, Hugo climbs the highest peak in the Yucatán, pleading for a sign from God. Though the creator never speaks, Hugo and the notebooks perish in a providential lightning strike, leaving his archaeological sponsor "saddened and perplexed."

Three science fiction short stories published in *Amazing Stories* in the late 1930s and early 1940s turn to additional strategies for accelerating and maintaining growth: irradiation, hormone treatment, and an early form of genetic engineering. The social and biological effects of these interventions in growth, including their gender implications, are as diverse as their strategies, demonstrating that the meaning of growth is not inherent but socially situated. In Ed Earl Repp's "The Gland Superman," the overt issue is what we would now call "doping," the use of steroids and growth hormone in athletics. Biochemical researcher Dr. Harland Gale exposes the pituitary gland to a "para-Roentgen ray," "infinitely more penetrating than the Roentgen ray—or the X-ray, as it is called," to remedy the growth defects caused by the "ductless glands" (Repp 1938, 12, 11).[30] The strategy is manifestly effective with animals, as the directors of the Mellon Institute learn when Professor Gale leads them past some of his creations: "He led them past cages crowded with unearthly creatures that brought startled gasps from the Directors. Rabbits as large as sheep; mongrel dogs the size of Saint Bernards; canaries whose beaks could have crushed walnuts; and a large, glass-lidded box full of such things as a foot-long grasshopper, a flea as big as an armadillo, and mice the size of puppies" (14). But with human beings, Dr. Gale has hit a snag. As he explains to the visiting board members from the Mellon Institute, his funding agency, "I can do nothing to remedy the horrible distortions man is subject to because of unruly glands. I can cause growth . . . but that is all" (13). When the Mellon visitors refuse to fund his experiment, he is desperate to find a source of support and finds his inspiration in the sports page of the evening newspaper. Able to cause, but not to direct, growth, he lets technology determine his strategy, vowing, "I'll build my own champion! A gland champion!" (15). He uses the para-Roentgen ray technique to enhance the build of a lightweight boxer, and after some technical setbacks, the scientist is successful. The boxer wins his bout, and Dr. Gale explains: "There's three-quarters of a million dollars waiting in the bank for you. . . . And Henry—you've made it possible for me to give humanity one of the greatest gifts it has ever received" (29).

Ross Rocklynne's "Big Man" (1941) continues the narrative experimentation with endocrinological interventions into growth, combining the theme of improving the human race (through intervening in growth) with the theme of addressing growth abnormalities. The story concerns a six-foot-tall, eighteen-year-old boy whose growth is accelerated with injections of "tethelin" until he becomes a "two mile giant" (Rocklynne 1941, 80). "Big Man" orders the president of the United States to put aside democratic rule and submit to his greater power to "solve every problem that puzzles a long-weary, long-unhappy world." "Henceforth," he declares, "I am the master of human destinies in America" (73). Yet as the story proceeds, we learn that Big Man was created by a scientist possessed of "a humanitarian instinct tinged with a lust for power," who developed a method to control the "master glands, and through them [to] make giants, giants whose glands would keep on working, past the limit of growth, in complete harmony" (75). Big Man is really a naive, nearly feral child, a slave whose massive growth was produced by the normal-size scientist, who, as Big Man's master, controls him from a collar-shaped control platform around his massive neck. In a cataclysmic ending, the story issues a clear critique of the notion that human growth can be controlled by the few to better the lot of the many: the powerful Big Man is blinded, and both slave and master face a devolutionary kind of destruction by walking into the sea.

In "The Test Tube Girl," by Frank Patton (1942), growth is accelerated by a primitive form of recombinant genetic engineering in order to preserve the last viable female embryo in a world rendered sterile after a nuclear disaster.[31] While the goal is to preserve humanity, the effect is not preservation but transformation. Accelerating growth through the combination of human and plant material, rehabilitating blood with "liquid plant chlorophyll," these scientists create a hybrid that harkens back to the plant-animal network explored by Wells in *The Food of the Gods* (Patton 1942, 16). Working with the "Lindbergh mechanism," a kind of ectogenetic facility like an artificial uterus, the scientists at

Eugenics Laboratories manage to induce "a month's growth in twenty-four hours" in the embryo they are culturing. Forced by a military accident to leave the Lindbergh mechanism unattended for five months, they return to find the ultimate liminal being: the "chlorophyll girl" "alive, and apparently developed to the stage, still in an embryonic environment, of a sixteen-year-old girl! By all the laws of nature she should have been walking and breathing now. Or she should be dead. But she's neither!" (24). The biologist who has created her marvels: "In all my years I've never dreamed of anything so biologically impossible as this. A full-grown human being, in less than seven months altogether. Twenty years, crowded into seven months" (24).

Like the giant creeper that provides one of the earliest indications that the Herakleophorbia has leaked into the environment, the test-tube girl has plantlike properties that seem to define her. As one of her creators explains, "The chlorophyll has so changed her that the human life span means nothing. She has the life span of a flower, growing with the spring, flowering with the summer, withering with the frost, and dying with the snowfall" (34). Yet we learn at the story's conclusion that not only do we need to rethink the norm for a life span because of her mingled plant and human origins, but we also need to make sure we have adopted the right botanical model. As she explains: "Did you ever hear of a perennial?" . . . Can't you understand. We *are* plants. As much plant as human. But we *aren't* going to die. We just change with the seasons" (53).[32]

"The Test Tube Girl" resembles those later works of science fiction, *Mr. Adam* and *The Children of Men*, in its tale of the gestation of the final human fetus before the human race succumbs to universal infertility. Through the story circulate themes we recall from earlier science fictions of growth by Rocklynne, Repp, and Rector: a focus on the way that rate of growth is determined by the earliest stages of life; an awareness of the linkages between the macro- and microprocesses of growth; and an appreciation of the differential growth rates not only characteristic of different species but present within each species. Since Minot's explorations of the relations between embryonic life and the end of life, Gilford's investigations of progeria, Rector's cautionary story of "crystals of growth," and the images of hypertrophic men and supple, perennial women, a significant shift has occurred in the attitudes toward intervention in growth.

"The Gland Superman." Ed Earl Repp,
"The Gland Superman," *Amazing Stories* 12,
no. 5 (October 1938).

(*facing page*)
"Big Man." Reprinted from Ross Rocklynne, "Big Man,"
Amazing Stories 15, no. 4 (April 1941).

"The Test Tube Girl." Reprinted from Frank Patton,
"The Test Tube Girl," *Amazing Stories* 16, no. 1 (January 1942).

BIG MAN

BY ROSS ROCKLYNNE

In maddened fury, the pilot drove his plane straight toward Big Man's face, guns spitting

What was the grim purpose of this collossal figure that stalked steadily toward Washington?

ON that night of June, 1978, A.D., across the miles of quiet water, from island to island, from coastal steamer to pleasure yacht, from ship to island to radio stations on the Florida mainland, flashed mad, coded messages:

"I saw it, I tell you. It was a man. I saw it with my own eyes. He came right out of the horizon. He filled the whole horizon and threw a big shadow down onto the water. I couldn't see the Moon. . . . For God's sake, tell me, did you see him?"

From a coast guard station on the Floridan coast:

"Are you crazy? If this is a gag! Listen—"

"But I *saw* it!" Wildly. "I *think* I saw it. . . . Maybe I didn't. My head aches. Wait till tomorrow. Maybe I'm sick. I *must* be sick."

But verifications skittered madly across that stretch of quiet sea. People had seen a man, a *big* man, coming up out of the sea!

"Did you see it, WX31D?"

"Did I! Thank God *you* saw it! If you hadn't—I don't know what . . ."

The sun came up, bathing the world in the hideous light of reality it must

70 71

The TEST TUBE GIRL

By Frank Patton

"Look at her! That's what she needs! A sun bath! She's like a flower in the sun!"

Nowhere in the world any more children—except one baby in a test tube! Was mankind doomed to die?

"LOOK, Allan, my boy, how beautiful it is—" the man in the soiled laboratory smock waved a trembling hand toward the ghostly, moonlit city spread far below the tiny veranda high in the tower of Eugenic Laboratories "—and in a few more minutes we will know whether or not it will all vanish from the Earth. . . .

"And in a few minutes, I, Henri Varrone, the man those desperate millions down there believe to be the greatest of all biologists, will know whether they are right, or horribly wrong. . . ."

"No! Wait—don't do it!"

Allan Sutton's hoarse shout interrupted the biologist's sombre tones.

Varrone whirled about, bewildered.

"What. . . ." he began, then, as Sutton stared upward in horror, his gaze went up the facade of the building beside the veranda to a window ledge.

A white-clad figure was outlined in the moonlight, standing on the very edge of the stone sill. It was a woman, her face pale, tragic, drawn.

"Myra!" called Sutton in a stricken whisper now that carried weirdly through the still night air. "Don't jump. . . ."

Then, as though released from a momentary paralysis, he began to edge

8 9

RETHINKING *GROWTH* AND GROWTH HORMONE

This chapter opened with a quotation taken from the first issue (in 1937) of an interdisciplinary journal dedicated to the investigation (regardless of the discipline used in its study) of its titular term, *Growth*: "a fundamental and common property of nature." The twentieth-century science fictions of growth that we have surveyed exemplify the qualities integral to the journal's intellectual position. The interaction between human, animal, vegetative, and mineral growth that we have seen in the science fiction echoes the journal's principled attention in its earlier years to scholarship on any kind of growth, whether human, animal, vegetative, or mineral. The attention to the chemical, physical, and hereditary aspects of growth dramatized by Wells and Wylie, like the mathematical, statistical, chemical, and physical measurements enacted in each of the works of science fiction, seem to function in an uncanny way as preemptive narrative enactments of the mission of the new journal: "To uncover these laws and to test them" ("Introduction to Growth" 1937, 1).

When in 1988 the magazine renamed itself *Growth, Development, and Aging*, its reorientation indicated the increasing integration of growth with the overall conception of the life span, making interventions in growth more relevant to those other interventions at the beginning and end of life. While still limiting itself to basic biology, *GDA* now invited submissions incorporating three new areas of investigation:

> (1) *Relationships among patterns of growth and development and patterns of aging*. To what degree are these all part of the same process, so that patterns of development predict subsequent patterns of growth and aging? (2) *Genetic and environmental effects on growth, development, and aging*. The modern techniques of molecular genetics promise to provide new types of information. Do some genes affect development, growth and aging in the same ways? (3) *Methods of measuring aging changes in different biological systems and evaluating effects of treatments to retard aging*. To study aging, its symptoms must be defined, to develop treatments that benefit these symptoms.[33]

The questions raised in this journal's new mission statement reflect three changes in the understanding of human growth that took place over the course of the twentieth century. First, we can see the increasing awareness that growth is an integral part of the life span, implicated in the process of aging just as the rate of aging is indicative of the future process of growth. Coupled with that is the awareness that aging occurs, as Minot had suggested more than a half century earlier, even during embryonic development. Second, we can see the slide from investigation and measurement of biological processes to interventions in those processes in order to treat the "symptoms" that are now seen to characterize them, whether those symptoms are spatial (being too short) or temporal (being too old). Third and finally, we can see those interventions on the threshold of being recategorized from medicine to cosmetics, or from therapy to enhancement technology. This latter term, which I am borrowing from bioethicist Carl Elliott, refers to "the use of medical technologies not to cure or control illness and disability but to enhance human capacities and characteristics."[34]

An example of this transformed understanding of human growth appears in the changing medical and social attitude toward the substance known as growth hormone, or GH. GH, a water-soluble neurotransmitter that can pass in the blood to all parts of the body, is produced by the placenta and later by the pituitary gland. Growth hormone was initially thought only to be important during the prenatal and childhood years, during which it causes skeletal growth, including the over- and undergrowth called acromegaly and achondroplasia, that gives us the giants and dwarfs we know from popular culture as well as fiction.[35] Similarly, the progeria and ateliosis investigated by Dr. Hastings Gilford we now know to have been the effect of excessive or inadequate GH.

While it lags somewhat behind them, the medical time line of significant growth hormone discoveries parallels the major science fiction treatments of growth. Yet while the stories move from an embrace of the intervention in growth on eugenic grounds (Wells) to a resistance to that intervention on social grounds (Wylie) or at least an articulation of the way such an intervention irreparably alters the human species (Patton), the actual course of GH discoveries and treatments takes the reverse course, moving from the narrowly medical to the broadly social. In 1921, Herbert Evans of UC Berkeley "was the first to postulate a specific growth-stimulating substance released by the anterior pituitary" (Schulman and Sweitzer 1993, 58). Isolating that substance from the pulverized pituitaries of cattle, Evans injected it in rats and produced excessive growth. UC Berkeley researcher Choh Hao Li, a student of Evans and a member of his Institute of Experimental Biology, succeeded in isolating GH from animals (1945) and from human cadavers in 1955. Although this prepared the way for growth hormone replacement therapy in 1957, when Maurice S. Raben of Tufts University Medical School "used pituitary-derived hGH to make an abnormally short, GH-deficient boy grow six times as fast as he had been growing," it also ushered in the new and frightening era of trans-species illnesses. "By mid-1970s," Schulman and Sweitzer report, "hGH had been collected from 82,500 human pituitary glands and was being used to treat about 3,000 very short children with proven GH deficiency" (61). Then, as Gina Kolata (1986) reports in *Science*, in 1985 "a catastrophe occurred": three people who had been treated with GH in the previous two decades died of Creutzfeld-Jakob disease, the rare brain disorder that has more recently been known as "mad cow disease." "The most likely reason that these people developed Creutzfeld-Jakob disease is that they received growth hormone extracted from pituitaries of other Creutzfeld-Jakob victims and that these pituitaries contained infectious slow virus particles [prions]."[36] Although the FDA immediately banned all use of GH from cadavers, a replacement was found. This new hormone was biosynthetic, produced in *E. coli* bacteria by genetic engineering. Because this new product was nearly the same as pituitary GH, it was given the name "somatrem" and was approved in October 1985 by the FDA for use with GH-deficient children.[37]

What did it mean to be GH deficient? The classic criteria for such a deficiency before 1963 were observational, "based on height, appearance, and medical history" (Schulman and Sweitzer 1993, 122). By those criteria, GH-deficient children usually exhibited a failure to grow at the normal rate; their growth curve generally departed dramatically from the expected familial growth pattern; and they retained the chubby cheeks and delicate bones of babyhood. Both passive and active tests were developed in 1963 to check for the production and efficient use of GH. In the "spontaneous test of GH production," blood is drawn over a period of every twenty minutes for four hours, or by sampling over a twelve-hour night, and the fluctuating rate of growth hormone production is monitored. Another kind of test measures GH production by stimulating it, by administering a range of treatments from beef broth through exercise to injection with substances known to trigger GH production: in this case, the amount of GH secreted will be measured (123–25). In both kinds of tests, however, GH deficiency is usually diagnosed if the test subject does not secrete the amount of growth hormone thought to be normal: "generally defined as between 8 and 10 ng/ml, on two successive tests using different sources of stimulation" (125).[38]

Neither the test itself nor the circumstances of its use were as absolute as such definitions suggest, however. As Lantos et al. (1989) point out, "The Food and Drug Administration's criteria blend two indications—the morphological abnormality of growth failure and the biochemical abnormality of inadequate growth hormone—that are both relative rather than absolute" (10–21). Moreover, in the early years of GH replacement with cadaver-derived GH, GH deficiency was in reality defined by the supply of GH rather than the demand for it. As Schulman and Sweitzer (1993) point out: "In some ways the question of GH deficiency was moot because whether there was deficiency or not, there was not enough pituitary GH to go around to all the GH-deficient children. Only a tiny amount could be gleaned from each human pituitary, and donated tissues became increasingly hard to come by. As a result, treatment was not extended to all those who might have benefited. Only children with no detectable GH production were given treatment. No partial deficiency was acknowledged. *It was this lack of* GH for treatment that originally defined GH deficiency" (61–62; italics mine).

In light of this ambiguity in the diagnosis of GH deficiency, Lantos et al. suggested in 1989 that GH therapy be compared not with treatment for a disease but with pediatric treatments given for non-life-threatening problems in three areas: preventive well-child treatment, treatment designed to improve performance in school and psychosocial well-being, and cosmetic therapy (Lantos et al. 1989, 1022). The question of repositioning GH therapy from medical to social treatment was still under debate a decade later. In a 1999 article, bioethicist Ruth Macklin turned the focus from the disease/prevention dichotomy to the more profound dichotomy underlying it, between normalization and enhancement. Considering the "questions surrounding the administration of synthetic growth hormones to short normal children," Macklin concludes: "It could be argued that treatment for short stature . . . is not truly an enhancement but is rather an attempt to normalize the child's condition. Children who are candidates for this treatment are considerably shorter than statistically average children: the purpose of therapy is to bring them into the range of normal height. Although there is no known organic cause for their short stature, from a statistical point of view these children are deficient in height" (Macklin 1999, 1, 9).

Macklin's analysis divides the ethical questions raised by such uses of growth hormones into four categories: (1) whether treatment is appropriate in the absence of "disease or biological deficiency," (2) how best to proceed when research results are "ambiguous or uncertain," (3) the difficulty of engaging in a risk/benefit assessment, and (4) how to decide what person or people should make the treatment decisions (Macklin 1999, 1). Macklin's intended audience is the bioethics community within medicine, as is made clear by her closing observation: "If and when sufficient data demonstrates the safety and efficacy of growth hormone treatment for short normal children, physicians, parents, and children themselves will have to make the difficult benefit-risk calculation to determine whether to embark on a course of therapy" (10). The implicit hierarchy of agency here puts physicians at the top and the children themselves at the bottom.

By the summer of 2003, however, the FDA had essentially taken the decision away from that entire hierarchy of agency, granting the pharmaceutical manufacturer Eli Lilly and Company the right to market their drug Humatrope for children between ages seven and fifteen who are abnormally short not because of disease or GH deficiency but simply because of unknown natural causes. As the *New York Times* reported: "Children who are healthy but abnormally short will be able to have injections of growth hormone in hope of gaining one to three more inches of height, the Food and Drug Administration said Friday, deciding an emotionally charged issue. . . . 'This is not cosmetic use,' said Dr. David Orloff, the agency's chief of endocrinology."[39] Although for sixteen years GH has been administered *only* to children who have medical conditions deemed to require it (and, because of GH shortages, to but a subset of them), the FDA's decision was the first to include in this GH-eligible group simply those whose growth curves predicted shorter-than-normal height: "boys predicted to be shorter than 5-foot-3 as adults, and girls shorter than 4-foot-11." While Lilly pledged to maintain "tight restrictions on Humatrope's availability," controlling the specialists who can prescribe it and the pharmacies where it is available, the fictional precedent provided by H. G. Wells's *Food of the Gods* calls this pledge into ironic question, sensitized as we are to the effect of social pressures, as well as human technological fallibility, on the hope to keep Humatrope (any more than Herakleophorbia) under strict control.

Note that while the theme of the interference with human growth rates has remained the same, the nature, strategy, and implications of that interference with growth rates have changed dramatically. If we return for a moment to Rector's 1926 story "Crystals of Growth" and recall how Brontley's sneering determination to show Jameson whether "or not" he needs the crystals of growth stands there as fully adequate narrative proof of his medical overreaching, we will realize the magnitude of the change that has occurred in our understanding of human growth. By 1999 the very acceptance of "normal variation" has ceded to an unquestioned process of normalization, no doubt because statistical representation trumps personal testimony. In the second half of the twentieth century, GH has increasingly taken on biomedical significance throughout the life span. The notion of a GH deficiency shifts from being something that is medically established to something that is clearly socially, economically, culturally, *and* medically constructed. The possible interventions with GH extend from gaining height in childhood to gaining bulk, strength, more lean muscle mass, and so on, in adulthood and into old age. Thus the representation of, and intervention in, growth in the twentieth century calls into question the oppositions that have been integral to the human life span: youth and old age; birth and death.

The shift in our understanding of GH from being a medical treatment to an enhancement technology interestingly interacts with our shift in the understanding of hormone replacement therapy (HRT) from being an enhancement technology (for women and men in their middle years) to being increasingly understood as a medical necessity. Can we assume the same thing will happen with GH? And what are the stakes of this for our understanding of the relationship of science fiction and medical science? Or to put it another way, what relationship exists between the representation of growth and development in scientific and science fiction representations, chosen as they are from both American and British literature in the twentieth century?

We have seen a significant departure from the defeat of the fanatical growth specialist in "Crystals of Growth," Gilford's understanding of the socially constructed nature of normal and abnormal growth, or Wells's exploration of the social implications of accelerating growth with the "food of the gods" to a statistically based intervention that recalls Minot's work with guinea pigs. Indeed, I am tempted to argue that humans now share with guinea pigs and all other research animals the condition of being conceptualized in the aggregate. And with the shift to a statistical, aggregate model, the whole issue of whether or not to intervene in human growth has been channeled through a set of medicalized assessments, and the liminal point shifted from individual preference ("I'm tall enough to suit me") to the norm within a population.

UNDERSTANDING THE CHANGED CONCEPTION OF GROWTH

Wells's 1904 novel *The Food of the Gods* urged a focus not on containment but on permeability, not on boundaries but on processes of dissemination. Thirty-three years later, *Growth* also proselytized for an interdisciplinary perspective, even if bounded by biological science. Inspired by Wells and the other science fiction writers, as well as by the scientists who founded *Growth*, to take a transdisciplinary perspective, we have seen how a shift in the conceptualization of growth occurred in the twentieth century across a range of different discourses, from literature to medicine to bioethics. Here again, the social and ethical issues appear more fully fleshed out in fiction long before they are debated in the medical literature. But they also encourage our willingness to read across the divide, to read both literatures in relation to each other so that we can get the full picture. From food of the gods to crystals of growth, through irradiation and hormone treatment to genetic engineering, science fiction has imagined methods of intervening in growth that have migrated from fiction to fact in the course of the twentieth century. Medicine has moved from the discovery of growth as an area of study (including its normal and abnormal facets), through the isolation of animal and then human growth hormone, to the synthesis of synthetic growth hormone, then to the notion of applying growth hormone to abnormally small children, and finally to an increasing acceptance that growth hormone is important for the maintenance of health, muscle tone, vigor, and so forth, throughout the life span.

What do the parallel investigations of growth that I have traced in this chapter tell us about the status of our epistemological and disciplinary categories for self-understanding? Recently several science studies scholars, plotting the trajectory of the contemporary shift in our modes of self-understanding and self-representation, have argued that it will extend to the digital and the informatic.[40] "The claim [is] . . . this: two co-occurrent, synergistic transformations—the ongoing move to parallel and distributed computing and the explosive growth in visualization—are reconfiguring contemporary technoscience. The effects of this emergent parallelism as it circulates through cultural space are being felt at every level from how we read, write, and see to the ways we understand ourselves as 'selves.' And that, in the process of its unfolding this parallelism is giving rise to a hitherto unavailable, and yet to be adequately identified, serialism" (Rotman 2000). (That is to say, the effect of changes occurring simultaneously, and distributed across populations, will be to shift the very nature of how human individuals grow and change, one by one.) While this argument is persuasive, in my view it doesn't go far enough, because it underestimates the extent to which this move to parallelism and then back to a new serialism is already well under way. The move to parallel and distributed computing, like the growth of visualization, has a prehistory: the medical and cultural reformulation of the human being made possible by a range of technologies from statistics and graphing to science fiction. Significant cultural work was being done even before the transformations attributed to contemporary technoscience. Dating as far back as the seventeenth century, synergistic transformations in literature and science took place in parallel at a number of liminal positions ranging from the boundaries of the human, to species boundaries, and finally to the structural social liminality we think of as disciplinary fields. These parallel points of intervention on living members of the community, reshaping the meaning of both the individual and the group, both performed and enforced a significant structural shift whose material and symbolic effects on the way we live out our lives we are only beginning to appreciate.

5

Incubabies and Rejuvenates

THE TRAFFIC BETWEEN TECHNOLOGIES OF REPRODUCTION AND AGE EXTENSION

On 16 August 1926, *Time* magazine carried the following story: "Physiologists convening in Stockholm all but forgot other topics in a furore created by Dr. Serge Voronoff, famed gland-grafter. . . . To his Swedish hosts he revealed that he had grafted within Nora, a mature female chimpanzee, the sex organs of a human female. Then, with assistance from Dr. Elie Ivanoff of Moscow, he had artificially impregnated Nora with human sperms. She was to bear her baby in January and it would be, biologically, a human child."[1] Voronoff's pregnant chimpanzee joins a long list of experimental subjects leading to the birth of Louise Brown, the first IVF baby, in 1978, and to the late-twentieth-century innovation of surrogate birth. This medical news item also interweaves modern literary and social history, for Nora is a whimsical echo of Ibsen's Nora, her name an ironic (perhaps even vengeful?) response to the feminist, suffragist agitations of the early twentieth century. Moreover, her "ape-child" invokes preoccupations with degeneration and race that gripped Britain in the years before World War II.[2] Yet while the tale of Nora the chimpanzee embodies the converging discourses of medical science, feminism, and racialization in the project of reshaping reproduction, it also embodies another preoccupation of early-twentieth-century science and culture: the project of scientific rejuvenation. In what follows, I will excavate some of the links between those two projects—now more familiar to us as assisted reproduction and age extension strategies—and I will suggest what we might learn from that forgotten history about our approach as a culture and society to the liminal human moments at both ends of the life span.

The "Nora experiment" was actually part of a far-larger project of gland grafting carried out in the 1920s and 1930s by the French-Russian scientist Dr. Serge Voronoff. Its overall goal was not to induce pregnancy but rather to mitigate the painful symptoms of aging.[3] While Nora shared with the rest of the graft recipients the condition of being postmenopausal, the rest of the recipients were human women, who had chimpanzee ovaries grafted into them as a rejuvenation treatment. This gland grafting for women was conceptualized as a medical intervention designed not only to ward off aging but in so doing to enhance "normal" familial ties, social relations, and gender roles.[4] Like the other scientific and parascientific therapies for rejuvenation carried out between the 1860s and the 1930s, Voronoff's gland grafts were motivated by a surge of cultural interest in rejuvenation in Great Britain, Europe, and the United States. Beginning before the turn of the century when Charles-Edouard Brown-Sequard injected himself with the extract of canine testicles, rejuvenation experimentation continued in the first three decades of the twentieth century with Voronoff's so-called monkey gland grafts; Viennese physiologist-endocrinologist Eugen Steinach's eponymous experiments with vasoligature (tying off the sperm ducts) and x-raying the ovaries; the "reactivation" treatments of New York physician Harry Benjamin, which added "diathermy," or electrical current treatment, to the surgical, radiological, and hormonal treatments already in existence; and the cellular treatments (injections of prepared fetal sheep cells) of Swiss glandular surgeon Paul Niehans.[5] Frequently represented through mythic tropes such as the elixir of life, the fountain of youth, or the Holy Grail, these "rejuvenation therapies" were situated at the lively intersection of fact and miracle, science and culture.[6] They generated controversy in the popular press; became the focus of science fiction short stories, bestsellers, and canonical novels; provided the theme for a smash Hollywood movie; and exercised a shaping impact on our attitude toward aging that persists to the present day.

The story of Voronoff's Nora suggests that early-twentieth-century rejuvenation therapy was closer than we have previously realized to that other major twentieth-century intervention in human life: the set of techniques for scientifically shaping conception, gestation, and birth known as reproductive technology or, most recently, assisted reproduction. Both projects bore the stamp of several closely linked discourses emerging or achieving prominence in the first three decades of the twentieth century: the eugenics movement, the new fields of chemical embryology and "sex endocrinology," the sexology movement, and the life span reconceptualization fundamental to the emergence of gerontology.[7] That these new fields would intertwine, influencing the ways medical science intervened in the beginnings and endings of human life as the century went on, is not surprising. Despite their different foci, they all shared an interest in improving the bodily bases of human life, in the male and female human life course, and in the efficacies of surgical and chemical intervention. What interests me is the way those scientific and technical strategies for reshaping the human *body* functioned in consort with cultural and social conditions shaping the human *subject*. In what follows, I will turn to the kinds of cultural work—on the body, the subject, and on social relations—performed by reproductive technology and rejuvenation therapy. But first we must consider their debt to the new—or newly consolidating—discourses circulating in the early decades of the twentieth century. We can then move on to explore the fantasies that fueled (and continue to fuel) reproductive technology and rejuvenation therapy, as articulated through their representation in fiction and popular culture.

EUGENICS

From the founding in 1907 of the Eugenics Education Society, the eugenics movement was a point of convergence for medical practitioners, scientists, and social reformers interested in improving the human species by enabling "the more suitable races or strains of blood a better chance of prevailing speedily over the less suitable," in the words of the society's honorary president Francis Galton.[8] The saga of the eugenics movement, from its origins in Francis Galton's and Karl Pearson's statistical analyses of heredity, through its widespread albeit contested popularity in Britain and the United States, as a movement housing both emancipatory and notoriously racist social projects, to its disappearance in the late 1930s with the revelations of Nazi eugenics, is too complex to detail here. In fact, the saga does not end in the 1930s, as Daniel Kevles and others have observed; while the eugenics movement went underground during World War II, it reemerged in the 1950s as human genetics. In 1954 the principal publication of the Galton Laboratory at University College, London, changed its name from the *Annals of Eugenics* to the *Annals of Human Genetics*, and a new field of genetic counseling emerged, to carry on the reform eugenics ideal under a kinder, gentler name.[9]

Two specific aspects of early-twentieth-century eugenics are worth highlighting, because they influenced the linked projects of reproductive technology and rejuvenation therapy. First, since eugenics addressed the improvement of the human species through negative and positive means, the point of intervention was the beginning of a human life: the processes of conception, gestation, and birth. Eugenicists were willing both to discourage or even forbid reproduction that they understood as disadvantageous to the human species as a whole (negative eugenics) and to provide incentives for reproduction that they deemed advantageous (positive eugenics). Of course, the very concept of what was "disadvantageous" was profoundly subject to ideological influence. Thus my second point: when eugenicists weighed desirable traits, longevity was frequently included. Although few scholars of the eugenics movement emphasize this point, the eugenics writers of the 1920s and 1930s included the timing and nature of the end of life in their assessment of the eugenically desirable and the "dysgenic."[10] So for eugenicists, the process of embryological development and the process of aging were crucial to the project of improving the species: they took as a prominent goal the longest possible deferral of the latter, and they took as their strategy intervention in the former.

CHEMICAL EMBRYOLOGY AND ENDOCRINOLOGY

From its nineteenth-century emphasis on morphology, or the structural development of the embryo, embryology began to shift, with the opening of the twentieth century, to an interest in the chemical processes that triggered embryological development. At the turn of the century, embryologists took a mechanical and experimental approach to the problem of how embryos developed: "variously compressing, constricting, and centrifuging eggs, and killing or separating the cells of early embryos, all directed to finding out at what stage cells become committed to specific pathways of differentiation."[11] In a series of classic experiments, Hans Spemann divided embryos with loops of hair, producing "[quite] different results . . . depending on the orientation of the constrictions. If the constriction was in the median plane, dividing the embryo into left and right halves, normal embryos resulted. If the constriction divided the embryo into dorsal and ventral halves, only the dorsal half developed normally" (Witkowski 1987, 259). Spemann continued this analysis later, using transplants and grafts. Continuing this emphasis not on naturalist observation but on experimentation, a group of scientists at Cambridge University, led by Joseph Needham and including Dorothy Needham and colleagues C. H. Waddington, J. D. Bernal, Dorothy Wrinch, and Joseph Woodger, began in 1920 to explore the "biochemical basis of embryonic development" (Witkowski 1987, 247). From Hans Spemann's turn-of-the-century work to the Cambridge group's work in the early 1930s, this interest in the chemical basis of embryological development led experimenters to investigate the effects of dividing embryos at early stages of growth, and (later) implanting centrifuged embryonic cells into host embryos.

The interest in the chemical bases of embryological development was paralleled by a new interest in the chemical basis of sexual development. As Nelly Oudshoorn has documented, the discovery of sex hormones in the first decade of the twentieth century led to an explosion of experimental laboratory endocrinology. In a series of experiments remarkably parallel to the work in embryology, scientists turned to the surgical removal and transplantation of gonads to investigate the chemical substances present in the sex glands. "In this surgical approach, scientists removed ovaries and testes from animals like rabbits and guinea-pigs, cut them into fragments, and reimplanted them into the same individuals at locations other than their normal positions in the body. With these experiments scientists tested the concept of hormones as agents having control over physical processes without the mediation of nervous tissue."[12] Organotherapy, as this procedure was called, was also used on human subjects, although as Dr. H. Lisser observed in 1925, "It is curious that far less interest has been manifested in ovarian implantations than the corresponding procedure in the male sex, despite the fact the ovarian castrations are infinitely more common than testicular castrations."[13]

Early-twentieth-century endocrinology anticipated to a considerable degree the issues that would be raised by the nascent field of reproductive technology, as is apparent in a "remarkable" case of ovarian grafting reported by Lisser in the journal *Endocrinology* in 1925: "In this case amenorrhoea had set in after labor and symptoms of absence [*sic*] developed. The small cystic ovaries were removed and a foreign ovary implanted. The woman again menstruated four months after the operation, and four years after the operation she conceived and bore a normal child. This case was earnestly discussed before the Edinburgh Obstetrical Society, and can not well be denied. *There is, indeed, in this case no doubt that if there has been no error of observation, this woman bore the child of another woman.* Serious objections might be raised against such procedures on ethical and forensic grounds."[14]

The surgical approach—with its manifest complications—was superseded once the so-called sex hormones were chemically isolated and identified in the 1930s: now rather than surgically transplanting gonads containing hormones, the chemical isolates themselves could be administered (Oudshoorn 1994, 48).

The developmental courses of embryology and sex endocrinology have been traced by a number of scholars, most notably Jane Oppenheimer, Evelyn Fox Keller, and Nelly Oudshoorn, and I will not rehearse their findings here.[15] Rather, I will simply delineate the shaping impact of these two fields, developing parallel to each other in the first three decades of the twentieth century, on the linked RTs. Having begun with a grossly surgical approach, both embryology and sex endocrinology moved in the early years of the twentieth century to a chemical approach, searching for the chemical properties that explained development, whether of the embryo in the former case, or of the sex organs and sexual characteristics of the individual in the latter case.

SEXOLOGY

To the surgical and chemical approaches to developmental and social/behavioral questions, the new field of sexology added psychological and psychoanalytic approaches as well. Catalyzed by the work of a number of German physicians, sexology achieved its first public presence in 1908 with the appearance of the *Journal of Sexual Science*, founded by psychiatrist Dr. Magnus Hirschfeld. The field attained institutional consolidation with the establishment of the Institute for Sexual Science, in Berlin, in 1919, and with the Second and Third International Congresses of the World League for Sexual Reform, in Copenhagen (1928) and London (1929).[16] Concerned with a range of issues including "sexual biology, sexual pathology, sexual ethnology and sexual sociology," the sexology movement taxonomized sexual behaviors (particularly homosexuality and the psychophysiological condition sexologists designated "intersexuality"), considered surgical and hormonal solutions to sexual difficulties, and helped to set the stage for medical scientific interventions in birth and aging.[17] Some of these interventions were given prominence in 1929, in *Sexual Reform Congress*, the volume compiled after the Third International Congress. As well as articles on eugenic sterilization, abortion, and contraception, the volume included two essays specifically on age extension: Dr. Harry Benjamin's essay "The Reactivation of Women," reporting on the range of "reactivation" or rejuvenation therapies in clinical use with female patients, and Dr. Peter Schmidt's "Seven Hundred Rejuvenation Operations: A Nine Years' Survey," providing his clinical assessment of seven hundred vasoligature cases he had performed in the nine years since the publication of Eugen Steinach's *Rejuvenation*.[18] An interdisciplinary or multidisciplinary field itself, sexology lent to the parallel and successor projects of rejuvenation therapy and reproductive technology a characteristically modern technoscientific optimism and reliance on a broad arsenal of medical scientific treatments for biosocial difficulties, as well as a proto-postmodern skepticism about normalizing categories.

GERONTOLOGY

Zoologist Elie Metchnikoff coined the term "gerontology" in 1903, and in a career mingling embryology and experimental medicine, he went on to lay the foundation for a new, multidisciplinary field, "the scientific study of old age."[19] Personally motivated to understand aging by an encounter with serious illness at the age of fifty-three, Metchnikoff modeled in his own life the convergence of disciplinary involvements—in embryology, zoology, and immunology—that would characterize all these emerging fields of study. The story of gerontology's emergence as a (multi)discipline from Metchnikoff's pioneering work onward has been compellingly told by W. Andrew Achenbaum and, more recently, Stephen Katz.[20] But there is one specific point about the relationship between gerontology and the dual projects of reproductive technology and rejuvenation therapy that merits emphasis. Preparatory to the emergence of gerontology in the twentieth century, a shift occurred in the construction of the human life span that provided the crucial context for a renewed interest in the perennial project of retarding age or regaining youth. As Stephen Katz has shown, before the mid-nineteenth century, Enlightenment-based views mingled science and the supernatural to construct the life span as an infinitely extendable period capped by old age as a philosophically significant but natural part of the life continuum. By the mid-nineteenth century, in contrast, a new modern construction of the life span reflected the taxonomizing, quantifying, and hierarchizing impulse of modernity, positioning old age as a discrete, separate, medically classifiable and finite developmental stage.[21] While hitherto the boundary between miracle and fact had been blurred—so that accounts of marvelous longevity mingled with Baconian-influenced treatises on life prolongation through proper diet, exercise, and avoidance of harmful emotions—with the emergence of modern medicine, "medical, demographic and insurantial investigations proved the life-span to be fixed, thus disbanding premodern images of excessive longevity as fanciful," and old age was constructed not as a site of miraculous possibilities but as a medical problem (Katz 1995, 62).[22] These gerontologically motivated changes in the construction of the life span both reflected and catalyzed the modern impulse to tinker with the life course scientifically, the impulse that—I am arguing—undergirds both reproductive technology and rejuvenation therapy.

SHARED TRAITS OF THE TWO RTS

The discourses of eugenics, embryology, endocrinology, sexology, and gerontology left the following imprints on reproductive technology and rejuvenation therapy, or what I am calling the two forms of RT: the scientific and social focus on a liminal subject (the embryo or fetus, and the aging person or neo-mort); the use of other species as research reservoirs (as in Voronoff's use of Nora the chimpanzee and Steinach's experimentation with hormone injections in cattle, and the more recent use of hamster and mouse ova as crucial aspects of the development of in vitro fertilization);[23] the rationalization and Taylorization of human bodies and of medical scientific interventions into bodily processes (beginning with the separation of sexuality from reproduction, foundational to both reproductive technology and rejuvenation therapy); the reliance on visualization technologies, from endoscope and microscope to the cinematic apparatus, as both a measurement tool and a rhetorical device (ranging from the use of laparoscopy in the development of in vitro fertilization to before-and-after films of rejuvenation subjects and the hit motion picture made from a novel about rejuvenation); the existence of national rivalries for scientific and medical discoveries (as representatives of different nations vie for control of the new rejuvenation techniques, just as different nations vie to be first in certain reproductive technologies); a linkage to the broader modern agenda of exercising medical scientific control over the body (as a decline in religious certainties and a breakdown of community, accompanied by an increasing medicalization of the body, has led people to take refuge in viewing the body as a project);[24] a gender-asymmetrical social uptake of the medical scientific innovations (as women's bodies become the primary focus of scientific and medical interventions in Western society since the eighteenth century, as fleshly objects, subjects of anatomical dissection, and gendered skeletons, and the primary source for research materials).[25] Finally, and most significant for my purposes here, both forms of RT are the subject of a fantasmatic investment whose dimensions and implications can be gauged by looking at their representation in imaginative literature.[26] These fantasmatic investments play out in the production of bodies, subjects, and societies, as I traced in the case of reproductive technology in my study *Babies in Bottles*, and as I will go on to detail in the case of rejuvenation therapy.

FICTIONS OF REJUVENATION

Two novels that explore the personal and social impact of successful rejuvenation therapy will give me my point of entry to some of these issues, particularly to the desire to participate, in fantasy, in techniques for warding off or rolling back the onset of aging. These novels had very different cultural positions: Gertrude Atherton's 1923 *Black Oxen*, about a woman who has received the Steinach treatment (the x-raying of the ovaries), displaced Sinclair Lewis's *Babbitt* on the U.S. bestseller list; and C. P. Snow's *New Lives for Old*, about the discovery of a technique for rejuvenation through hormone injection, was published anonymously and to little fanfare in Britain in 1933.[27] These novels reveal the contestation over the terms on which rejuvenation therapy was practiced, the meaning that it would have in the cultural imaginary, and how those meanings were inflected by the gendered project of modernity.

Rejuvenation therapy was big news in 1923, the year *Black Oxen* appeared. If we trace the networks leading out from the different methods of rejuvenation therapy, as evidenced by the coverage in the *New York Times* and *Time* magazine for that year, we see they extend from cities to small towns, join nations around the world, connect disciplinary colleagues from different countries, and even involve members of different species.[28] Thus when a visit to the United States by Eugen Steinach was impending, the *Times Sunday Magazine* enthused: "New men for old! Within a few months the name of the Professor of Biology at the University of Vienna has become the talk of cities and has penetrated to the furthest hamlets [of the United States]."[29] Scientific, aesthetic, civic, and personal contexts converged in this new interest in rejuvenation. On 21 July 1923, Serge Voronoff was reported to have galvanized "700 of the world's leading surgeons" at the International Congress of Surgeons in London, where he spoke on "the success of his work in the 'rejuvenation' of old men," relying on film both to document his scien-

C. P. Snow. Photograph by Mark Gerson. Reprinted from John de la Mothe, *C. P. Snow and the Struggle of Modernity* (Austin: University of Texas Press, 1992). Reprinted by permission.

tific successes and to persuade his audience: "At London, Voronoff presented moving pictures showing the transference of monkey glands to human beings, with 'before-and-after' effects on three specimen cases—men aged 65, 74 and 77 respectively, in more or less advanced stages of decrepitude. Within periods of four to 20 months after the operations, the films showed them as hale and active, apparently in middle age, riding horseback, rowing and doing other athletic feats."[30]

By 13 October the *Times* reported that Voronoff had so impressed his fellow physicians at the French Surgical Congress that he was scheduled to perform his gland graft operations on eight of them right after the congress adjourned. The new frenzy for gland operations had even enhanced the international value of the chimpanzee, Voronoff's preferred graft source, resulting in new government regulations and international collaborations. As the *Times* reported, "So scarce have chimpanzees become that the Governor of French West Africa has issued an ordinance protecting them and the Pasteur Institute has sent Dr. Vulbert to Africa to establish a farm to raise them."[31] As the year wound to a close, and the *Times* continued its frequent reports on rejuvenation-related items, even the rare bit of negative press was embedded in a modernist discourse of instrumentality, mechanism, and Taylorized bodies. In a two-hour address to the International Congress of Comparative Pathology in Rome, a Chicago surgeon attacked rejuvenation therapy but claimed that "further developments in surgical science would make it possible to patch up all human maladies by the use of spare parts, just as one did with a motor car."[32]

This was the climate, then, in which Gertrude Atherton's *Black Oxen* appeared. While clearly inspired by newspaper coverage, the novel had the extra impetus of Atherton's own Steinach treatment in 1922 and took its title from the work of another Steinach patient, W. B. Yeats, who wrote in "Countess Cathleen," "The years, like Great Black Oxen, tread the world, / And God, the herdsman, goads them on behind."[33] An "instantaneous best-seller," *Black Oxen* was also instantaneously controversial. By October 1923, the mayor of Rochester, New York, had ordered every copy of Atherton's novel "removed from the city's public libraries" on the grounds that the novel was "unfit for young minds."[34] Prohibition proved a spice to sales, and by the following year, *Black Oxen* had become a hit movie, produced by Frank Lloyd and starring Corinne Griffith, Conway Tearle, and Clara Bow.

The novel's plot can be summarized fairly quickly: Mary Ogden Zattiany, a legendary beauty, undergoes Steinach's treatments shortly after World War I and is dramatically rejuvenated: at age fifty-eight, she now looks a young thirty. When she returns, incognito, to New York City, to settle her financial affairs and raise money for her Vienna Fund for Austrian war orphans, she provokes curiosity and confusion in her friends (who think she is her own daughter), resentment once her secret is out (from her age cohort *and* from women a generation younger who are competing for male suitors), and finally love in the thirty-four-year-old aspiring playwright Lee Clavering. In short, the news of her rejuvenation treatment provokes a social and scientific tempest: "Although she would not consent to be interviewed, there were double-page stories in the Sunday issues, embellished with snapshots and a photograph of the Mary Ogden of the eighties: a photographer who had had the honor to 'take' her was still in existence and had exhumed the plates. Doctors,

"At Dunwiddie's Mountain Lodge." Movie still from *The Black Oxen*.

biologists, endocrinologists were interviewed. . . . When it was discovered that New York actually held a practicing physician who had studied with the great endocrinologists of Vienna, the street in front of his house looked as if some ambitious hostess were holding a continual reception" (214–15).[35]

Yet if Mary's recaptured youthful beauty has caused a younger man to love her, that does not mean that she can unproblematically return his love. Atherton's novel is explicit that rejuvenation therapy has an asymmetrical meaning for the two sexes. Women are sterile, even if rejuvenated, and social constraints limit their enjoyment of even nonreproductive sexuality: "That is where men have the supreme advantage of women. . . . If these [rejuvenated] men indulge occasionally in the pleasures of youth, or even marry young wives, the world will not be interested. But with women, who renew their youth and return to its follies, it will be quite another matter. If they are not made the theme of obscene lampoons they may count themselves as fortunate" (325–26). No doubt in response to this social reality, but also in response to Mary's experience as a nurse at the front during World War I, the novel's conclusion retrospectively constructs Mary's motivation for undergoing rejuvenation—and thus the motivation of all women—not as a desire for youth, sex, and love but as an attempt to achieve agency.[36]

Mary decides *not* to marry the young man she loves. Rather, she accepts the proposal of an older suitor, who is soon to be the chancellor of Austria, because he will provide her with the position she needs to help the orphaned children of Vienna recover from the ravages of World War I.

C. P. Snow's *New Lives for Old* shares with *Black Oxen* the sense that men and women will have a range of different reasons for being rejuvenated, which they can gratify to a greater or lesser extent depending on social conventions. However, Snow's novel differs in the protagonists' position in relation to rejuvenation treatment, as well as in the *kind* of treatment it represents. While Atherton wrote her novel from the perspective of the *object* of medical scientific intervention (since she had actually undergone the Steinach treatment), Snow wrote his novel from the perspective of a *subject* of science. A celebrated participant in the traffic between science and literature, and author of the well-known study *The Two Cultures,* C. P. Snow took a Ph.D. in physical chemistry at Cambridge University, and *New Lives for Old* reflects the excitement of that scientific involvement. But in its initial anonymous publication, the novel also reflects Snow's experience of the more troubling side of science, for it replays "the biggest embarrassment of [Snow's] young scientific career," when he and his research partner Philip Bowden mistakenly announced that they had successfully synthesized vitamin A, and their work was sharply criticized by two senior scientists, leaders in the field of vitamin chemistry.[37] In an instance of fictional reparation, Snow's autobiographically tinged protagonists, Billy Pilgrim, professor of biophysics in King's College, London, and his junior research partner David Callan are successful where Snow and Bowden failed. They synthesize a human sex hormone, collophage, which can prevent the physical atrophy of the body associated with aging. As Billy Pilgrim explains: "The machinery of the body could go on for years and years after old age begins . . . except that with the years less and less collophage is being made, and somehow the body dries up and dies before its time. . . . And if you provide collophage artificially, you prevent this decay happening for a long time. . . . To-night is nothing more nor less than the beginning of the science of rejuvenation" (19).

When Pilgrim informs the prime minister of their accomplishment, he attempts to block the public announcement of the discovery on the grounds that it may be too destabilizing. But when Pilgrim threatens to release the news first to the Russians, Britain implements rejuvenation therapy. The consequences are not wholly surprising—even at the moment of discovery, the two scientists quarreled bitterly over whether their new substance should be offered to the public—but Snow documents them extensively. They include the creation of two classes of people (those who are biologically or economically able to be rejuvenated and those who are not); the disruption of the structure of employment (as older workers fail to retire, and younger workers are unable to advance); the dislocation of the family and of the generational system on which it is founded (as rejuvenated wives leave unrejuvenated husbands for younger lovers, and rejuvenated mothers compete sexually with their daughters); competition over access to rejuvenation therapy (as classes and nations vie for access, resulting in misapplication and needless deaths); overpopulation (as rejuvenated men continue to father children with young women); and the creation of new economic markets among the rejuvenated (for art, for cosmetics and cosmetic surgery).[38] Whereas Atherton's novel stressed the intimate personal benefits and costs of rejuvenation to one woman, Snow has given us a balance sheet of its implications for a whole society, men and women. He paints a dire picture for his privileged protagonists: what began as a scientific discovery ends as a revolution, when the poor—unable to afford the costly rejuvenation treatment so widely used by the rich—riot and seize power in Great Britain.

In contrast to Snow's predictions that rejuvenation therapy would have revolutionary consequences, the medical practices developing from rejuvenation therapy in fact rather than fiction seem anything but revolutionary. Instead, between 1935 and 1985, society quietly assimilated hormone replacement therapy for women into normal medicine, framed less as a *rejuvenation* strategy than as a "treatment" for the female menopause. Although significantly less publicity has been given to male hormone treatment, whether as therapy or as a rejuvenation strategy, it too has had its advocates.[39] This developmental trajectory from rejuvenation treatment to hormone therapy for the "disease" of menopause reflects changes in the new discipline of endocrinology as well as the traffic between popular culture and science, as I understand it. Several factors channeled the development of sex endocrinology, and thus hormone therapy, to a predominately female population in the 1920s and 1930s. First, as we have already seen, the identification and classification of the sex hormones in the 1920s led to a shift from a focus on the gonads as agents of sexual difference to a stress on a chemical approach to sex endocrinology, and on hormones as keys to both human sexuality and human health. Then the existence of disciplinary sites for intervention in women's bodies (the gynecological and obstetric clinic) narrowed the global focus on glands from a symmetrical focus on both "male" and "female" sex hormones, to an asymmetrical concentration of scientific, medical, and pharmacological attention on ovaries, on "female" sex hormones, and on women as the site of intervention.[40] A smaller market existed for male sex hormone therapy, reflecting the much-smaller number of intake sites—or medical occasions—for men to be treated, whether by "organotherapy" or hormone injection. Urological problems did not bring men into clinics as regularly as gynecology and obstetrics did women, and to reiterate Dr. Lisser's point in 1925, "ovarian castrations are infinitely more common than testicular castrations" (14). Finally, a recoil occurred away from the therapeutic uses of male hormones, in response to the exaggerated claims made in the popular press for their uses in rejuvenation therapy. Popular culture thus collaborated with the disciplinary structures of clinical and research medicine to focus hormone therapies on the female body and to shape medical claims away from overt talk of rejuvenation.

A similar recoil away from exaggerated popular press claims for rejuvenation powers shaped gerontology, according to Achenbaum. Reflecting this, a feature in the first issue of the *Journal of Gerontology*, in 1945, took pains to distinguish the work of responsible scientific gerontologists from "alchemists, charlatans, con artists, or uninformed adventurers. 'What kind of fountain of youth are we seeking: a fountain that will miraculously erase the wrinkles of age, or a fountain that will make the later years of life a healthy and intellectually occupied period?' " (Achenbaum 1995, 128). When we consider these novels in relation to their era, in which gerontology and endocrinology were struggling for scientific legitimation against the backdrop of a rejuvenation-mad popular press, what we make of the whole will depend, to some extent, on our position. Those of us committed to understanding the drive to "fix" aging, as if it were a disease rather than a naturally occurring process, may be struck by the movement from external to internal fixes, from the macro to the micro, from grafting to gland therapy, to injections, and finally now—with the discovery of the gene for Werner's syndrome of premature aging—to genetic manipulation.[41] Those of us interested in the intersection between popular culture and scientific culture may notice the powerful influence that negative popular press can have on medical scientific development, as well as the enhancing impact of positive press.[42] Those of us interested in literary representations of aging will focus on the way that the two novels' plots differently present the meaning of rejuvenation, demonstrating that the discourse of rejuvenation therapy is as fissured and contested as that of reproductive technology.[43]

Finally, for those of us interested in the interplay between science and culture, Atherton's *Black Oxen* and Snow's *New Lives for Old* register a change of direction not only in early-twentieth-century endocrinology but in the culture at large, in their representation of different methods of rejuvenation therapy.[44] If Snow's *New Lives for Old* portrays a class-based rejuvenation revolution, in a sense that may be because a prior sex-based revolution had already been squelched: the potential revolution embodied by Mary Zattiany's choice of rejuvenation not for sex or love but for agency and power. Through a combination of structural, ideological, and cultural forces, both within and beyond the field of medical science, the nascent field of rejuvenation therapy was being redirected: from the feminist goal of providing aging women with a vigorous life free from the entanglements of (hetero)sexuality and reproduction, to the scientific goal of (re)shaping aging women's bodies to the standard of male delectation, so that they stay visually pleasing and reproductively functional.

SYNTHETIC WOMEN

Another short story from the same era, also expressing the fantasy of constructing a woman wholly for male delectation, can return us from a concentration on rejuvenation therapy to attention to the interplay between the two forms of RT in the twentieth century. Jep Powell's 1940 story "The Synthetic Woman" concerns not rejuvenation but synthetic conception and gestation: reproductive technology, in short.[45] Yet exemplifying the intertwined nature of these two life span interventions, Powell's story of reproductive technology also explores some of the central issues raised by rejuvenation therapy: discrepancy between age and generational positioning, disturbance of the life course, discrepancy between physiological and psychological aging, and the problem of the ethical, forensic, or ideological implications of technoscientific interventions in the life span. The tale concerns Vivian, an "incubaby" formed by the combination of chemicals and ectogenetically gestated, whose aging process has been accelerated by the injection of a growth-enhancing substance, "oxydyne," so that at chronological age eight she has all the physiological appearance of a "beautiful, full-blown bud of womanhood" (Powell 1940, 117, 102). A technoscientific anomaly, legally underage yet the result of scientifically accelerated development, Vivian wants to marry the wealthy young man who has kissed her in the laboratory, although she fails to understand the first thing about marriage, gender differences, and sexuality. Yet her creator, the pseudomystical Dr. Shaiman (the pun is clearly intended), is saving Vivian for another: the racially and intellectually devolved, hirsute, and remarkably "dark" incubaby Bruno, who has been damaged in (artificial) utero by an excessively large application of testosterone. The tale ends happily; Vivian is saved from the devolved Bruno, marries her wealthy suitor, and has a "naturally born" son, catalyzing a dispute between Dr. Shaiman and the attorney-narrator about who is the rightful grandfather (124). Yet as is so often the case with mass-market science fiction, the romance plot and the optimistic conclusion are shadowed by some troubling questions about the social implications of science: "Could she, after all, marry Daniel Laird? What was her legal age? Would the law hold her to her calendar age, or would it recognize her biological age? Was she a legal entity at all? . . . What would be the legal interpretation of 'born'? . . . Surely Vivian looked, and acted, and seemed like a normal person. But, in following out the very letter of the law, must she be labeled a 'processed person,' a synthetic? The thought was ghastly" (108).

"The Synthetic Woman" draws on the embryological and endocrinological experiments of the first three decades of the twentieth century to imagine a woman whose sexuality is out of synchrony with her chronological age, troubling her legal and social status, and indeed her very identity. Nearly forty years before the birth of the first test-tube baby, this tale of Vivian the "incubaby" forecasts the profound sociolegal and psychological disturbances that would result from scientific

Cover illustration of "The Synthetic Woman."
Reprinted from Jep Powell, "The Synthetic Woman,"
Amazing Stories 14, no. 9 (1940): 100.

interventions in reproduction and aging. The story attempts to recontain the threat of these interventions in its concluding section, "All's Well," where we learn of the birth of Vivian's baby. "'Born naturally,' Shaiman cackled triumphantly. 'And I'm a grandfather!' 'You?' I snorted. '*I'm* the grandfather. Didn't I adopt her legally?'" (124). Destabilized by technoscientific interventions, biological and generational relations can only be sutured, now, by another form of expert intervention: the legal process. Now that birth and aging are subject to (re)construction, a cultural shift has occurred that requires the full force of the ideological state apparatuses to remedy.

The true story of Voronoff's Nora, a decade and more before the publication of "The Synthetic Woman," registered an early tremor in the large cultural shift predicted by Powell's science fiction story. Some months after Voronoff announced Nora's impregnation with "human sperms," she was found not to be pregnant after all. The experiment in surrogate mothering in a postmenopausal female subjected to hormone-induced rejuvenation had failed, and scientists would have to work for another sixty plus years before they succeeded. Finally, in 1993 and 1994, the Western world saw a sudden rash of postmenopausal pregnancies following hormone treatment and the implantation of embryos (from donor ova). From the project of gland grafting into a chimpanzee that gave us the (falsely pregnant) Nora, through the fiction of a synthetically created baby subjected to rapid aging by hormone injections, to the actual hormone treatments and ovum transfer that gave us the (successfully pregnant) sixty-one-year-old Rossana Dalla Corte, we have arrived at a troubling new situation for women.[46] A convergence of the two major technologies to be exerted on women's bodies in the twentieth century—reproductive technology and what used to be known as rejuvenation therapy—has the potential to produce a drastic narrowing of a woman's arena for agency and intervention, from the wide world of political and social relations to the far more limited territory of her own body.

My point is not that either of these two technologies is *inherently* constructed to limit women's options; at earlier moments, each could have been shaped to be emancipatory. Now, however, they are being deployed to reinforce each other, making each a matter of coercion rather than choice, technoscientific intervention rather than personal action.[47] Synthetic woman, indeed. Yet to conclude in the mode of modernist paranoia, satisfying as it is, neglects the more complex coda to this story. If in its popular cultural image, between 1935 and 1985, hormone therapy seemed to cling to the asymmetrically gendered control project of modernity, with its persistent focus on a female patient to be "normalized" or synthesized by a male medical profession, a subterranean discourse of male rejuvenation through hormone therapy has always also existed, practiced out of the public eye, in exclusive Harley Street surgeries, Manhattan clinics, and Swiss spas. In the last decade, mainstream medicine has demonstrated a more postmodern and complex attitude toward both hormone treatment and the nature of maleness and femaleness. The rise of evidence-based medicine has definitively challenged the efficacy of hormone replacement therapy to prevent heart disease, osteoporosis, and Alzheimer's disease.[48] While medicine has retreated from intervention in the female menopause, the interest in what is being called the "male menopause" has been growing. A new medical specialty, andrology, has emerged, marked by a new publication, the *Journal of Andrology*.[49] After a study by the University of Pittsburgh Medical School assessed the medical impact of a scrotal testosterone patch, and other studies by the National Institute on Aging, the Medical College of Wisconsin, and the University of Pennsylvania School of Medicine explored the replacement of hGH, testosterone, and other less-well-known hormones, testosterone replacement therapy has begun to receive the marketing push previously dedicated to Premarin and other estrogen replacement products.[50]

From the original goal of improving the human "product," rejuvenation therapy and reproductive technology may have shifted to a broader goal: reconfiguring the human life span. Dedicated to blurring its constitutive categories—those fixed biological life stages of parenthood and generationality—both projects may now serve a new construction of birth and aging, as exemplified in the notion of the postmenopausal mother. These conjoined narratives—of reproductive technology and rejuvenation therapy, of *Black Oxen* and *New Lives for Old*, of synthetic women and postmenopausal mothers—suggest five questions that we should ask about the ways we approach (theoretically and personally) the experiences of giving birth and of aging. I have argued elsewhere that reproductive technology has shifted our perspective on reproduction, so that "from conceptualizing both gestation gone right and gestation gone awry as natural (because both outcomes were found in nature), we have come in our era to policing the outcome of gestation medically."[51] Adapting one of the findings of Dr. Peter Snyder's Pittsburgh study of male hormone replacement therapy, we should ask:

1. "Whether these bodily changes [infertility, aging] represent a condition in need of treatment or whether the changes are physiologically normal aspects of [life] that men [and women] just have to accept."[52]

2. If we "fight" aging through what I am calling rejuvenation therapy, as we "fight" infertility through reproductive technology, for whom or what are we doing so? For love? Or, like Mary Zattiany, for agency? Do

we even want to accept that either/or? If it is agency, where is that agency located? Only in our bodies, or in the world?

3. What can bodies or babies be if they are not our products?

4. What collective responsibility are we overlooking when we conceptualize both sorts of RT—rejuvenation therapy as well as reproductive technology—as individual choices?

5. How will we negotiate the boundaries of identity and subjectivity as both the beginnings and the ends of life become increasingly characterized by technoscientific collaborations with the human?

Not only do these intertwined narratives raise questions about the medical interventions into women's bodies that characterize modernity and postmodernity, but they also suggest some of the forces that come into play when a new medical field is struggling for legitimation. Between the falsely pregnant chimpanzee Nora and the successfully pregnant Rossana Dalla Corte, the two RTs have emerged as asymmetrically situated in medical practice and research. Reproductive technology—or assisted reproduction, as it is now called—has forged a solid and profitable interdisciplinary home for itself in contemporary medical practice. In contrast, "rejuvenation therapy" has moved from a central position in the work of Voronoff, Benjamin, and Steinach in the early years of the twentieth century to a marginal position at present. It lingers mainly as an unacknowledged goal or a welcome by-product of more or less legitimate medical endeavors (from the increasingly marginalized medical use of hormone replacement therapy to the increasingly widespread dermatological uses of Retin-A) and as the continuing recipient of a strong fantasmatic investment expressed in mass-market tabloids and popular books on how to "stop the clock" and "beat aging" forever. Yet powerful cultural linkages exist between reproductive technology and rejuvenation therapy, between fantasies of incubabies and of rejuvenates. As we will see, with the growth of regenerative medicine from the late 1990s into the first decade of the twenty-first century, new strategies are emerging, applied with increasing efficacy, that resituate the marginalized project of rejuvenation within orthodox medical practice.

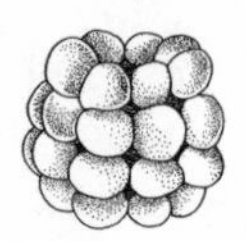

6

Transplant Medicine and Transformative Narrative

An English artist steals human bodies and body parts to make plaster casts of them, which he decorates with precious metals.[1] A German anatomist "plastinates" cadavers, immersing them in cooled liquid acetone, infusing them with a secret polymer preparation, arranging them in often stylized scenes, and drawing ten thousand people in one day to his exposition of "anatomy art."[2] A Canadian artist exhibits jewelry made from twelve-week-old fetuses and solicits townspeople for donations of body parts to incorporate into his art projects. A Dutch pathologist collects the organs and bodies of young children, keeping the head of an eleven-year-old child in a jar, and labeling one nine-week-old fetus "Inflated monster. Humpty Dumpty."[3] LifeGems, a company based in Chicago, Illinois, sells alternative "personalized" memorials to grieving people: synthetic diamonds of between .25 carat and 1.3 carats made from the carbon remains of a loved one, whether human or animal. As the company's Web site explains, "The proprietary LifeGem creation process creates diamonds from the true essence of our loved ones, the carbon. Our families receive the ashes as all others do when choosing cremation, except our families also receive a certified, high-quality LifeGem created diamond to memorialize their loved one's unique and wonderful life."[4] These disturbing manipulations of the human body, the subject of media and judicial attention at and beyond the beginning of the twenty-first century, raise questions about the way we understand human organs and body parts.

All these interventions manipulate the human body, are meant to be consumed by viewers (whether the private collector, the public audience of an art exhibition, or the bereaved visiting or viewing a memorial), and invoke a set of aesthetic, economic, legal, and temporal tensions: between art and science; between worthless plaster and precious metals; between the socially central and the marginal, the criminal and the celebrated; between the perishable and the everlasting. In short, these episodes provide me with another perspective on the liminal lives that are my subject, because they participate in a central aspect of the reevaluation of human life at the beginning of the twenty-first century: the transformation of the value and duration assigned to human organs and body parts.

Although they occur in the seemingly very divergent realms of art and medicine, they share a preoccupation with the manipulation of the human body that has been, since *Frankenstein*, a central trope in science fiction. As such, they draw our attention to the role science fiction has played, both bridging and boundary breaking: raising questions about the epistemological impact of the disciplinary divide as well as registering shifts (both semiotic and material) occurring across the entire field of nature-culture. In what follows, I will move from exploration of four instances of organ and body part manipulation that occurred around the turn of the twenty-first century back to three science fiction representations of the same phenomenon whose publication dates span the twentieth century. I have selected the science fiction short stories I discuss not for their distinctiveness but as representative expressions of the biomedical imaginary of their era. I assume that many other SF works express similar issues and concerns; I have chosen these stories as case studies of that broader phenomenon.

My goal here is to engage in an exploration of the foundations of the contemporary field of transplant surgery in the biomedical and cultural imaginary. Anthropologist Nancy Scheper-Hughes has observed, "Transplant surgery has reconceptualized social relations between self and other, between individual and society, and among the 'three bodies'—the existential lived body-self, the social, representational body, and the body political. Finally, it has redefined real/unreal, seen/unseen, life/death, body/corpse/cadaver, person/nonperson, and rumor/fiction/fact."[5] The various ways that human beings in Anglo-America have imaginatively negotiated relations between the "three bodies" itemized by Scheper-Hughes tell us much about the role of culture and society in the normalization of transplant surgery. In particular, science fiction has functioned as a pivot point, the zone of the in-between, the uncategorizable, even the abject, rather than fitting securely in either of the seemingly secure zones of "art" and "science." As I will argue, because of its status as an in-between or liminal zone (freed from the epistemological constraints of disciplinary knowledge), science fiction is able to perform an imaginative transformation of the body that can predate and, in fact, enable its biomedical transformation.

ORGAN ART

Anthony-Noel Kelly, a part-time teacher of sculpture and cousin of the Duke of Norfolk, was tried in 1998 for "theft of anatomical specimens from the Royal College of Surgeons" and received a three-month prison sentence.[6] Having worked as a butcher and in a slaughterhouse, Kelly used "body parts in his art, explaining his procedure of molding plaster casts from the remains and gilding the plaster copies in gold and silver. He then disposed of the body parts themselves, which were not incorporated into the finished artwork in any way" (Trull 2003, 2). He obtained some of the bodies he needed for his work from mortuaries, relying on a friend who worked for the Royal College of Surgeons, but he also worked on family members, "reportedly craft[ing] one of his pieces from the body of his grandmother, a former diplomat" (2). The fact that Kelly's grisly sculptures were based on actual bodies was discovered when a visitor to one of his exhibits recognized the face of the sculpture before him: "an old man who had recently died. The head was coated in silver and part of the exposed brain had been removed" (1).

While Kelly's case of a scientist gone bad has a lineage extending back to Barnes and Hyde, the Victorian "body snatchers," as well as the fictional Victor Frankenstein, it is most interesting for the way it unsettles our understanding not of science but of art. Kelly's conviction was based on the 1994 Anatomy Act of Great Britain, which outlaws any use of human remains other than medical research and any disposal of those remains other than burial (Trull 2003, 2). The defense in the Kelly trial explicitly challenged the division between art and medical science, arguing that—as one commentator put it—"the heart of what's at stake in the Anthony-Noel Kelly case [is]: should individuals have the right to donate their bodies to art, in the same way we can donate our bodies to science? The Anatomy Act doesn't give Britons that choice, and that doesn't seem right" (2). Not surprisingly, the groups that found Kelly's sculptures most offensive were "funeral directors and medical people: the former were concerned about the respect due to the dead, while the latter feared that the supply of bodies would reduce if the public felt that hospital staff could not be trusted to look after them" (Walker 1999, 225). For both groups, the offense taken was not aesthetic but professional and pecuniary.[7] The Kelly case defense reveals that our understanding of human organs is based on the foundational distinction between medicine and art, a distinction that is enforced both socially and economically (226).

We will come back to the question of the "supply of bodies" and, more importantly, of organs in a moment. First we need to consider the question of the difference between medical and artistic access to organs and bodies. If Kelly's activities were punished because they transgressed the tacit division between art and science, as Jonathan Meades has argued, the activities of another artist may have escaped such punishment in part because he possessed medical credentials. In 1998, the same year as Kelly's trial, "Body Worlds," the show of the German anatomist and artist Dr. Gunther von Hagens, opened in the Berlin Post Office and was attended by huge crowds. Von Hagens's art relies on a technique he invented himself, plastination, which enables him to preserve, pose, and exhibit organs, tissues, even entire flayed corpses, by injecting them with synthetic polymers so that the color stays true to life while the tissue becomes dry, odorless, durable, and hard.[8] Photographs on von Hagens's Web site document his "sculptures": flayed chess players; trifurcated Giacometti-like bald men; thoughtful or transfixed heads whose play of muscles or tendons is exposed directly to the viewer; a skinless Adam musing, after the model of Leonardo, on the drooping length of his own skin he holds up before us.

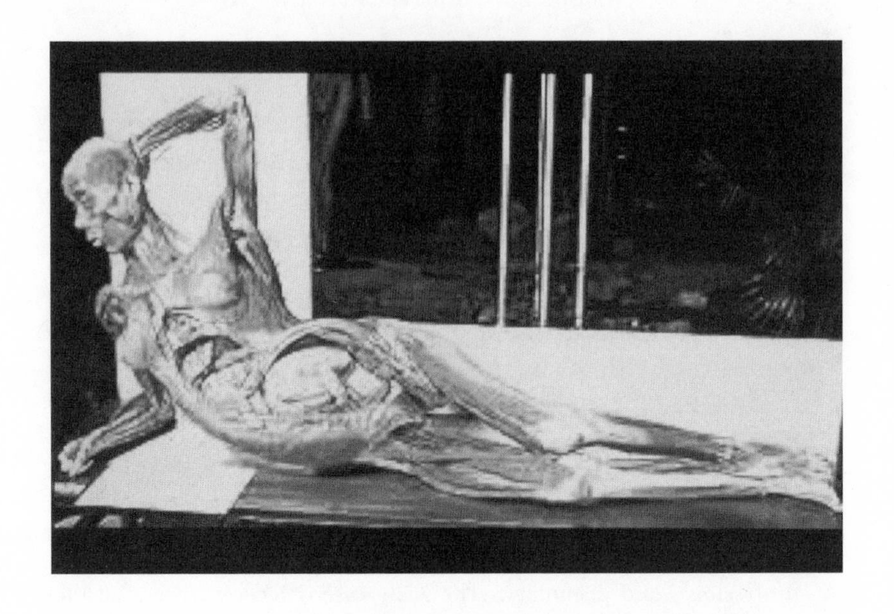

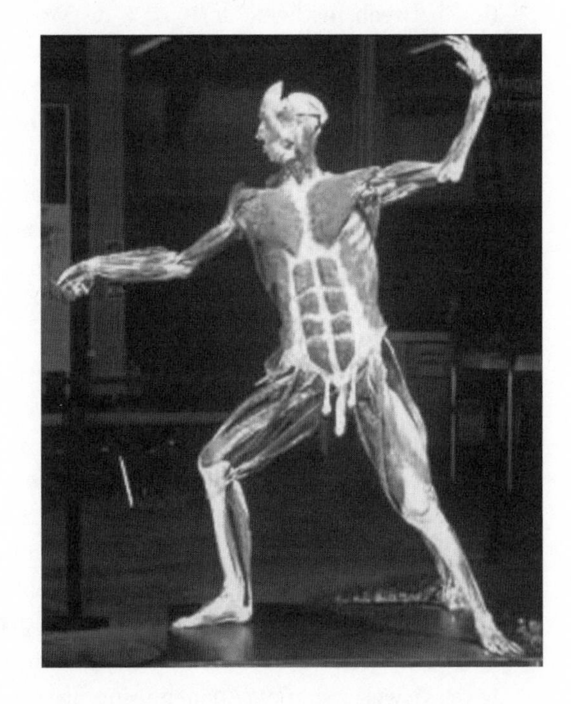

(*above*) *Odalisque*,
by Gunther von Hagens.
(*left*) *Fencer*,
by Gunther von Hagens.
Both reproduced
from Body Worlds
Web page, http://
www.bodyworlds.com.

Eerie as these sculptures are, what is even more striking is the conglomeration of discursive strategies—the mixture of aesthetic, educational, demographic, and commercial discourses—with which von Hagens promotes his exhibit. As the sculptor explains on his Web site: "The democratization of anatomy by the Koerperwelten exhibition expresses itself above all in the fact that the layman in the evaluation of the exhibition behaves differently than predicted of experts: The high numbers of visitors reflect the need of the broad population to want to know more about the structure and function of their own body."[9] Although speaking as an artist assessing his art, von Hagens sounds more like a scientist or social scientist. Evoking the distinction between layman and expert, he frames the exhibit almost as an experiment, in that it confounds expert predictions, appeals to a broad "population," brings a specialized epistemological tool (anatomy) to a wide public, and approaches the human body from a perspective not of aesthetics but of "structure and function." As he explains to an interviewer, "Ninety five per cent of the visitors in the Heidelberg exhibition in the summer of 1998 said that they were satisfied by what they saw. On a relevant questionnaire they replied that the exhibition was either 'good' or 'very good' and that it provoked very positive feelings."[10]

That von Hagens believes the scientific and commercial values of the technique trump the aesthetic value of its creation becomes apparent when we learn that he has also established an "Institute of Plastination" in Heidelberg.[11] If we follow a link on the Koerperwelten Web site, we come to another site that explains "Plastination—A Teaching and Research Tool." As that site explains, "Plastination is a unique technique of tissue preservation developed by Dr. Gunther von Hagens in Heidelberg, Germany, in 1978." Photographs of different tissue specimens demonstrate that different choices of polymer and different kinds of plastination will be appropriate for different kinds of tissue: "*Silicone* is used for whole specimens and thick body and organ slices to obtain a natural look. *Epoxy* resins are used for thin, transparent body and organ slices. *Polyester-copolymer* is exclusively used for brain slices to gain an excellent distinction of gray and white matter." After addressing the four main steps of his technique (fixation, dehydration, forced impregnation, and hardening or curing), the Web site addresses the bottom line: the ever-growing market for the technique. "Plastination is carried out in many institutions worldwide and has obtained great acceptance particularly because of the durability, the possibility for direct comparison to CT- and MR-images, and the high teaching value plastinated specimens have."

Von Hagens's dominant commitment is clearly to technoscience. As he explains his technique, the organs are less important than the revolutionary material that preserves them. Ironically, however, the artistic commitment of another contemporaneous artist results in a similar overvaluation of technique over organs. In 1989, the Canadian artist Rick Gibson and the owner of the Young Unknowns Gallery were found guilty of "outraging public decency for creating and displaying" earrings made of twelve-week-old fetuses. As early as 1984, Gibson had held a show at the Cuts Gallery in Kensington entitled "Dead Animals," which incorporated animal and human body parts (Walker 1999, 150). Next Gibson walked around Reading, England, wearing a placard reading: "Wanted: legally preserved human limbs and human foetuses" (150). When he placed the advertisement in a gallery window, he received two three-inch-long human fetuses from a lecturer in pathology. But (like von Hagens) Gibson only created the legally disruptive fetus earrings when he found a way to preserve them permanently: a technique for freeze-drying used by the Natural History Museum. He intended the resulting "human earrings" to raise questions about the materials appropriate for art and self-adornment, as well as the ethics of using such materials. Although his undergraduate degree was in psychology, Gibson felt that the implications of his art reached beyond the species boundary, testifying to the fact "that humans had once been or still were part of the animal kingdom" (Walker 1999, 151). How do we explain Gibson's conviction for turning the discarded fetuses into "human earrings," when we consider the legally sanctioned enterprise LifeGems, which turns the cherished cremated remains of human or animal loved ones into synthetic diamonds? In the contrast between the art made from Gibson's discarded fetuses (whose protolives have been cut short by abortion) and the luxury product produced by LifeGems from the expensively harvested carbonized cadaveric remains (which persist posthumously because "diamonds are forever"), we find raised issues of valuation, property rights, and duration to which I will return because they illuminate the changing nature of the human body in the twenty-first century.

ORGAN STRIPPING

The cases of art using human organs (Kelly, von Hagens, and Gibson) all amplify questions also raised by a scandal that occurred in the world of medicine: questions of "professional interest, commerce, and . . . ownership."[12] In 1999, Dirk van Velzen, a Dutch pathologist who worked at the Alder Hey Children's Hospital, in Liverpool, England, was found to have "harvested" thousands of organs, without parental consent, from children who had been autopsied at his hospital. A shocked British public was fed vivid details of the scandal by the daily press. An "archive" of human and fetal organs had been discovered at the Alder Hey Hospital, including a heart collection containing more than two thousand hearts; a fetal collection containing around 1,500 fetuses, and an additional collection that by December 1999 had accumulated more than 445 partial or full fetal remains. This collection, kept by Dr. van Velzen as a private museum, included one jar holding the head of an eleven-year-old child and another containing a nine-week-old fetus.[13] While the hospital was found to have retained more than 1,500 fetuses ("miscarried, stillborn or aborted without consent"), the scandal extended beyond Alder Hey. A committee of inquiry led by the United Kingdom's chief medical officer, Professor Liam Donaldson, soon revealed that the practice of taking the organs from deceased children was a common one.[14] In 1999 investigators found that the Bristol Royal Infirmary had collected hearts and other organs for decades. The Donaldson committee found that "105,000 organs" were retained at medical schools and hospitals throughout Britain.

The response to the Alder Hey scandal occupied a prominent position in British newspapers and their Web pages in 1999 and 2000. After the hospital admitted that it had retained the organs of more than eight hundred children who had been autopsied between 1988 and 1995, the parents involved were contacted and given the opportunity to retrieve the organs and tissues that had—without their knowledge—been removed and retained.[15] For Paula O'Leary, whose eleven-month-old son Andrew had died of SIDS eighteen years before the scandal broke, the process of retrieving her child's remains was painfully complex. First she was told that his heart had been retained by the hospital. Then after

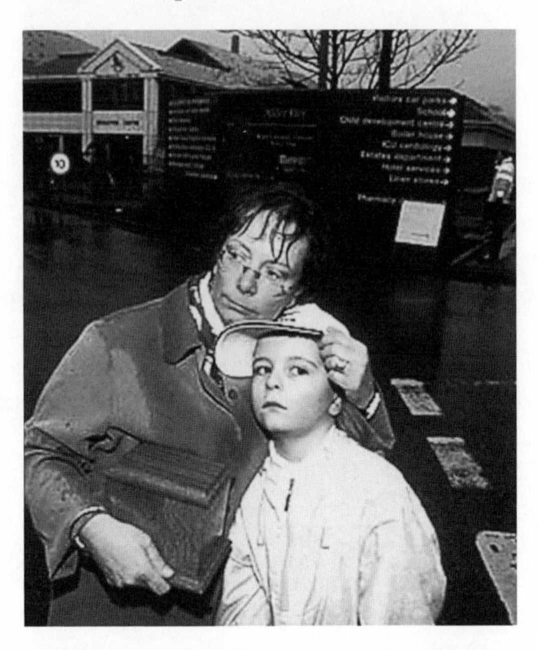

Carol and Joseph Wainwright at Alder Hey with her child's body parts. Photograph by John Giles, PA. Reprinted from *Guardian Unlimited,* 30 January 2001.

the family buried the heart in a small white casket, they were told that some tissues were still uninterred. When Mrs. O'Leary and her lawyer went to the hospital, "She was shown a box full of Andrew's organs preserved in glass—including his liver and gall bladder, spleen, adrenal glands, thymus gland, kidneys, skull, brain and spinal cord. His pancreas and thyroid gland had disappeared, she was told."[16] Not only the body of Andrew O'Leary but also the memory of him that his mother retained were grotesquely dismembered by the Alder Hey scandal. "Now when she closes her eyes she finds it hard not to see 36 different parts of his body eerily suspended in glass blocks, laid out on a table before her at the offices of the solicitors acting for Alder Hey Children's Hospital" (Barwick 2001a).

The disappearance of the pancreas and thyroid gland (according to the hospital) points to another important way in which the scandal triggered a reevaluation of the status of organs. The Donaldson inquiry established that it was accepted practice not only for hospitals to remove organs without parental consent but also to trade in them for profit—in short, to *sell* them. In January 2001 a reporter for the *Independent* reported that healthy thymus glands, removed from children during heart surgery without parental consent, were sold to Aventis Pasteur "for research on a drug treatment for aplastic anaemia," in return for a ten-pound donation to the hospital's cardiac department for each gland received.[17] Indeed, a quick check with other British hospitals suggested that such practices were common: "Birmingham Children's Hospital admitted it had given organs to drug companies in the early 1990s in return for cash" (Laurance 2001, 2). As the shadow health secretary Dr. Liam Fox explained: "What we are effectively talking about here is cash for organs. . . . It brings many ethical issues into play here, and I think we need to know the full details . . . because if it has happened in Alder Hey we have to ask if it has happened elsewhere."[18]

What issues are raised by the Alder Hey scandal and its press coverage? It exposes the profit-driven, even grotesque, aspects of a supposedly benevolent institution (the children's hospital). More than that, this scandal of unlicensed organ retrieval and sale reveals the fluctuating, contested, and constructed—in short, liminal—position of human organs at the turn of the twenty-first century. It raises a number of crucial questions: the possibility and locus of ownership of organs, the professional use of organs, the [relative or absolute] merits of organ gift and organ sale, and the question of informed consent for all of the above. If we round back from the Alder Hey organ-stripping scandal to the instances of organ art with which I began, we are led to ask: To which kinds of commerce can human organs legitimately be subjected? Artistic, including the purchase of organs and body parts for artistic use, the sale of organ-derived art objects, and the mounting of organ-related gallery shows?[19] Scientific or technical, including the development of new models for anatomical illustration, new forms of cadaver preservation and memorializing, new tissue preservation techniques, and new strategies for diagnosis and testing? Curatorial, including the accumulation of organs and body parts for private or public museums? Then, what kind of ownership of organs and body parts is possible or acceptable? Ownership of the organ by the human being in whom it is found, the medical scientist or artist to whom it is transferred, or the sanctified ground in which it ultimately rests? Ownership of the fetus by the gestating woman, the woman who has decided to abort it, the abortionist, pathologist, or anatomist who has retrieved it, the artist who employs it as part of his intellectual property, or the state that controls the access to, and outcome of, the process of fetal development? Ownership of the cadaver's cremated remains by the grieving survivor(s), the funeral director who supplies the cremated remains to LifeGem, the LifeGem franchisee who subjects the remains to the patented carbon retrieval process from which the material of the colored synthetic diamond is formed, or the person who becomes the recipient (and owner) of the resulting LifeGem "memorial"?

While plastinated chess-playing corpses, freeze-dried fetus earrings, diamonds made of human or animal carbon, and gilded casts of partly dissected corpses are brutally different from the scandal of the Alder Hey hospital, with its painful tales of parents clamoring to return their children's organs to their already interred bodies, these different artistic and medical instances of organ and body part manipulation demonstrate the same fact. The value and significance of a human organ or body part is no longer self evident, but rather is *produced* through a complex set of institutional negotiations involving medicine, art, society, the legal system, human emotions, and economic calculations. From the initial distinction between worthless organs and valuable ones, we have now moved to distinguishing between the various reasons for which organs are valued: as metonymic representation of a loved one, as the actual/memorial essence of the deceased, as information, as replacement parts, as exchangeable commodity.

The "Report of a Census of Organs and Tissues Retained by Pathology Services in England" (2000), issued in response to the Alder Hey scandal, portrayed a medical culture in which organs and tissues fluctuated between being valueless and valuable in several different ways. "Human tissues, organs and body parts have not been regarded as sensitive materials for disposal, even those of children. Relatives were rarely asked for views and the majority of such tissues and organs were disposed of as clinical waste" (5). Noting that "this [disposal as clinical waste] may be seen as unacceptable by many people," the report suggests that tissues and organs are viewed as valuable for one of several reasons: because they were part of a loved one, because they advance knowledge, or because they can be sold to companies extracting materials from them for profit. The changing valuation of organs has also affected a core medical institution: the autopsy.

ORGAN OF INFORMATION: THE DEMISE OF THE AUTOPSY

The general public's response to the revelations from Alder Hey was horror at Dr. van Velzen's ghoulish museum of fetal remains, the hospital's flagrant disregard for principles of informed consent for postmortems, its ethically dubious practice of selling healthy thymus glands to a pharmaceutical company, and the callous manner in which the hospital informed parents that organs thought long buried with their children had been retained and were now available for "reclamation." But medical experts were disturbed by quite another aspect of the whole affair. As Dr. Ian Bogle, chairman of the British Medical Association Council, explained: "I am deeply shocked that . . . even worse, no real research was conducted on these children's organs so that there was no possible benefit to patient care."[20] Following a national summit on the retention of organs and tissues following postmortem examination, the Royal College of Pathologists forwarded a statement to Professor Liam Donaldson, chief medical officer, England and Wales, stipulating the crucial value of organs as purveyors for information. The statement revealed that to physicians, the pressing need for information mitigated the offensive impact of the Alder Hey scandal. While the public found the museums of human remains disturbing, the pathologists reasserted the pedagogical importance of "pathology museums comprising extensive collections of preserved organs or substantial portions of organs" while offering the reassurance that the provenance of the organ is customarily withheld.[21] The physicians' neglect to obtain informed consent for autopsies was interpreted as reflecting the desire to spare families anguish and distress—both the families whose children were to be autopsied and the "families [and] future patients" whose welfare would benefit from the autopsies done "to improve medical knowledge and to make reliable diagnoses" (RCP 2001, 12).

A comparison of the public's and the pathologists' responses to the Alder Hey scandal suggests that the informational value ascribed to a human organ fluctuates depending on whether the medical response to mortal illness is curative or palliative, and whether the physicians involved are focused on increasing medical knowledge or on caring for patients and families, including the deceased and their next of kin. Recently, a surgical resident in the United States suggested that the latter view is winning out: "The autopsy is in a precarious state. A generation ago, it was routine; now it has become a rarity. Human beings have never quite become comfortable with the idea of having their bodies cut open after they die."[22]

Despite the discomfort with organ harvesting resulting in a decline in autopsies, the procedure of organ transplantation has become an accepted part of Western medicine. Recently, a friend who is suffering from incurable lung disease said to me, "I'm thinking more seriously about the idea of a lung transplant. Not now, but in six years, say, when the techniques have matured." Indeed, throughout the world, people with means increasingly take for granted the notion that transplantation procedures will continue to be refined, making it ever easier to replace diseased organs with healthy ones and thus prolong life.[23] Whether we think of this as the acceptance of a cadaver organ, excised quickly from the body of a brain-dead accident victim, as the "gift" of a living organ, surgically removed from a healthy sibling or close relative, or even as the theft of an organ from an inadequately informed and vulnerable organ "donor," this procedure is increasingly routine. Curiously, however, science fiction is still frequently invoked as a gauge for the progress of organ transplantation, as an excerpt from one transplant hospital Web site reveals: "Before the 1950s, organ transplantation belonged more to the realm of science fiction than to medical science. Although references to tissue and organ transplantation date back as far as 2000 B.C. in Egyptian manuscripts, it was not until revolutionary progress in the world of medicine during the later half of the 20th Century that organ transplantation became a reliable treatment for the thousands of people suffering from organ failure."[24]

The curious role of science fiction in the normalization of this biomedical practice is the point of departure for what follows. Instead of taking as self-evident the difference between science fiction and this particular area of medicine—a difference that the hospital's comment relies on for its impact—I want to inquire into its status, to investigate the nature of the relationship between the fields of science fiction and organ transplantation. Can it be that the transformative procedure of transplant medicine is *enabled* somehow by the transformative narrative that is science fiction? And if so, can we abstract from this instance a model for the relations between science fiction and biomedicine? Can we generate, beyond that, a model for the relations between the realms of the symbolic (art) and the material (biology and medicine)?

Memoirs by leading transplant surgeons Dr. Paul Terasaki and Dr. Thomas E. Starzl, as well as surveys of the field of transplant surgery by health professionals in related fields, reveal that the central barrier to the acceptance of organ transplantation is the immune system's hostile response to the new organ.[25] Two kinds of immune response had to be overcome to produce a successful organ transplantation, which has been defined "as the function of a graft from a genetically nonidentical donor for at least 6 months after transplantation": the initial rejection of the organ's cells by the immune system of the recipient, and the somewhat later process by which the *recipient's cells* were rejected by the donor organ's cells, a process known as graft-versus-host disease.[26] In both cases, the field of modern transplant immunology played a major role in making organ transplantation successful. From the early reliance on blood group typing, to the short-lived reliance on total body irradiation (TBI), the uses of the immune suppressant drugs 6-MP (6-mercaptopurine) and azathioprine (Imuran), and the use of HLA (human leukocyte antigen) matching for donor selection, to the discovery in 1972 of the immunosuppressive drug cyclosporine, "a watershed for both solid organ and bone marrow transplantation," and the later discovery in 1988 of FK 506 (FR900506), transplant immunologists worked throughout the twentieth century to manage the self-other interactions central to the organ transplantation process.[27] Despite twenty-first-century improvements in immunology, this issue continues to be crucial to the success of organ transplantation.[28]

The complex physiological transactions within the transplant recipient are echoed by transplant technology's complex process of institutional and social expansion. From its start in mainly Western medicine, organ transplantation has grown to be a global business, transacted in "transnational spaces with surgeons, patients, organ donors, recipients, brokers and intermediaries, following new paths of capital and technology in the global economy" (Scheper-Hughes et al. 2000). This field, which began with the transplantation of cadaver organs and continued with the transplantation of gift organs donated by healthy, living family members, has in recent years progressed to the transplantation of organs not merely donated but purchased or in some cases even stolen from unrelated donors. Contemporary sociologists, ethnographers, medical ethicists, and anthropologists have responded differently to the political, cultural, and ethical implications of the global trade in transplant organs, some (like Swazey and Fox) choosing to leave the field out of unease at the personal and social costs of the technology, others (like the members of the Bellagio Task Force on Organ Transplantation) banding together to examine "the ethical, social, and medical effects of the commercialization of human organs" (Scheper-Hughes et al. 2000, 2). A number of rumors and urban legends circulating in popular culture may articulate local resistance to this sweeping global trend. While Benjamin Radford described as an urban legend the tale of a grandmother trying to sell her five-year-old grandson to a man who would kill him for his kidneys, later discussions of such reports have tended to emphasize the ways they condense a number of actual events: the existence of a global organ traffic, the vulnerable position of street urchins in a number of cultures, and the imbalance (racial, social, agential, economic) between organ donor and organ recipient (Scheper-Hughes et al. 2000; Radford 2001). Most recently, Stephen Frears's film *Dirty Pretty Things* (2002) dramatizes the place of illegal organ procurement in a complex global economy in which work, nationality, race, sex, identity, and even health are alienable, subject to commodification, and available to anyone *for a price.*[29]

Yet on both the scholarly and the popular levels, the issues that are being raised about organ transplantation, and the questions generated by it, are not new. Long before concerns about organ transplantation appeared in the writings of ethicists, public policy writers, and anthropologists, they were aired in works of science fiction. Anthropologist Donald Joralemon has observed that "organ transplantation seems to be protected by a massive dose of cultural denial, an ideological equivalent of the cyclosporine which prevents the individual body's rejection of a strange organ. This dose of denial is needed to overcome the social body's resistance to the alien idea of transplantation and the new kinds of bodies and publics that it requires."[30] In particular, live-donor organ transplantation requires particular psychic effort, on a personal and social level, to endorse. As one reporter put it, "the live-donor transplant represents an important resource for people with liver disease—and a dramatic simultaneous surgical procedure that can seem, at times, like science fiction."[31]

I will argue in what follows that throughout the twentieth century, science fiction writing was as crucial in the cultural realm as immunology was in the realm of medicine in bringing about public acceptance of organ transplant technology. Indeed, we might think of science fiction as functioning as a kind of ideological cyclosporine. Science fiction does not just purvey "a massive dose of cultural denial," inoculating us against the terrible desires fueling organ transplantation, though arguably it does that. Rather, in science fiction, we find articulated and negotiated issues integral to the normalization and institutionalization of transplant technology: the relation between body, body part, and identity; the role of race, class, and age in the constitution of that identity; and the notion that even the death of the self is subject to social and scientific construction.

I will consider these questions by way of three science fiction short stories, all published after 1912, when Alexis Carrel received the Nobel Prize for the techniques of suturing blood vessels essential to organ transplantation, and before 1983, when the immunosuppressive drug cyclosporine was approved for general use as part of clinical organ transplantation.[32] Two of these short stories are fictional narratives of transplant operations that have gone on to be realized in fact, while the third story projects an institutional structure for all organ transplantation that is, as yet, only hypothetical. "Transplant technology fulfills the human desire to live longer," T. Awaya has explained, and it is this human desire that the third story addresses.[33] However, life extension itself holds meaning only in relation to a larger social structure, from which come the values that lead people to turn to organ transplantation. Concerned with the implications of organ transplantation for human identity, these stories all articulate specific resistances to the practice of organ transplantation framed in terms of class, ethnicity, race, and age, as well as different motives for undergoing the procedure. Read chronologically, they reveal changes in our understanding of human life. No longer is human existence defined by its unique temporal and spatial coordinates: one body, one life, in a specific space and time. Instead human life is increasingly defined by the agential, instrumental deployment of resources for bodily renewal, both its temporal and spatial context subject to extension or translocation. Like the organ-art with which I began, these science fiction short stories too both enact and articulate this profound transformation in our understanding of human life.

"NEW STOMACHS FOR OLD"

"New Stomachs for Old," W. Alexander's story of rejuvenation through organ transplantation, was published in 1927 in Hugo Gernsback's classic science fiction magazine *Amazing Stories*. Founded in 1926, the magazine's aim was not merely to reflect but to jump-start new scientific developments. Gernsback promised to deliver "extravagant fiction today . . . cold fact tomorrow."[34] In keeping with that crusade, an editorial box provides the scientific context for this work of fiction: "Several years ago the German Professor, Dr. Walter Finkler, amputated the heads of various insects and transplanted them on others. Strange to say, the insects with the transplanted heads, after the new ones had grown, managed to get along the same as with their original heads. So the operation of exchanging your old stomach for a new one may, after all, not be an impossibility, but you may get the surprise of your life if you ever make such an exchange. At least one millionaire who bought himself a new stomach found this out rapidly in totally unexpected results."[35] Even as it introduces the notion of live-donor organ transplantation, this short story also draws attention to what would become highly contested aspects of contemporary organ transplantation practices: the issue of compensation for the donation; the potential for a coercive relationship between donor and recipient; the social discrepancies (ethnic or racial, economic, geographic, or class based) between donors and recipients; and the impact of the transplant on the health of the donor as well as the recipient.

The organ recipient in this short story is Colonel Seymore, a dyspeptic millionaire whose digestion has been ruined by years of ill-treatment, so that his very nature seems to be changing. As his doctor counsels him, "I would earnestly advise you, Colonel, to undergo this operation at once, for your stomach trouble is seriously affecting your disposition. From being the most amiable rich man of my acquaintance, you are rapidly becoming one of the most grouchy" (Alexander 1927, 1039). With his doctor's help, Colonel Seymore obtains a new

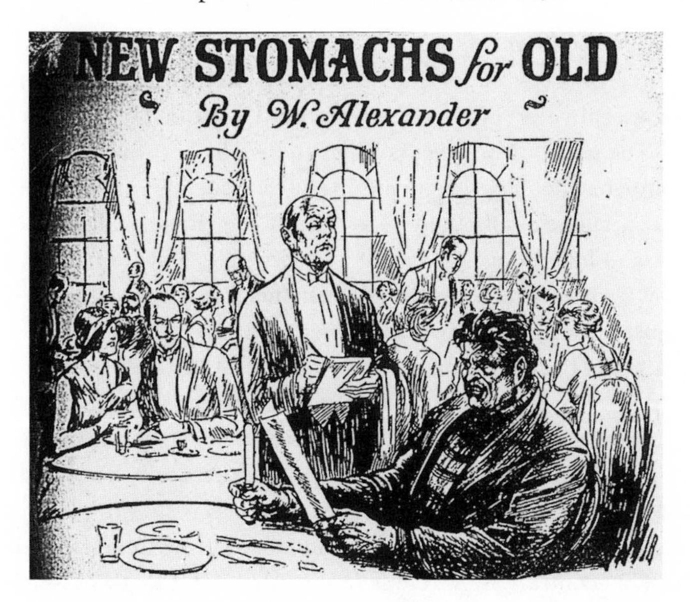

"New Stomachs for Old." Reprinted from W. Alexander, "New Stomachs for Old," *Amazing Stories* 1, no. 22 (February 1927).

stomach from a "strong young man" who is willing to make the stomach exchange in return for "all hospital fees and . . . ten thousand dollars" (1039).

When the colonel stipulates to his physician that the organ donor he has obtained for him must suffer no injuries from the exchange, the doctor reassures him: "I will stake my professional reputation on it, that your stomach will function perfectly in his body, for there is nothing organically wrong with it. You have merely weakened it by your method of living. I trust you will change your habits when you have acquired this new stomach and take the out-of-doors exercises I have long and unsuccessfully advocated for you" (1040). This exchange anticipates a current area of hot debate—the ethics of live-donor transplants—as well as marking for contemporary readers the ethical distance we have traveled since the story was published in 1927. The way the story frames issues of informed consent, donor risk, donor coercion, and donor reimbursement illuminates—by contrast—the contemporary landscape of organ donation. Since June 2000, the donor protection requirements for any such transaction have been stiff, as stipulated by a consensus statement on the live organ donor, drafted by an executive group convened by the American Medical Association: "The person who gives consent to be a live organ donor should be competent, willing to donate, free from coercion, medically and psychosocially suitable, fully informed of the risks and benefits as a donor, and fully informed of the risks, benefits, and alternative treatment available to the recipient. The benefits to both donor and recipient must outweigh the risks associated with the donation and transplantation of the living donor organ."[36]

As this consensus statement indicates, current practice mandates great care to avoid donor coercion, which can take the form of psychological, social, or economic pressure. Only very recently has live-donor organ transplant been considered medically acceptable, in response to the perceived "increasing shortage and long waiting times for cadaver organs." In contrast, in Alexander's story, Dr. Wentworth's assurances that the donor will not be harmed are sufficient to justify performing the transplant operation. The unequal social positions of the donor and recipient in "New Stomachs for Old" not only means that they have different levels of *choice* about the transplant operation, and different *rights* in relation to it, but also that they are given different amounts of information about the parties involved and the procedures to be anticipated. The moment when the millionaire decides to proceed with the transplant is dramatized, but not the moment where the donor agrees to the donation. Current practice would require that the young donor not be reimbursed for his donation, putting in question Colonel Seymore's offer of ten thousand pounds *in addition to* "all hospital fees." In Alexander's short story, the millionaire is able to see his intended organ donor—"a swarthy young man sitting stiffly in a chair" in the hospital waiting room, but the donor is unaware of the recipient's identity (1040). With that glimpse, the colonel remarks to his doctor: "I shudder to think of coming in contact with any part of that Italian's anatomy, let alone making it my own, but I suppose I should count myself lucky that you found some one willing to make such an exchange" (1040). The doctor's response weighs the colonel's class assessment against the only index that counts in organ transplantation: organ health. "You are indeed lucky. . . . That young man is as near a perfect specimen of the *genus homo* as it has ever been my privilege to examine. I am not introducing you two as it might later be a source of embarrassment to you. He does not know of your identity" (1040). The one-way glimpse between Colonel Seymore and the "swarthy young man" in Alexander's story encodes four interanimated kinds of difference between them: difference in class position, difference in ethnicity, difference in race, and difference in access to information. In contrast, contemporary transplant protocols affirm equality of donor and recipient, who would most likely never be informed of each other's identities, since the assumption is that both individuals are possessed equally of rights to privacy and autonomy. "At some transplant centers, the policy is not to share donor information with the recipient, respecting the autonomy and confidentially of both the donor and the recipient" ("Consensus Statement on the Live Organ Donor" 2000).

When the transplant surgery has been completed, the vigorous young working-class man and the elderly millionaire part, each with the other's stomach. More than an exchange of organs has taken place, however. The two men discover they have changed *tastes* as well. The millionaire now scorns the food at his exclusive club as "tasteless and insipid" and haunts dubious tenement coffeehouses where he eagerly devours platefuls of spicy Italian food. The young Italian finds himself for the first time unable to resist fine food, although still without the means to pay for it. For each, the consequences of the transplantation are ruinous. The millionaire's culinary slumming threatens his income. The board of directors of the bank refuses to support his planned business merger because they "cannot afford to be associated in any enterprise, with a man who is known to find his business or pleasure in questionable joints of the slums" (Alexander 1927, 1041). The young Italian is hauled into court, charged with larceny, "accused of enjoying a large meal at the Ritz, and then refusing to pay for the same, claiming to have no funds" (1031). Speaking in his own defense, he explains to the judge, "Me, Tony Moreno, eata one beeg meal, have no da mun. Dees stomach I getta from reech mans, no will eata da speeget, maka me spend mucha da mun. Now no gotta da mun" (1041). Once transferred, the stomach has brought with it the tastes and desires that went with it, to the dismay of both men, who find those tastes and desires impossible to harmonize with their social situations.

The story concludes with the colonel proposing that "it would be best for each of us to get our own stomachs back again." With that in mind, he pledges to cover all the court costs for the organ donor's larceny case, as well as all hospital fees for another surgery "to re-exchange [their] stomachs, and pay Tony another ten thousand dollars" (1073). The second surgery is accomplished speedily, and the medical note on the whole case suggests that its results were not only reparative but actually rejuvenating: "The Colonel was never again afflicted with stomach trouble. Dr. Wentworth's explanation as given in the May issue of the Medical Journal, when stripped of medical phraseology and technicalities is this: 'The Colonel's stomach having lain for months in the powerful, vigorous body of the young Italian, was built up by nature to function properly in its new environment and so was returned to the Colonel, a powerful organ, just as an athlete might return in the pink of condition after undergoing a course of training' " (1073).

In its articulation of the notion that a classed and racialized identity (producing distinct tastes and desires) inheres in specific organs, and can be transferred with the organ from one person to another, "New Stomachs for Old" recalls ancient medical practices in which desired attributes were obtained by ingesting the organ in question. But this story also addresses a theme that would be prominent in the second half of the twentieth century as organ transplantation was globalized: the enabling relationship between inequities of race, ethnicity, gender, and class, and the commodification of body parts. The donor in "New Stomachs for Old" is a "swarthy" working-class Italian pauper, whose healthy body is his only capital (even if to the colonel it also serves as a social liability because of its "coarseness"). The recipient, an upper-class white man, is possessed of both economic and social capital (despite the fact that he suffers from physical illness). Perhaps the *wrongness* of the stomach transplant lies precisely in its transgression of the identity categories of race, class, and ethnicity. Yet with the normalization of organ transplantation, death itself has been reconstructed as a process of nature/culture. Once the medical definitions of death were changed to include the notion of irreversible coma (at the end of the 1950s) and brain stem death (at the end of the 1960s), "death became an epiphenomenon of transplantation," according to medical anthropologist Nancy Scheper-Hughes (Scheper-Hughes et al. 2001, 18). As the complications of cadaver organ donations increased the interest in live-donor organ transplantation, and as advances in immunosuppressive medicine made it possible to use organs from an unrelated donor, the role of race, ethnicity, gender, and class in organ transplantation became increasingly clear. Scheper-Hughes has given one of the most important early formulations of the principles governing global organ traffic: "In general, the flow of organs follows the modern routes of capital: from South to North, from Third to First World, from poor to rich, from black and brown to white, and from female to male" (6). Yet Stephen Frears's portrait of organ trafficking in 2002 demonstrates that with the globalization of capital and transportation (and thus the widespread availability of medical interventions), that formula for "the flow of organs" is too simple in its neatly paralleled power hierarchies. All the participants in the illegal organ trade in *Dirty Pretty Things* are so-called third-world people, and all of them either black or brown. Yet one of the would-be organ recipients is a young Indian girl, and the two coerced organ donors are, respectively, Somalian and Turkish.

"THE BLACK HAND"

The rich, dyspeptic colonel's desire for a new stomach, which leads to his retreat from an operation that gave him more than he bargained for, has its parallel in another science fiction tale of organ transplantation, Charles Gardner Bowers's 1931 "The Black Hand."[37] This story, too, begins with the interview between a potential organ recipient and his doctor, and here too the doctor acts as advocate for an operation that the recipient initially resists. But in this case, the potential recipient is the well-born and well-connected artist named Van Puyster, portraitist of the prince of Siam, who learns that he must have his infected hand amputated. Desperate at the thought of the amputation, unwilling to accept "one of your infernal leather and metal contraptions," he learns from his doctor that a hand transplant may be possible. "There is a condemned man in the state prison who has agreed to sell you his arm before he dies" (Bowers 1931, 910).

The themes of racial, social, and economic coercion, and of the medical motives for involvement in transplant surgery, introduced in Alexander's story, return even more vividly in this story. As the doctor explains to his patient Mr. Van Puyster, "The negro wishes $10,000 to go to his estate and he wants an impressive burial. I shall charge no more than my regular fee for amputations, *as I greatly desire the honor of*

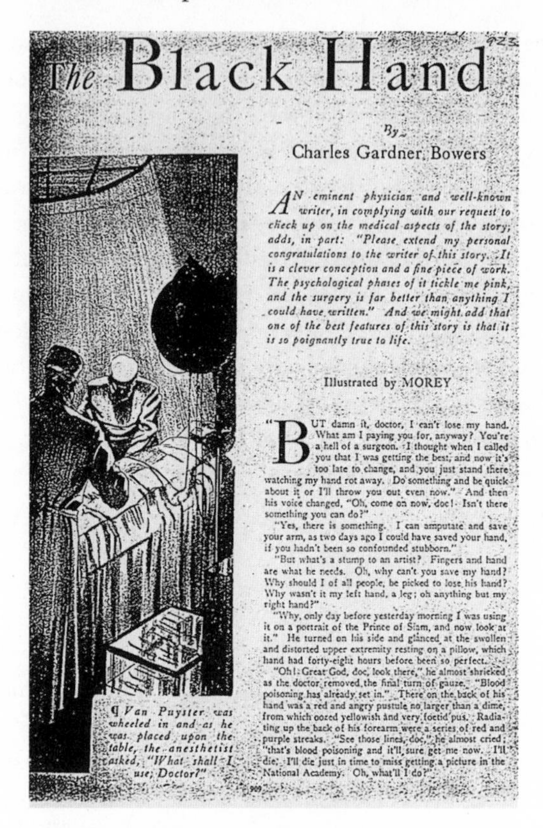

The Black Hand

By

Charles Gardner Bowers

AN eminent physician and well-known writer, in complying with our request to check up on the medical aspects of the story, adds, in part: "Please extend my personal congratulations to the writer of this story. It is a clever conception and a fine piece of work. The psychological phases of it tickle me pink, and the surgery is far better than anything I could have written." And we might add that one of the best features of this story is that it is so poignantly true to life.

Illustrated by MOREY

"BUT damn it, doctor, I can't lose my hand. What am I paying you for, anyway? You're a hell of a surgeon. I thought when I called you that I was getting the best, and now it's too late to change, and you just stand there watching my hand rot away. Do something and be quick about it or I'll throw you out even now." And then his voice changed, "Oh, come on now, doc! Isn't there something you can do?"

"Yes, there is something. I can amputate and save your arm, as two days ago I could have saved your hand, if you hadn't been so confounded stubborn."

"But what's a stump to an artist? Fingers and hand are what he needs. Oh, why can't you save my hand? Why should I of all people, be picked to lose his hand? Why wasn't it my left hand, a leg; oh anything but my right hand?"

"Why, only day before yesterday morning I was using it on a portrait of the Prince of Siam, and now look at it." He turned on his side and glanced at the swollen and distorted upper extremity resting on a pillow, which hand had forty-eight hours before been so perfect.

"Oh! Great God, doc, look there," he almost shrieked as the doctor removed the final turn of gauze. "Blood poisoning has already set in." There on the back of his hand was a red and angry pustule no larger than a dime, from which oozed yellowish and very foetid pus. Radiating up the back of his forearm were a series of red and purple streaks. "See those lines, doc," he almost cried; "that's blood poisoning and it'll sure get me now. I'll die; I'll die just in time to miss getting a picture in the National Academy. Oh, what'll I do?"

¶ Van Puyster was wheeled in and as he was placed upon the table, the anesthetist asked, "What shall I use, Doctor?"

909

"The Black Hand." Reprinted from Charles Gardner Bowers, "The Black Hand," *Amazing Stories* 5, no. 10 (January 1931).

being the first to accomplish this operation" (911). That the organ donor is a condemned criminal raises some potential complications for the surgery, Dr. Evans explains: "I intend to make a direct transfer of his arm to your stump, as I do not think you would relish being bound by the side of a condemned criminal for ten minutes, much less ten days, furthermore, he is doomed to execution before that time would have expired. . . . The negro being condemned to the lethal chamber shall, instead, die under the anaesthesia" (911).

While grafts are customarily carried out by suturing the recipient and the donor tissues together until the graft "takes," clearly such a union is problematic for moral, physiological, and psychological reasons. Morally, the donor is socially untouchable. Physiologically, the donor will be dead (and a danger to the recipient) before the graft is completed. And psychologically, body part grafting is more comfortable to the potential recipient, in this work of fiction as it also is in fact. Realism and aesthetic appeal (a graft) frequently win out over pragmatic ease and function (a prosthesis). The story also anticipates the debate around the practice of using convicted criminals as organ donors, either as cadaver donors (as was the case in a phase of Western transplantation) or as unwilling living donors, as has been reported at the turn of the twenty-first century by anthropologists working in India and China (Scheper-Hughes et al. 2000; L. Cohen 1999). Finally, the story invokes the troubling relationship between death (induced or confirmed) and the process of organ transplantation. The organ donor is a convicted criminal sentenced to death for his crime, just as—with the new acceptance of brain death as the requirement for organ harvesting—the "beating-heart" organ donor (or "neo-mort") would die as a result of that act of organ harvesting.[38]

The most striking resemblance between these stories, however, is the symmetrical response of the potential organ recipients to their donors. Both men feel revulsion, the one based predominately (though not exclusively) on class, the other on race. Colonel Seymore confesses with a grimace that the very notion of being in contact with any part of the donor's body makes him "shudder," and Van Puyster demonstrates a reluctance to undergo the transplant operation, since as an artist he is especially sensitive to the visual impact of such a transformation: "The shock to his aesthetic mind was almost overpowering. The thought of a black hand was revolting, but the thought of no hand at all was like death itself. Would the hand be large and awkward, or would it be slender and sensitive? Was it coal black, or only a light mulatto? Could he ever return to his society with such a stigma?" (911). Despite his reluctance, Van Puyster accepts the graft of "the arm that has murdered an unarmed man," an arm that (to make matters worse) "is black" (910). The minutely described operation ends with the application of a cast, and the separation of the two individuals: one to recover in the surgical ward of the hospital, the other to meet his fate. "The negro, still under anaesthesia, was turned over to the prison officials" (923). As the narrative concludes, we learn that the operation has been a success: "Two weeks later and the arm was healing rapidly. Two years later and complete sensation had returned. Five years later and the black hand was painting masterpieces, but Van Puyster always wore gloves" (923).

If the story ended here, we would comment on the author's accurate prediction that donor organs would, more often than not, come from racial "others," while the recipients would most frequently be white. We would remark that Van Puyster's habitual gloves echo that power imbalance through the symbolic assertion that the painting of masterpieces and the possession of a "black hand" are incompatible. We might also note the metonymy that (ironically) makes that same black hand the agent of genius, as distinct from Van Puyster who wields it.

But as the story goes on, our attention shifts from the donor to the recipient, and we move into the genre of psychiatric case history to read the case of Van Puyster the artist. "H.V.P., a native born American of Dutch and English ancestry, age 46, unmarried, white male," has been admitted to the "Psychopathic Hospital" after the judge committed him, "as the jury had found him insane when on trial for a series of homicides of Negroes" (923). As the case history stipulates, H.V.P. was the recipient of "the world's first arm-grafting operation . . . [and] the arm was taken from a condemned negro criminal" (923).

After the surgery, Van Puyster undergoes a transformation in character and conduct: though he is still painting marvelous pictures, they have begun to verge on the fantastic, and then the grotesque. He avoids encountering "negroes" at any time, discharges his Negro valet, and is manifestly uncomfortable even at the mention of "a negro." Socially increasingly reclusive, his behavior begins to mimic that of the ghastly Dr. Hyde: "It was during this period that he started going out during the night and on several occasions he was known to have stayed away from home for as many as three days at a time and to have returned at the end of that period with his clothes mud-spattered and torn and in an altogether disreputable state. His man also noticed that these events corresponded with a series of brutal murders" (923). The case report charts Van Puyster's arrest and incarceration, during which he is assailed by hallucinations that he is "being pursued by a negro, who was attempting to cut off his arm. . . . He grew more morose and solitary and under no conditions would he permit an attendant to remove his glove" (923). Just as physicians have diagnosed paranoia and decided that "his right arm again be amputated, this time above the region of the graft," Van Puyster is "found dead in his cell, having bled to death from a self-inflicted wound which had severed his *right* radial artery" (923). In its understated rationality, the case summary that concludes Bowers's story manages still to suggest its uncanny alternative. The judgment is suicide ("A patient with a negative psychiatric history became criminally insane following a graft of a negro's arm"), but the implication lingers that the deed was really murder, as the Negro criminal has come at last to reclaim his donated hand.[39]

In form and outcome, these early-twentieth-century narratives anticipate two important shifts in the social climate surrounding organ transplants. First, the overt attention to racial difference in early-twentieth-century transplant narratives has been succeeded, in the early twenty-first century, by a covert and *enabling* racial imbalance between organ recipients (frequently, though not always, white and "first world") and organ donors (frequently, though not always, black or brown and "third world"). Second, most likely in response to that growing local and global inequity, there has been a shift in the psychiatric perspective on organ transplantation cases. Whereas the early-twentieth-century stories explore the psychiatric and social effects of being the organ *recipient*, by the early twenty-first century, it would become standard medical procedure to assess the psychiatric and social fitness for, and effects of, being a live organ *donor*.

FROM SCIENCE FICTION TO SCIENCE FACT: TWO HAND TRANSPLANTS

In his contribution to "an ethics of organ transplantation," Lawrence Cohen has observed: "As they are cut out from the flesh, organs reconstitute the spaces of bodily analysis." We can extend his observation to the second half of the operation: "As they are stitched into the flesh, organs reconstitute the spaces of analysis: bodily, social, scientific, cultural and political.[40] On 23 September 1998, more than half a century after Bowers's story appeared, newspapers reported that Dr. Jean-Michel Dubernard of Lyon, France, working with an international team of surgeons, had performed the world's first successful hand transplant. To be specific, "they had attached the right hand and forearm of an anonymous donor to the arm of a 48-year-old Australian man whose own hand was amputated after a logging accident in 1989."[41] As one reporter observed, "It sounds like something right out of a campy, late-night horror film: eager doctors working to attach the hand of a cadaver to the arm of a living person."[42]

Initially the operation produced not only the standard celebratory publicity but quite a bit of criticism. Newspaper articles condemned the team of surgeons for the secrecy with which the procedure was carried out and for their poor planning. Robert Beasley, M.D., director of New York University's hand surgery department, accused the French team of "a grossly unethical act of showmanship."[43] Other physicians voiced their concern that the small likelihood of regaining hand and arm function meant that the patient would be trading the great risk posed by long-term use of immunosuppressant medications for what was largely a cosmetic gain. For example, Dr. Vincent Hentz of the Stanford University School of Medicine was reported as saying, "I can almost guarantee that you will have a shorter life, on average, in order to have this transplant done. That is the real ethical issue here."[44] Even surgeons such as Dr. Jerome Kassirer, editor of the *New England Journal of Medicine*, who defended the surgery, observing that "our society is comfortable" with the notion of organ transplantation, still acknowledged the underlying motivation: "Doctors are a competitive lot. They would very much like to be first."[45]

Yet if the operation took the world by surprise, even more surprising were the revelations that emerged in the following weeks. It seems that Clint Hallam, the transplant recipient, was not an Australian businessman who had lost his hand in a logging accident with a circular saw (as he had claimed) but a criminal who had lost his arm while serving a two-year sentence for fraud in a New Zealand prison. The revelation that the transplant doctors had been unaware of Hallam's real criminal identity was initially used to criticize the care with which the surgeons had chosen him as a transplant recipient. In response, Dr. Dubernard, "who had been Hallam's doctor for more than two years, said he was 'duped' into believing the logging-accident story."[46] More interesting still, the surgery was soon repositioned not as life-extending therapy but as cosmetic experimentation. As *Time* magazine breezily summarized, "a criminal past is no reason to deny someone medical treatment—even a treatment that is purely experimental and has nothing to do with saving lives."[47] "Embarrassed as they might have been, the surgeons had no grounds for canceling the operation—especially given how badly Hallam wanted that arm. He was so eager to be a guinea pig, in fact, that he'd also registered with a U.S. group that hoped to be the first to transplant a hand. The winning team insisted they were not in a race with the Americans or anyone else, but they couldn't help crowing last week."[48]

As the saga continued, the news stories it generated increasingly seemed less like "a campy, late-night horror film" than a sting, and finally a farce. Initially Hallam was elated with the surgery and experienced the challenge of recovery as a problem of internal communication: "'It's now a question of my mind saying, "I've got my hand back,"' Clint Hallam, 48, of New Zealand said at a news conference at the Edouard Herriot Hospital."[49] As the recovery dragged on from months to years, things began to go wrong with the transplant. Although for a year after the surgery, "his new hand, which had previously belonged to a motorcyclist killed in an accident, had functioned well," Hallam then was troubled by "pockets of rejection" that covered the grafted hand with scabs.

As Hallam was struggling through his postoperative complications, in January 2000, Dr. Dubernard and his team operated again, upping the ante by performing a "*double hand-and-forearm transplant* . . . on a 33-year-old Frenchman who lost his hands in 1996."[50] Unlike Hallam, this transplant recipient's name (Denis Chatelier) was kept secret for several years, although his morphological, political, and reproductive status was announced.[51] Reporters learned that he hailed from the Atlantic coast of France and was the father of two children, "a man of small build who lost his hands when a homemade rocket exploded as he was preparing to launch it" (Altman 2000). According to his surgeon, he was also "strong-willed, tenacious . . . with the spirit of a marathon runner" (ibid.). In a curious coda, it was reported that he had received his graft from a "brain-dead 19-year-old man who had fallen off a bridge," and whose heart, kidneys, liver, and arms had been donated to different recipients. After the arms were removed, it was reported, "the surgeons attached prosthetic arms to make the donor's body look normal."

While Chatelier's double-hand-transplant donor's body could be made to "look normal" with the use of prosthetic arms and hands, the transplant recipient Hallam was still struggling to live with his scab-encrusted grafted hand and arm, which were far from normal looking.

Finally, three years after the surgery, it was reported that "the world's first recipient of a hand transplant has had the hand amputated at his request."[52] The ninety-minute operation was necessary, doctors told reporters, because the transplant recipient had for two months failed to take the essential immunosuppressant medications "prescribed as part of the anti-rejection treatment, his surgeon said. 'His life was in danger,' he added." Although the French surgeon who led the team performing the transplant operation in 1998 declined to comment, Hallam, the grant recipient, was vocal on the subject. He denied that he had refused to take the immunosuppressant, and instead explained his choice of amputation using the same mind/body distinction that he employed in 1998, but now to opposite effect: "Mr. Hallam told the British Broadcasting Corporation . . . that his body and mind had said 'enough is enough' and the hand should be amputated."

In its trajectory, the Hallam hand-and-arm transplant story, like Alexander's "The Black Hand" a half century earlier, illuminated the range of tensions produced by organ transplantation: between disability and mortal injury, between medical care and medical coercion, between donor and recipient, criminal and free citizen, hand and body, psychiatric illness and health, as well as between scientific and political communities, public relations and journalistic practice.[53] When we add to it the tale of Denis Chatelier's successful double hand transplant, the two actual stories trouble the categories of self and other, prosthesis and transplant, amputation and surgical attachment, morphological normalcy and abnormality. Whether the "pockets of rejection" endangered Hallam's health or merely his self-image, his hand transplant that only a year earlier was hailed as a success and "a triumph for immunology more than surgery," was now clearly a failure. In contrast, Denis Chatelier's doctors affirmed the success of his double hand transplant by comparing the functioning of his grafted hands to the prostheses that they replaced. "By one year, the patient could perform the same daily activities that were possible with myoelectric prostheses used before transplantation, and he could also perform additional activities including holding a pen, a glass, or a pair of scissors; shaving; and other personal hygiene tasks" (Barclay 2003).

We see a remarkable shift in the nature and locus of identity if we compare both the Hallam hand-and-arm transplant *and* Chatelier's double hand transplant to Bowers's "The Black Hand." With the move from Bowers's story to Hallam's transplant, although the race-based motive for rejection articulated by the fiction may have disappeared, difficulties in negotiating the self/other boundary remain, with the mind rejecting the hand as no longer part of the body. In contrast, when we compare Chatelier's double hand transplant to the work of fiction, there is no sense in the former instance that the provenance of the organ endangers or contaminates the organ recipient. Rather, it seems that the organ recipient is somehow able to alter the status of the grafted organs. An interview with a member of the surgical team in 2003 reveals that Chatelier was able to make the mental adjustment that Hallam could not, coming to view the grafted hands as self rather than other. "Ongoing psychological support and evaluation revealed that during the first three months, the patient was worried . . . troubled by seeing the transplanted hands. However, by three months, he considered 'the' hands to be 'his own' hands" (Barclay 2003). Most remarkably, this mental shift or reconceptualization appears to have been accompanied by a change in brain structure and function as well:

> Functional magnetic resonance imaging (fMRI) at 2, 4, 6 and 12 months showed cortical reorganization, with a progressive shift of cortical hand representation from the lateral to the medial region in the motor cortex. (Barclay 2003)

> The same basic problem of uncertainty persists that has been central to organ transplantation since its inception: "the innate and unrelenting intolerance of individuals to grafts of other people's tissues and organs." (Swazey and Fox 1992, 10)

> Among the preconscious sentiments that these [organ] recipients seem to share is the belief that some of the psychic and social, as well as physical, qualities of the donor are transferred with his or her organ into the person into whom it is implanted. (Swazey and Fox 1992, 36)

Let's return for a moment to a comparison of the two science fiction short stories of organ transplantation, to see what they share and how they differ. Both short stories make the point that transplantation in humans relies on clinical trials that bridge differences of species. Recall that "New Stomachs for Old" features an informational box describing how "the German Professor, Dr. Walter Finkler, amputated the heads of various insects and transplanted them on others" (Alexander 1927,

1039). "The Black Hand" includes the description of an operation grafting the posterior end of one frog onto the anterior end of the other, so that "the composite frog lived until its untimely death, three weeks later, by an accident" (Bowers 1931, 910). Similarly, both short stories address what we might call the internal management of ethnic and racial difference: the former by stressing the healthy physique of the young Italian whose stomach will rejuvenate the elderly, dyspeptic colonel; the latter by dramatizing the horrifying effect on the artist of receiving a donated "black hand." In each of these short stories, a technical and surgical success is followed by a social failure: the wealthy Caucasian dyspeptic finds that his career and personal life are ruined as he takes on the habits of the poor Italian debtor, and the brilliant white artist becomes a pathological, racially motivated murderer after the grafting of a black hand. In short, these stories share the sense that identity is linked to the transplanted organ. That identity (as lower-class lover of spicy Italian food; as murderer) travels with the transplanted organ and comes into conflict with (or dominates) the recipient.

When these stories were written, organ transplantation was not a reality. Voronoy first reported the transplantation of a kidney from a human cadaver donor (which was unsuccessful, since the organ never functioned) in 1937, six years after the publication of "New Stomachs for Old."[54] I have argued that science fiction acts as a kind of ideological cyclosporine, enabling the acceptance (on an individual as well as a collective level) of organ transplantation. For transplantation to become a reality, the notion of organs as being identity containing and identity conferring clearly had to be dislodged, or at least moved from a socially central to a socially marginal belief. The final story I want to consider helps us to see that shift in the process of occurring, because there the grafted organ is conceptualized not as identity containing and conferring but rather as identity maintaining.

FROM THE MILITARY DRAFT TO THE ORGAN DRAFT

Robert Silverberg's "Caught in the Organ Draft" ([1972] 1983) appeared during the height of the Vietnam War, and it captures Scheper-Hughes's point that the flow of organs follows the modern routes of capital, from the dispossessed to the affluent, but with a twist.[55] Here we follow the flow of *physical capital* from the vigorous and disenfranchised young to the frail but institutionally dominant old. An organ draft law, "put through by an administration of old men," has given rise to a system providing a never-ending supply of healthy organs for the elderly pillars of society. Through this new organ draft system, young men and women register on turning nineteen, and those who are high in "organ reservoir potential" receive an "organ draft" notice calling them to report to Transplant House for their ritual physical exam, after which the draftee goes "on call," and within an average of two months they are "carving [him] up" (Silverberg [1972] 1983, 143). The result is a society in which the young feel victim to the old, much as they did during World War I, but the nature of their victimization is medical rather than military. As the protagonist thinks when he eyes a pair of "splendid seniors" as the story opens:

> We can guess at their medical histories. She's had at least three hearts, he's working on his fourth set of lungs, they apply for new kidneys every five years, their brittle bones are reinforced with hundreds of skeletal snips from the arms and legs of hapless younger folk; their dimming sensory apparatus is aided by countless nerve grafts obtained the same way; their ancient arteries are freshly sheathed with sleek Teflon. Ambulatory assemblages of second-hand human parts, spiced here and there with synthetic or mechanical organ substitutes, that's all they are. And what am I, then, or you? . . . In their eyes I'm nothing but a ready stockpile of healthy organs, waiting to serve their needs. (142)

The story follows a young man from his draft notice through his organ "donation," tracing his evolution from draft resistance to cooperation and finally to collaboration with the system of organ draft once he himself stands to gain from its fund of endlessly procurable organs. The system of organ transplantation that Silverberg's fiction institutionalizes embodies the two central elements of medical practice in late modernity: the adaptation of a statistical, rather than individual, approach to human life, and the tendency of medical technology to create the need it satisfies. As the protagonist explains, "Nobody escapes. They always clip you, once you qualify. The need for young organs inexorably expands to match the pool of available organpower" (144).

As he works through his own emotional and strategic responses to the organ draft notice, the protagonist raises a variety of prominent issues in contemporary organ transplantation debate: whether human beings have the right to bodily integrity; whether it is ethical to practice live-donor organ transplantation; whether body enhancement is an appropriate medical practice; whether body parts (including tissues and DNA samples) can be possessed; whether there *is* really a "shortage of transplantable organs," or whether that reflects broader power shifts in society that privilege longevity of the powerful over quality of life of the relatively powerless; whether there are limits to the medical fight against death. The protagonist traces the progress in organ transplantation in two registers: medically, where it takes the form of an increasing victory over the immune system's tendency to reject the transplanted organ, and socially, where it takes the form of an increasingly institutionalized system of interlinked greater and lesser coercions that shape the populace toward more and more widespread involvement with the donation system. While "eventually everybody will have a 6-A Preferred Recipient status by virtue of having donated," the protagonist still struggles against another kind of resistance: psychic rather than physical: "Drugs, radiation treatment, metabolic shock—one way and another, the organ rejection problem was long ago conquered. I can't conquer my draft rejection problem. Aged and rapacious legislators, I reject you and your legislation" (154). Yet after a night spent weighing the options, the protagonist makes his decision: "When the time comes, I'll surrender peacefully. I report to Transplant House for conscriptive donative surgery in three hours" (155). His own likely future need for transplant organs has changed his mind. When the story concludes, the transplant operation has been completed. "I've given up unto the powers that be my humble pound of flesh. When I leave the hospital . . . I'll carry a card testifying to my new 6-A status. Top priority for the rest of my life. Why I might live for a thousand years" (156).

Written more than thirty years ago, Silverberg's story clearly references the social security system in its carefully calibrated relationship between present financial contribution and future receipt. But it does so, strikingly, by anticipating the emergence of a bioeconomy: a market (and a futures market) in human body parts and products. The story articulates all the contemporary critiques of the organ transplantation system, including the problematic developments of organ commodification, the live-donor system, and the inequitable distribution of obligations and gifts. In the time since the story's publication, we have continued on to build the institutional system of organ transplantation that Silverberg warned us against, so that by now his science fiction short story seems more parable of the present than prediction of the future.

The three science fictions of organ transplantation by Alexander, Bowers, and Silverberg reflect the changing view of the human body during the era in which organ transplantation was naturalized. Alexander's tale of the stomach transplant and Bowers's tale of the hand transplant figure the human organ as the carrier of identity. The transplanted organ is rejected in each story precisely because it transfers troublesome identity categories (class and race) to the recipient. In the 1920s and 1930s, the human body is, implicitly, transient, and no amount of wealth, power, or artistic talent can render it durable. Yet by the time Silverberg writes, in the 1970s, identity is vested no longer in the transplanted organ but in its recipient: the powerful senior citizen whose life span is extended by the replacement organs obtained through the organ draft. Transplant technology has been socially and psychologically naturalized: "we" are now comfortable with the notion of being a transplant recipient, which also means we are unalarmed by the notion of *someone else* being a transplant donor. Of course, "we" and "they" are terms shaped by social, economic, racial, geographic, and class coordinates. Silverberg's interesting wrinkle is to suggest that with the demographic shift to an aging society (in the West at least) may come an inversion in the meaning of age as one of those coordinates. In this era of globalization, this normalization of organ transplantation has taken place quietly, aided in part by the effects of social, geographic, and racial/ethnic distance: the political, structural, and legal institutional mechanisms by which it has occurred remain to be discovered and debated.

In "Immortality for Achievers," cartoonist Tom Tomorrow brings Robert Silverberg's idea up to date, in what might now be called a graphic short story. In a frame headed "The President unveils a modest proposal!" President George W. Bush is pictured in front of the American flag, explaining, "So you see, by harvesting vital organs from the poor—we can prolong life for the wealthy! I call it my immortality for achievers initiative!"[56] Democrats respond with a "slightly watered down alternative" entitling "anyone who is forced to donate an organ under the President's plan" to "a modest tax credit." Criticism of the

THE BACK PAGE BY TOM TOMORROW

IMMORTALITY FOR ACHIEVERS

88 THE NEW YORKER, JULY 2, 2001

"Immortality for Achievers." Cartoon by Tom Tomorrow. Reprinted from the *New Yorker*, 2 July 2001, 88.

initiative isn't totally absent. A TV pundit ventures the timid criticism: "I—I don't want to go out on a limb—but doesn't this plan seem slightly tilted in favor of the rich?" while "a few crazy extremists insist that both parties are somehow 'in thrall' to their 'wealthy contributors'—but nobody pays attention to them!" Finally, the plan achieves bipartisan consensus. "And so, under the MANDATORY ORGAN DONATION ACT of 2002, next of kin will receive a gift certificate worth five dollars off their next purchase at any government surplus facility!" In the last frame of the strip, a flushed, sweating Richard Cheney commands, "Now somebody get me a new ticker—and I mean Pronto!" Like Silverberg's 1972 science fiction story, this futuristic comic by Tom Tomorrow shows us a society in which control of the bioeconomy is more important than access to the cash economy. Those who have access to replacement organs have the option of longevity, whereas those who have surrendered their organs in return for cash, tax credits, or (most absurdly) government surplus gift certificates only have more options in the present. In short, this new economic order neatly illustrates an observation by the social anthropologist Michael Thompson: "The people near the top have the power to make things durable and to make things transient, so they can ensure that their own objects are always durable and that those of others are always transient."[57]

TRANSPLANT MEDICINE: WOMB OR "ROTTING PIECE OF MEAT"?

If the core life-sustaining organs are no longer conceptualized as carriers of identity, what of those organs whose transplantation is a question not of life preservation but of lifestyle? And what of organs that might, *literally*, be carriers of another's identity? A recent transplant procedure both recapitulates and reverses some of the issues I have traced through fact and fiction in the twentieth century. On 17 March 1999, Cynthia McCulloch wrote to Med Help International's "Maternal and Child Health Forum" with the following question: "enlight [*sic*] of the incredible miracles of science, I'm wondering if it would be at all possible to achieve pregnancy after a 'complete' hysterectomy? I'm thinking if a person can receive a heart transplant, maybe there could be a medical procedure that could implant a uterus."

McCulloch's question was answered by one of the anonymous "professionals" from the "Henry Ford Health System": "Uterus transplant is not a doable operation. All transplants require drugs to suppress the immune system: most are harmful to early pregnancy development."[58]

The possibility of rejection may have troubled the professionals of the Henry Ford Health System, but it does not seem to have worried doctors at King Fahad Hospital and Research Center in Jeddah, who just a year later performed the world's first uterine transplant, on 6 April 2000. The uterus was taken from a forty-six-year-old woman "with ovarian cysts who had been advised to have a hysterectomy," and transplanted into a twenty-six-year-old woman who had lost her own uterus when she hemorrhaged during a cesarean section.[59] It survived for ninety-nine days, through two menstrual periods, until its blood supply was compromised and it had to be removed.

While the Saudi doctors announced the uterine transplant as a success, opinion elsewhere varied.[60] Jeremy Laurence, health editor of the *Independent*, described the surgery as "the first step towards restoring the capacity to bear children in women who are prevented from becoming pregnant because of uterus defects," and quoted the Royal College of Obstetricians and Gynaecologists (U.K.) describing the report as " 'very exciting.' If the surgery could be refined and the transplanted wombs made to function normally, it opened up the possibility of mothers donating wombs to their daughters, or sisters to sisters, the college said."[61] And Dr. Richard Smith, consultant gynecologist at the Chelsea and Westminster hospital in London, praised the Saudi team for having "achieved a lot . . . [and] shown it is technically feasible to perform the operation in a woman, which is a world first."[62] Smith reported that his own work in uterine transplants was stalled for lack of funding, but he reported that the danger of medications to suppress organ rejection would not be a major impediment. "Our view always was that the uterus would go in and the woman would have one or two babies and then the uterus would come out—she would only be on immuno-suppressive therapy for a few years."[63]

In contrast to Smith's breezy optimism, Britain's Lord Winston, the "fertility expert," was scathing in an interview on the BBC: "The truth is that it was a complete failure, because of course blood clotting is exactly what you would expect. . . . A rotting piece of meat in the pelvis which is what will happen with a womb transplant will endanger the life of the recipient of the transplant. . . . The idea of a uterus slowly being rejected with a pregnancy inside it with a live baby gradually dying in consequence is actually quite horrific."[64] And Dr. Jonathan Bromberg, a transplant specialist at New York's Mount Sinai School of Medicine, pointed out that "the immune-suppressing drugs used to protect the new organ have significant rates of serious complications, including certain cancers. . . . You've got to differentiate between life-saving procedures and lifestyle procedures. . . . As important as pregnancy is, getting a new uterus is not like getting a kidney or a heart or a lung. It's not even close" (Marcus 2002, 1). Bromberg questioned why "the Saudi doctors [would] have performed the procedure in a woman when there's so little evidence that it could work even in animals. . . . It's still very experimental and very much science fiction."

Once again, simply the passage of time has demonstrated that research science and science fiction share the same liminal territory, bounded (and linked) by the zest for extrapolation. As this book was going to press, in July 2003, another womb transplant made the news. "Womb Transplant Baby 'within Three Years' " was the headline to the BBC News Online story drawn from the European Society for Human Reproduction and Endocrinology annual meeting in Madrid. Swedish researcher Mats Brannstrom, of Sahlgrenska Academy at Gothenburg University, reported that fertile, healthy mouse pups had been born to female mice that had received womb transplants. Having claimed on the base of this mouse experiment, which he reported in the *Journal of Endocrinology* in 2002, to be able to surpass the Saudi doctors' unsuccessful womb transplant attempt by using "a different surgical technique," Dr. Brannstrom challenged boundaries of gender *and* generation in the text of his 2003 announcement.[65] "The Swedish expert behind the research says that one of the best candidates to be an organ donor would be the patient's own mother—raising the prospect of carrying your children in the same womb that carried you. He says that it may be even technically possible to transplant a womb into a man, and use hormone injections to allow pregnancy to succeed" (Hutchinson 2003).

What is the ethical status of such a womb transplant? Brannstrom observed that there was likely to be debate about whether such major surgery, followed by the mandatory lifelong reliance on immunosuppressant drugs, was ethically acceptable, since a womb transplant "would be the first organ transplant which is not needed to cure a life-threatening illness" (Hutchinson 2003). Yet while there have not been what might be called "lifestyle" organ transplants, we do have the precedent of the elective hand transplants. And we discover that we have entered a zone that is liminal not only ethically but also generically if we begin to explore other nonlifesaving transplants, such as the first successful transgender hair transplant, or the first successful human tongue transplant, in 2003.

Comparing Alexander's 1927 story "New Stomachs for Old" to reports that a forty-two-year-old man with a malignant tumor of the jaw and tongue had been the recipient of what was presumably a cadaver tongue, we realize that our changing definition of medical success influences (and changes) our notion of ethics. While in the Alexander short story, the stomach transplants were considered unsuccessful because they drastically altered the recipients' sense of taste, now in 2003, Viennese physicians Dr. Christian Kermer and Dr. Franz Watzinger say that they will consider this inaugural tongue transplant a success "if the patient . . . regains his ability to eat and speak . . . [although] it's very unlikely he'll regain his sense of taste."[66] The ethical boundary is a slippery one, it appears: while last year womb transplantation seemed ethically unthinkable, this year it seems just around the corner. And that leaves us with one final example of the shared liminality of genre and ethics: a CNN.com report, "Face transplants not just science fiction: Doctor: Procedure technically feasible but ethically ambiguous."[67]

RUBBISH THEORY AS A LIFE STRATEGY

How do the different genres of visual art and science fiction contribute to the revaluation of the human body and its constituent parts (organs, appendages)? Michael Thompson's analysis of the process by which objects acquire or lose value and durability has as its motor a "region of flexibility" in which economic, aesthetic, and temporal values are (as yet) undetermined.[68] In his book *Rubbish Theory* (1979), Thompson argues that we categorize objects in the physical and social world in two very different ways: as *transient* objects that "decrease in value over time and have finite life-spans" and as *durable* objects that "increase in value over time and have (ideally) infinite life-spans" (7). These two kinds of objects are seen hierarchically in relation to each other, the durable being clearly superior to the transient, both economically and aesthetically (6). Moreover, our treatment of objects reflects the category within which we place them and the value (both economic and aesthetic) we attribute to them. So we hang on to objects that seem to us to have value, we discard objects that seem valueless, and we take greater care of durable objects than of objects with a finite and brief life span.

Thompson argues that there exists a third realm in which objects have neither an inherent life span nor a fixed and determinate value, either economically or aesthetically. He proposes that this "third *covert* category" of cultural value, which he calls *rubbish*, "is able to provide the path for the seemingly impossible transfer of an object from transience to durability" (9). Thompson implies that rubbish operates as a transformative category: a machine through which objects acquire new values and properties. Thompson draws on the body to explain this third category. He explains that we can distinguish between body products that are rubbish ("excrement, urine, finger- and toe-nail clippings, pus, menstrual blood, scabs and so on") and those that aren't ("milk, tears, babies . . . and, sometimes, sperm"). Body parts or products that can be *used*, in other words, are no longer rubbish. Thus while rubbish is something we customarily discard, the concept itself is culturally constructed. We may decide *not* to discard an object in the rubbish category, as when we decide to save milk to feed a baby later, or to make a deposit to a sperm bank. "If rubbishness were self-evident and derived from the intrinsic physical properties of objects then this division of body products into rubbish and non-rubbish items would be fixed and unchangeable. Yet, in recent years, some body products have crossed from one side to the other" (10). To put this in terms closer to my own, Thompson's rubbish category is a realm of pure potential. A liminal zone, it functions as the machine or enabling space for cultural and material transformations.

While there are limitations to Thompson's notion of the rubbish category—in particular his tendency to think of it as working one way only, as a category that moves things from the transient and valueless to the durable and valuable—I find it useful as a way to think about the cultural function of science fiction as a genre. In its stress on the two indices of time span and value, Thompson's rubbish theory enables us to see that science fiction may be uniquely possible to provide that liminal zone, that transformative space, precisely because its own value is in the process of changing. From being a genre that has been relegated to the literary margins, it is taking on increasing centrality. Aesthetically, its movement from the margins to the center signifies the greater currency of science fiction as a genre, both in the academy and in the marketplace. Furthermore, the very transient topicality—the genre's preoccupation with the new, or the *novum*, as Darko Suvin has it—which was once the cause of its devaluation has become the reason for the genre's increasing appeal.[69]

Rubbish theory may be particularly able to illuminate the cultural function of science fiction because there is a generic and discursive diversity that is inherent to rubbish theory. As Thompson explains, "the rubbish theorist has to deal in different discourses simultaneously. And since they cannot be mixed they must be juxtaposed. The joke, the paradox, the shock technique and the journalistic style, far from being unscholarly devices to be avoided at all costs, become rubbish theory's inseparable accompaniments" (Thompson 1979, 5). The rubbish theorist, looking at the relations of science fiction to other forms of literature, will obey the imperative to deal in different forms of discourse, to juxtapose them, to welcome what seems *unscholarly.* He or she will of necessity have to problematize questions of value and life span, rather than accepting them as given: this holds true both for the value and life span of texts and for the value and life span of living things—among them the human body.

This set of oppositions grounds Zygmunt Bauman's striking metaphor for life in the postmodern era: "Life is not a novel with a finite set of characters, a plot and a denouement; it is instead a railway-station bookstand packed to overflowing with the latest best-sellers. And the shelf life of a best-seller, as one of their authors observed, is somewhere between milk and yogurt" (Bauman 1992, 190). On a first reading, this comparison between a novel and a shelf of bestsellers seems to favor the novel, valued because it is unique and irreplaceable, unlike the bestsellers that are indistinguishable from each other. But if we think of life not as a state but as a journey, then those bestsellers will be much more useful. They can help us pass the time without running out of characters or plot; we won't have to face the denouement, much less the ending; and though they won't achieve immortality, they'll last long enough to make the trip.

As Bauman's metaphor suggests, science fiction, like the shelf full of railway station bestsellers, falls into the category of perishables. Its very appeal lies not in timelessness but in its timeliness; not in its singularity but in its fungibility. The genre of science fiction is like the body in our postmodern era: capable of endless renewal. This brings us back to the postmodern *life strategy* (the term is Bauman's) of Silverberg's "splendid seniors." By claiming the organs of their juniors, they enact Jean-François Lyotard's observation that we live "in an *open space-time*, in which there are no more *identities*, only *transformations*" (Lyotard, cited in Bauman 1992, 184). With the incorporation of drafted organs, the space of the human body is expanded to include organs from elsewhere, and the time of the human body incorporates another's life span. The term "selective service" takes on an eerie new meaning in Silverberg's medicalization of the draft: "The allurements of survival offered by advancing medicine—survival itself, as defined by the medical 'state of the art'—become more and more selective—*socially* selective" (192). Focusing on the undoing of death, we value mobility, transience, ephemerality, inconsequentiality, over fixity, duration, eternality, significance.

In short, increasingly we value *rubbish*: like the organs that circulate in Silverberg's story, rubbish pleases because it circulates; and it circulates *because* it exists perpetually in the present, in the novum. Initially the trajectory of the transformation in "Caught in the Organ Draft" seems to be from transience to durability, from perishable bodies to eternally durable ones (with the provision of drafted organs). Yet that is only because we are failing to read both levels of Silverberg's plot, and to hear its echoes of the military draft from World War I through the Vietnam war. Not only does the story detail a surprising postmodern movement of old people toward everlastingly renewable life, but it also recalls the familiar modern story of young people entrapped in a system that codes them as powerless and perishable. The story turns on the difference between the old institution of a military draft and the new institution of an organ draft created in response to a drastic economic reevaluation: rejection of the belief that objects hold intrinsic value in favor of the belief that objects hold value in relation to their transience.[70] "Since the organ draft legislation went into effect," we learn, there have been no "field engagements" because "the old ones can't afford to waste precious young bodies on the battlefields. So robots wage our territorial struggles for us. . . . It's a subtler war than the kind they used to wage: nobody dies" (Silverberg [1972] 1983, 145).

In the new cultural zone articulated by Silverberg's science fiction story, the human body and its components have been (re)categorized as valuable precisely because of their transience. Human bodies have become both expendable *and* infinitely valuable, depending on your social location. In place of a military draft devoted to the goal of accumulating valuable territory and goods, we have an organ draft that provides valuable replacement organs to some, enabling an extended life span, at the cost of a truncated life span for others. Through its use of extrapolation, Silverberg's story thus demonstrates how the genre of science fiction functions as the liminal zone in which duration and value are transformed, or as the transmission machine, the *shifter* of duration and value.

SEPTEMBER 11 AND AFTER

Contemporary medicine is turning to transplantation not just of organs and body parts but of tissues as well, even embryonic tissues.[71] In a recent bulletin in *Reuters Medical News for the Web*, Dr. Curt R. Freed told a remarkable story to demonstrate the advantages of such transplantation for producing relief from the tremors, rigidity, and immobility resulting from Parkinsonism: "The advantage of transplant therapy is the consistent effect achieved for symptoms of Parkinson's disease, Dr. Freed said. One of the study patients, who had discontinued L-dopa therapy after receiving his transplant, was working on the 34th floor of one of the World Trade Center towers during the September 11th attack. Dr. Freed said the patient was able to walk down 33 flights of stairs and about 3 additional miles to reach public transportation to get home."[72]

These embryonic cell transplants recall the debate about the ethics of using discarded embryos (whether the product of miscarriages, abortions, or excess research embryos) as material for another sort of transplantation intended to prolong life and to enhance its quality. As in Silverberg's story, in Freed's story too we must address the issue of using the young (in this case the *embryonic*) to improve life expectancy and life quality for the very old (in this case, forty patients between thirty-four and seventy-five years old). Moreover, in his reference to the attacks of 11 September 2001, Dr. Freed's report returns us to questions about science fiction, the human body, and the definition of rubbish.

In an article entitled "Hauling the Debris, and Darker Burdens," published in the *New York Times* six days after the September 11 attacks on the World Trade Center, journalist Charlie LeDuff reported that a fine line "separates rescue from recovery."[73] For the rescue workers, toiling in the unimaginable chaos of what came to be known as "ground zero," the distinction between working to save living people or to recover bodies or body parts was subject to delicate shifts and fluctuations. " 'There is no clear demarcation where we are in rescue and suddenly we're into recovery,' an official from the Federal Emergency Management Agency said. 'The elements of the two can coexist side by side, but there will come a point when rescue operations will be over.' " In this attack, so minutely anticipated by science fiction and film that its factual form seemed at first redundant, none of the people trapped in the rubble of the collapsed towers had the option of life extension through organ transplantation. Indeed, as one rescue worker observed, the *haul* of human body parts and organs was pitifully small. Organs were no longer valued as life extenders but now welcomed only as the definitive mark of a life having ended. Moreover, journalists commented on how the accumulated posters bearing the faces of the deceased formed a new form of art installation. Here again, the contrast must be remarked: this new memorial art is as distant as can be imagined from von Hagens's Koerperwelten or Anthony-Noel Kelly's gilded corpses. Grounded not in the presence of aestheticized organs but in their absence, it celebrates a human body that is marked by a precious ephemerality and vulnerability.

The attack on the World Trade Center was also an assault on the carefully crafted illusion of risk management that is integral to our dominant contemporary life strategy, as columnist Maureen Dowd observed, echoing social theorists Zygmunt Bauman and Ulrich Beck.[74]

> It was always a delusional vanity, this fixation boomers had about controlling their environment. They thought they could make life safe and healthy and fend off death and aging. . . . They would overcome flab with diet and exercise, wrinkles with collagen and Botox . . . decay with human growth hormone, disease with stem cell research and bio-engineering. . . . After all these finicky years of fighting everyday germs and inevitable mortality with fancy products, Americans are now confronted with the specter of . . . hazardous-waste trucks spreading really terrifying, deadly toxins like plague, smallpox, blister agents, nerve gas and botulism.[75]

Writing of the uncanny resemblance of the terror attacks to blockbuster disaster movies, *New Yorker* columnist Anthony Lane speculated, "This ruination was the opposite of invention; of conjured worlds; whether it will also signal, in part, a death of invention is more difficult to call."[76]

We might raise the same question about both transplant medicine and the transformative function of science fiction narratives: will we see a death of invention in both areas, now that they have been surpassed by the events of 11 September 2001? Silverberg's character supports the organ draft because it promises to extend his life: "Why I might live for a thousand years!" (Silverberg [1972] 1983, 156). But investment in immunosuppressant medicine and organ transplantation technology requires funds that may now have been reallocated away from replacement and regeneration medicine to another sort of life extension project. And if we play the extrapolative game so central to science fiction, what we are likely to predict is not an organ draft but rather a military draft, as part of the remilitarization of ordinary life in the battle against terrorism.

If we recall the proposition that science fiction functions as *rubbish*—a culturally fertile zone of transformation and possibility—recent commentary on science fiction movies suggests that a shift may be occurring in our cultural valuation of this genre. Science fiction, for so long the fertile ground for imaginative transformation of life as we know it, now seems only a grim commentary on life as we must live it; no longer fantasy, but documentary. Ironically, now that science fiction's representations are moving from the ephemeral and fantastic to the durable and mimetic, that which was once rubbish (stigmatized for it by some, valued for it by others) has now attained cultural resonance. While its representational value (its factual accuracy, if you will) has increased, it is consequently valued less by (re)viewers. As one film correspondent wrote of *The Sum of All Fears*, a 2002 Paramount Pictures summer blockbuster in which "a president is sent to an underground bunker, in which an East Coast city is devastated by a bomb, in which there's a dusty, gaping Ground Zero," "The terrors depicted in 'Sum' have plausibility and authority, but forgive me if I like my technothrillers a little less real."[77] Or as another critic observed, "Watching this film, we can't shake our fears that we may be witnessing a version of our own previously unthinkable future."[78] These comments demonstrate the existence of an interesting new sort of genre policing, one that critiques science fiction not because it purveys unrealistic cultural fantasies but because its cultural representations are all-too-realistic. The imaginative emancipation offered by science fiction traditionally lay in its capacity to create magical new worlds. In contrast, science fiction now shocks us by its resemblance to fact. Unable any longer to inoculate us against our fears, science fiction sutures us to a destructive present that is all-too-real.

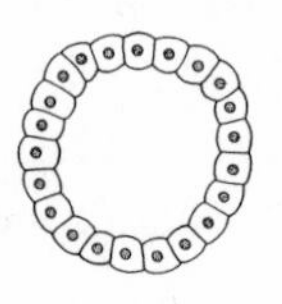

7

Liminal Performances of Aging

FROM REPLACEMENT TO REGENERATION

Doctors and patients alike say the new diagnosis holds one other bright spot: it is giving a new, first-person voice to Alzheimer's, allowing patients to talk about the disease before it robs them of the ability to understand that they have it. . . . Now, patients whose impairment is still relatively mild are starting to tell their stories [Dr. Alan Dienstag] said.[1]

In the summer of 2002, newspapers reported an increase in the numbers of early-onset Alzheimer's diagnoses, fueled both by new brain-imaging technologies and new drug trials. While the prognosis for Alzheimer's patients is still dismal—one physician likened it to having a sword of Damocles hanging over your head—still the possibility of making an early diagnosis gave some people reason for optimism. "If we in fact can intervene at this earlier stage and alter the course of the disease, that would have a big impact on quality of life," said Dr. Ronald Petersen of the Alzheimer's research group at the Mayo Clinic (Kolata 2002, 20). Yet while the new drug trials and the new scanning technologies will have their impact only in some hoped-for future, news reports suggest that early diagnosis is also making an impact on the quality of life right now. It has catalyzed the public emergence of a new voice and subject: the Alzheimer's patient, narrating his or her own experience. In this concluding chapter, I want to explore how we can understand this new kind of narrative, considered in relation to a new kind of sociomedical performance, whose implications I will explore.

Victor Turner's expansion of the concept of liminal play to include practices spanning the entire cultural field, from the arts to the sciences, helpfully directs us to the wide array of performative strategies with which we address the last liminal moment: aging. These performances, Turner points out, are both therapeutic and prophylactic; that is, they not only treat the afflictions that have arrived with aging but attempt to intervene before aging takes place, to prevent them from occurring.[2] Of course, "performative" here has layers of meaning, from the anthropological attention to cultural rituals as modes of negotiating human transitions (the "rites of passage" of Turner and Van Gennep), to the more specific notion of a productive, regulatory, and discursively motivated enactment of a way of being (in this case, *being old*) that simultaneously secures its meaning and fails to attain its ultimate expression.[3] Whether we are thinking of performance as the way we negotiate a natural passage to old age, or examine it as the process through which we both produce and regulate a culturally constructed entity, the elderly person, when we follow Turner or Butler, we generally understand performance as the cultural overlay on a natural foundation.

Yet as I suggested in the introduction, there is an important inconsistency in Turner's cultural theory, an inconsistency that is productive for any exploration of the performance of aging in the biomedical imaginary. Turner's notion of the *limen*, or threshold, which he defines as "a no-man's land betwixt and between the structural past and the structural future as anticipated by the society's normative control of biological development," is grounded in the social acceptance of an unchanging biological norm.[4] His implicit analogy—liminality is to culture as variability is to nature—posits a culturally accepted limit or boundary to the prophylactic performance of aging. In other words, Turner assumes that we can only prevent or ward off the discomforts of aging within the parameters of a "normal" body living a "normal" life span.

Since the late twentieth century, however, Turner's foundational opposition between stable biology and shifting culture has unraveled dramatically. Both biomedicine and culture more generally have challenged that notion of a naturally limited human variability. While the experience of aging has located human beings, often painfully, at a point of tension between social construction and materiality, between (technological, scientific, or intellectual) capacity and (physiological) incapacity, as we increasingly come to understand the liminality of aging as simultaneously biological and cultural, we have also begun to understand that our relation to aging is negotiable not only in cultural but also in biological terms. With the turn to a modern notion of the body seen as a project, the way each person experiences aging has shifted from being universal and ineluctable to being something particular and *chosen*, and the medical response to the experience of aging is increasingly grounded not in the notion of acceptance but in that of *cure*. At issue in this chapter is the question of just how that cure is attained.

Since the beginning of the twentieth century, Western medicine has increasingly attempted to forestall and even to cure aging by replacing failing organs and tissues, whether through organ and tissue transplantation or through the introduction of mechanical, or *biotic*, organs and tissues. Such a focus on replacement is characteristic of modernity across the range of discourses from medicine to literature. As one scholar has observed, "Modernist texts have a particular fascination with the limits of the body, either in terms of its mechanical functioning, its energy levels, or its abilities as a perceptual system. . . . Modernism is . . . characterized by the desire to *intervene* in the body; to render it part of modernity by techniques which may be biological, mechanical, or behavioral" (Armstrong 1998, 5, 6).

Whether these are deployed in biomedicine or in fiction, we can think of these replacement strategies as *prosthetic*. Negative prosthetics serve as compensation (by filling a lack or erasing a disability), and positive prosthetics provide a utopian extension of the body's potential (extrapolating from human abilities through replacing or extending human organs, or extending human senses) (Armstrong 1998, 6). Whether these technologies are replacement or enhancement oriented, however, they have a drawback. To the extent that we use our technologies prosthetically to perform, intervene in, and manage aging on the level of the individual by engaging in strategies of replacement, they enable us to avoid the meaning of our mortality, distancing human beings from our own embodied experiences and from each other. We have come to understand aging as an experience of the individual, to be remedied by the individual through manipulations of the individual body and mind.

This is not the end of the story, however. With the advent of the twenty-first century, we have seen the growth of regenerative medicine, a field characterized by the interest in generating new body parts in vivo rather than replacing them with parts from elsewhere. Are the drawbacks of replacement medicine compensated for with this new medical model? Does the new regenerative medicine of the twenty-first century offer a satisfying solution to the isolation, disempowerment, and existential avoidance characterizing our negotiations with the boundary of the life span? In this concluding chapter, I will consider these questions, moving from a survey of the early-twentieth-century performances of aging, in scientific fact and science fiction, to the biomedical and literary models of aging in the late twentieth century and beyond.

In making this transdisciplinary move, following the rules of method for exploring literature and science in action that I proposed in chapter 1, I am crossing a discursive boundary that is repeatedly invoked in the popular representation of regenerative medicine. To take just one recent example, in an Internet article from *Medscape General Medicine* entitled "Tissue Regeneration and Organ Repair: Science or Science Fiction?" a reporter summarized the "highlights" of the Twenty-seventh International Congress on Chemical Engineering, Environmental Protection, and Biotechnology as follows: "Indeed new options for organ/tissue repair might soon be available for application in the clinic, at least for certain organs and diseases. Does this mean that we will get to a stage where we will be able to order 'spare parts' for our body as we do for our cars or appliances? At the moment, *independently of all ethical consideration*, substitution of body parts, different from the first attempts at tissue repair, remains a science fiction scenario."[5] Refusing to accede to that generic and instrumental cordoning off of science from fiction, ethics from practice, in what follows I want to return to those first attempts at tissue repair in medical science and in science fiction, discussed earlier in this book, to tease out the ethical implications of the difference between *replacement* and *regenerative* medicine for our experience of the end of life.

THE FACT AND FICTION OF REPLACEMENT MEDICINE

In scientific practice and science fiction, the strategies of resisting aging available through tissue culture, tissue and organ grafting, and hormone therapy emphasized individuality and isolation. Honor Fell's description of the tissue culture experiments in the Strangeways Laboratory may have addressed themselves to a classroom audience (whether in person or in some projected future of educational television), but the implications were felt, dramatically, on an individual level. When she demonstrated the permeability of the boundary between life and death in a lecture to medical students, the implications were breathtakingly personal, even when expressed in her driest lecture manner.[6] Similarly, although they were shown to a group of surgeons attending the International Congress of Surgeons in 1923, Voronoff's moving pictures worked on the level of the individual. They *persuaded* the surgeons to submit to his surgery, each coming to it as an individual seeking a longer, more vigorous life. His gland grafting, too, focused on the effect of introducing the gonads of a chimpanzee into one specific man, whose individual response was a gauge for the operation's success or failure. These twentieth-century performances of aging were targeted to the individual, whether as consumer of the medical performance (the gland or organ transplantation; the act of culturing the tissue) or as spectator of the social performance (the movie watched by an individual who decides to have surgery; the televised experiment that persuades an individual to hope for eternal life). One of the most dramatic examples of that individualizing perspective was the special supplement entitled "Life and Death" published in July 1927 by England's preeminent scientific journal, *Nature.* The supplement featured Sir Humphrey Rolleston's lengthy essay "Concerning Old Age," a leisurely survey of the problem of aging.[7] Beginning with a statistical assessment of the legendary life spans of biblical times, the essay proceeds through discussions of "factors influencing longevity," the "physiology of old age," the "normal structural changes in old age," and "healthy and pathological old age," to round back to a consideration of old age as it is figured in Ecclesiastes 12. I will return to this important publication later, but for now I want to mark its participation in a conversation with science fiction, when in its introduction it inquired, "Are there any means by which the span of life may be prolonged or old age rendered less irksome, by which, in fact, the body may be *rejuvenated*?" (Rolleston 1927, 1). Rolleston's notion of a rejuvenated body as the response to aging, like the interventions of Voronoff and Fell, was founded on strategies characteristic of Western modernity since the nineteenth century, "in which individuals are isolated, separated, and *inhabit time* as disempowered."[8] Addressing the fear of decline and death at the level of the individual, they turn to medical science to supply a prosthetic vigor specifically to the individual, or to the individual cell or organ.[9] As we will see, this individualistic strategy is recapitulated in science fiction short stories of the same era.

Interest in the human life span was integral to *Amazing Stories*. In its first decade of publication, the magazine offered its readers a number of different dramatic performances of the assault on aging and death, incorporating both mechanical and electronic technologies. In seven different science fiction short stories published during that first decade alone, we can find electricity, organ transplantation, and tissue culture harnessed and put to work to lengthen life. The stories provide a fascinating glimpse of the biomedical imagination of the early twentieth century. They also anticipate some of the primary preoccupations of early-twenty-first-century antiaging medicine.

Organ transplantation joins telegraphy in "The Talking Brain," M. M. Hasta's 1926 story based on the tissue culture work of Leo Loeb and Alexis Carrel.[10] When his young student is severely injured in a car accident, the solitary Professor Murtha is seized by an impulse: "Perhaps he promised Vinton [the student] immortality in this world—freedom from the body's limitations, time without end for learning and thought and the creative activity the boy loved. . . . Vinton knew the things Murtha had accomplished; he knew of the work of Loeb and Carrel. If a heart could be kept beating in a bottle for years at a time, why should not a brain be kept thinking in a bottle forever? There seemed nothing impossible in the plan. How could he communicate? That was simple—for he knew his Morse and Murtha had solved the problem of efferent impulses" (Hasta 1926, 445).

Removing Vinton's brain, Murtha places it in the wax model of a head, perfusing it with blood and attaching the necessary electrical connections. But once that brain is held in its perfusion chamber, the portrait of disembodied life shifts from triumph to tragedy. Communicat-

"The Talking Brain." Reprinted from M. M. Hasta, "The Talking Brain," *Amazing Stories* 1, no. 5 (August 1926).

ing by Morse code, the wax head chatters: "This place is more terrible than you can know. Set me free. Kill me or let him kill me. Now. Now. Now. Now. Now. Now" (Hasta 1926, 478). The appalled protagonist contemplates the horror communicated not by touch or voice but by telegraph: the disconnected experience of "a disembodied mind, apart from the obedient creature of bone and muscle that served me" (478).

From January through April 1927, *Amazing Stories* followed Hasta's foray into the margin of life with a sustained encounter with the questions raised by Rolleston in *Nature.* One story in each issue addressed the challenge of aging and offered a different technological strategy for rejuvenation. In January, Dr. Miles J. Breuer's "The Man with the

Strange Head" represented aging as a problem of fatigue and timing in the body-as-machine, a problem that is solved by recourse to a stronger and technologically more advanced prosthetic machine-body.[11] As we have already seen, the February 1927 issue responded to the topic of rejuvenation strategies in W. Alexander's "New Stomachs for Old," the story of rejuvenation through organ transplantation that I discussed in chapter 6. The story explicitly gestures to the science of rejuvenation in its editorial sidebar summarizing Dr. Finkler's experiments with head transplantation in insects.[12] The insect head exchanges merely produce renewed vigor (no doubt because insects are viewed as having a distributed rather than unitary consciousness, as is demonstrated in Julian Huxley's *Ants*), while the stomach exchange in Alexander's story yields results not only unexpected but ultimately intolerable. Alexander's story dramatizes racial, ethnic, and class anxieties in its exploration of how the stomach transplant went awry, but contains them with its concluding notion that the working-class man has served as a valuable reservoir for upper-class physical renewal.

In addition to the important implications of this story for our changing attitude toward human organs, especially the commodification of body parts, "New Stomachs for Old" also embodies several additional important aspects of the modern performance of aging. Especially significant are the story's exploration of the unequal access to rejuvenation strategies; the resulting inverse connection between physical and economic vigor; and the technologically and sociologically determinant role of race and class in a person's longevity. The story suggests that whether we see these rejuvenation technologies as threatening or not depends on whether they seem to shore up, or to threaten, our sense of individual autonomy.

Electricity provides the technology for responding to senescence in the satiric tale "Advanced Chemistry," published in March 1927.[13] Professor Paul Carbonic, working with his ancient servant Mag Nesia, experiments in bringing dead rats back to life. When he succeeds, he erects a street sign "in brilliant letters": "DEATH IS ONLY A DISEASE—IT CAN BE CURED BY PROFESSOR PAUL CARBONIC." The professor quickly finds customers. First he revives little Sal Soda, who has fallen down two flights of stairs, by drilling a hole in her brain and injecting "the immortal fluid" that he has prepared from zinc oxide, copper sulfate, and sal ammoniac. The little girl comes back to life, and the profes-

"Advanced Chemistry." Reprinted from Jack G. Huekels, "Advanced Chemistry," *Amazing Stories* 1, no. 12 (March 1927).

sor accepts his $1.25 pay. The next customer is a gentleman whose friend has been killed in an accident. As they journey to the dead man's side, the professor explains to his interlocutor the nature of his discovery:

> Electricity is the basis of every motive power we have. . . . A body entirely devoid of electricity, is a body dead. Magnetism is apparent in many things including the human race, and its presence in many people is prominent. . . . Science has known for years that the body's power is brought into action through the brain. The brain is our generator. The little cells and the fluid that separate them, have the same action as the liquid of a wet battery; like a wet battery this fluid wears out and we must replace the fluid or the sal ammoniac or we lose the use of the battery or body. I have discovered what fluid to use that will produce the electricity in the brain cells which the human being is unable to produce. (Huekels 1927, 1128)

The professor revives the deceased man, Murray Attic, by drilling into his skull and inserting the fluid he has brought with him. No sooner does the dead man move than the professor himself falls victim to a heart attack. The stranger drills a similar hole in the professor's head and pours in one of the vials of liquid, until "a chemical reaction was going on in the professor's brain, with a dose powerful enough to restore ten men" (1128). The two revived dead men clasp hands, only to be electrocuted. When the story ends, the revivifying formula remains a secret. Huekels's story explores the attempt to create life by electric shock, its chemical agents anthropomorphized into human protagonists. The linguistic and onomastic wordplay provides another instance of the scaling effect that we saw in the imagination of the Strangeways researchers, but here the ground of being is not the cell but chemical and electrical impulse; the story suggests that life is not an essence but a process.

Finally, in the April 1927 story "The Plague of the Living Dead," A. Hyatt Verrill extrapolates from Carrel's famous experiment with the chicken heart—"immortality in the laboratory"—to its illicit social applications.[14] Dr. Farnham, a world-famous biologist, announces that he has discovered the technique to prolong life, only to be mocked by the public, harassed by the press, and shunned by his colleagues. Shocked, he withdraws with his laboratory assistants and his "supposedly immortal menagerie" to a secret Caribbean island, where he continues his experiments. Though he has announced his ability to prolong life, his true ambition extends beyond the indefinite life span of any individual currently existing. He hopes that "the young of these supposedly immortal individuals would inherit immortality" (Verrill 1927, 9). When he discovers not only the truth of that wish but also that "the serum with its new constituents would not only check the inroads of age but . . . would restore life," his research program seems to have been victorious. Of course, the tale turns on itself: the doctor realizes that he has literally done away with death. In a realization that anticipates Dr. Honor Fell's tissue culture lectures of the 1930s, when she observed that doctors must necessarily use the word "death" in "a restricted sense," Dr. Farnham confides his discovery to his lab notebook: "To sum up: It is impossible to define life or death in exact or scientific terms. It is impossible to state definitely when death takes place until decomposition sets in. It is impossible to say what causes life or produces death" (12).

The rest of the tale takes his discovery to its logical conclusions, as a volcanic eruption plunges the biologist out of the laboratory into the world, where like Sinclair Lewis's 1925 character Martin Arrowsmith, Dr. Farnham must use his untested knowledge in response to a crisis.

"The Plague of the Living Dead." Reprinted from A. Hyatt Verrill, "The Plague of the Living Dead," *Amazing Stories* 2, no. 1 (April 1927).

The results are as bad outside the laboratory as they were within it. Just as his terribly wounded, even decapitated, laboratory animals continued to live, so the human beings buried in volcanic ash or mutilated in a riot do not die but keep living though subjected to any amount of violence. "Doctor Farnham had a fleeting, instantaneous vision of the two fellows being chopped into bits or torn to pieces and each separate fragment of their anatomies continuing to live, or perhaps even reuniting to form a complete man again" (15). This final awful literalization of the Strangeways tissue culture experiments leads Farnham to acknowledge the wisdom of limits on the life span, if only in contrast with the horror he had brought about.

> Perhaps only the physical organism could be restored to life, and the mental processes remained dead. Perhaps, after all, there *was* such a thing as a soul or spirit and this fled from the body at death and could not be restored. . . . As the possibilities of this, as the probabilities of it, fully dawned upon Doctor Farnham, he cursed the day when he had first discovered his compound. Instead of benefitting mankind he had wrought destruction. He had produced immortal beings devoid of every spark of humanity, love, affection, kindness, intelligence, knowledge of right or wrong, self-restraint or any of the attributes of human beings. They could not be destroyed, they could propagate their kind, and gradually, but surely and irresistibly they would occupy the world to the exclusion of all other beings. (16)

Reopening the mechanist-vitalist controversy in response to the frightening implications of tissue culture experimentation, Verrill's story testifies to a rich vein of cultural anxiety about the biomedical assault on aging. Mechanistic survival of the body without the soul is the result of the engineered propagation produced by Dr. Farnham's compound. Moving beyond rejuvenation into immortality, Farnham's creations epitomize the alienating victory of technology. Unlike H. G. Wells's Herakleophorbia, which challenges the norms of human, animal, and vegetable size with its huge weeds, wasps, chickens, and even giant children while asserting the importance of human generosity and acceptance, Farnham's compound negates "the attributes of human beings," affirming instead the greater endurance of the soulless machinic body over the ensouled mortal mind-body.

Amazing Stories continued during the 1930s and 1940s to publish reports of new technologies that could delay aging or prevent death. Just to choose two representative ones: John Pease published a narrative of "man made immortality" inspired by the work of Professor Chichulin, of the Moscow Brain Institute, who established "that a decapitated head can live, that its eyes can blink and its throat swallow four hours after being severed from the body."[15] And Henry Gade details how a whole body is cryogenically frozen and kept in "suspended animation" so that life can be "suspended, to be revived later at the option of the scientist." While the first report was fictional, the second one, about an entire body in suspended animation, was factual. *Amazing Stories* contributor Henry Gade was reporting on a documentary film shown at the 1940 meeting of the American Medical Association. As summarized in

"Suspended Animation." Medical research receives a science fiction treatment. Henry Gade, "Suspended Animation," *Amazing Stories* 14, no. 1 (January 1940): 145.

the August 1939 issue of the *Journal of the American Medical Association*, the film recorded an experiment in which Doctors Lawrence W. Smith and Temple Fay packed cancer patients in cracked ice to produce a suspension of cellular growth. "Cracked ice was used, closely packed around the patients, and for five days they lay in a motionless state. . . . The result of this experiment was found to be beneficial, since the patients were improved. Their normal healthy cells were not impaired in their function, and restored themselves, while cancer cells were held in virtual hibernation, and were kept from growing. The experiment was carried on by Drs. Temple Fay and Lawrence W. Smith, in Philadelphia. Since then doctors have been anxious to try the treatment on other diseases, such as heart disease, tuberculosis, and infectious diseases of various sorts."[16]

Focusing on the strategy of replacement, modernist performances of aging in medical science and science fiction rely primarily on two procedures: *the cut* and *the exchange*. In the cut, the body part, or even the entire body in the case of Breuer's Mr. Anstruther, is removed from its context, isolated, and manipulated by the expert scientist, doctor, or technician. Ironically, even once the new mechanical body has replaced the failing biological one, a residue remains. The clockwork motion of Mr. Anstruther in his new mechanical torso strikingly recalls the diseased body it has superseded with its rigidity, tremor, and characteristic gait. In the exchange, the failing part is replaced: by a more vigorous organ (stomach, brain, head, mechanical torso), by greater chemical or electrical vitality, even by a different rate or portion of time itself, as when those "persons desiring a re-balancing of age differences" avail themselves of "the suspended animation machine" (Gade 1940).

Connecting biological with economic futurity, Gade's elaboration of the Smith and Fay experiment demonstrates how modern fiction and nonfiction both enact and test the limits of the modern response to aging, maintaining corporeal integrity at the price of context: spatial, temporal, and social. Though it begins by faithfully reporting on this actual scientific experiment, Gade's article in *Amazing Stories* moves by the end into fiction, departing from the disciplinary constraints of the *JAMA* article to offer a solution to problems not merely biological but economic as well: "It would even be possible to invest a sum of money, then sleep in suspended animation, to awake with a comfortable amount of security, enabling the investor to live out the balance of his natural life in retirement, without financial worry" (145). Gade's modernist narrative of replacement medicine contains an anticipation of what would come next as the initially fragmented body is reunified and revitalized, linked as a commodity to the circuit of desire.

POSTMODERN REGENERATION

While the modernist response to aging has been characterized by the strategy of the cut and the exchange, at the beginning of the twenty-first century, we are entering a system based not on replacement but on renewal. Although it is rooted in the central modernist life strategies that I have explored in this book—tissue culture, xenogenesis, hormonal and surgical intervention—the emergent field of regenerative medicine claims nonetheless to offer "a new therapeutic paradigm."[17]

Rather than replacing defective body parts (from other live or dead human donors, from animal donors, or with bioengineered mechanical parts), regenerative medicine aims to grow them anew. Part of a broader reconceptualization of the human body and human history as media, regenerative medicine enacts a movement from archival or mechanical storage to database, or textual/inscriptive storage. Unlike the former, which tends to be static and to function additively, database storage is flexible, subject to recombination and proliferation. As the Department of Bioartificial Organs at Kyoto University explains:

> Human body has functions to regenerate its own body. Cut wounds could be cured and broken bones could be retouched to the original conditions. But, not all the functions can be restored naturally like the cut tail of a lizard. Regenerative Medical Science applies newly invented methods, such as to provide conditions for the human body to be regenerated or to stimulate its regeneration in order to assist tissues or organs to regain their lost functions. In regenerative medical science, we are trying to develop a new treatment, which used to be impossible only with organ transplantation or implantation of artificial organs.[18]

Such regeneration occurs throughout the life span. Tissue exchange flows into tissue regeneration, with the commodification of all processes from conception to extinction: "Eyes (cornea), sperm, eggs, embryos, and blood have now been socialized, mutualized, and preserved in special banks. Deterritorialized blood flows from body to body through an enormous international network in which we can no longer distinguish the economic, technological, or medical components. The red fluid of life irrigates a collective body, formless and dispersed. . . . The collective body modifies our private flesh. At times it resuscitates or fecundates it *in vitro*."[19] However, physicians and researchers are perhaps naturally most excited about its implications for prolonging life. Dr. William Haseltine, the CEO of Human Genome Sciences, observed in 2000: "When we know, in effect, what our cells know, health care will be revolutionized, giving birth to regenerative medicine—ultimately including the prolongation of life by regenerating our aging bodies with younger cells" (Wade 2000).

Early in 2002, we saw a flurry of news stories about the promises of regenerative medicine to generate new therapies and offering startling, even disturbing, new visions of human growth. Even as they praised its revolutionary promise, these reports contained echoes—for those who can hear them—of modernist medical strategies. Thus, in the first week of the New Year, regenerative medicine recalled the modern techniques of tissue culture and transspecies interventions (what we would now call xenotherapeutics) when two major biotechnology groups, PPL Therapeutics PLC and Immerge Bio Therapeutics Inc., were reported to be racing to clone so-called "knockout pigs": swine that were cloned with a specific gene switched off, so that the immune reaction that would normally result with transplantation would not take place.[20] Also that week came a report focusing on the central strategy of replacement medicine, the organ transplant, but having exciting reverberations for replacement medicine. The *New England Journal of Medicine* contained the results of a study of heart transplantation carried out by a team headed by Dr. Pierro Anversa at New York Medical College, demonstrating that "a large number of primitive cells migrate from the recipient into the grafted heart."[21] Dr. Robert Bolli commented in an editorial on the topic that the Anversa team's study provided " 'incontrovertible evidence that endogenous mechanisms exhibit that result in the formation of new myocardium in the adult human heart,' 'a striking departure' from the widely held belief that cardiac tissue cannot be regenerated."

Finally, on 3 January 2002, attention to regeneration within a brain-dead patient propelled modernist experiments in hormone-based rejuvenation and organ transplantation across the life and death barrier. As reported by *Reuters Health*, a study by Dr. George C. Velmahos of the University of Southern California demonstrated that "administering thyroid hormone to brain-dead patients, who are potential organ donors, could lead to a dramatic increase in the number and quality of organs for transplantation."[22] What impressed me about these three news stories, as they came across my computer monitor in early 2002, was not their novelty but their familiarity. Here again we see the basic building blocks of replacement medicine: tissue culture, hybridity (of cell, of embryo, of whole organ, of whole being), organ transplantation, and hormone treatment. Though the media coverage it generated appears on the Web rather than in a tabloid, still it recalls coverage of the tissue culture research in the 1930s: "Carrel predicts, in the near future, a time when healthy hearts, kidneys, stomachs and other parts of the body will be kept alive and labeled, ready for inserting into anyone who may need them. . . . Already other parts of the human body have been grown in test-tubes."[23] The modernist context for Carrel's work led to a focus on the *replacement* (and, inevitably, the commodification) of the body part or body, as expressed most memorably by the mechanistic image of the body/car entering the hospital/garage for the installation of new (body) parts. While contemporary medicine has certainly maintained this focus on replacement medicine, as well as extending its for-profit implications, media coverage increasingly focuses on the drawbacks of such an approach: the unsupervised nature of tissue banks, leading to uneven, unorthodox, and even unsafe methods of tissue recovery, tissue banking, and tissue transfer, and consequently the potential for lethal infections or illnesses to be transferred to recipients with the tissues.[24]

As the dangers of replacement medicine are coming more into the public eye, another kind of medicine is emerging as well. And as we have seen, the popular representations of this new medical approach make a point of distinguishing it from the "science fiction scenario" whereby "we will be able to order 'spare parts' for our body as we do for our cars or appliances" (Armandola 2003). The poster child of this whole new kind of medicine is stem cell therapy, which exemplifies not only the frame-breaking, innovative nature of contemporary medicine but also its debt to what came before: the basic research in tissue culture exemplified by the work at the Strangeways Laboratory in England and in Alexis Carrel's Rockefeller University Laboratory in New York. As the National Institutes of Health explains in their June 2001 report, stem cell therapies involve taking primitive or undifferentiated cells, characterized by their plasticity, from embryonic, fetal, or adult tissue and transplanting them with the goal of inducing them to assume the characteristics of the cells in the transplant destination.[25] The NIH report lists several different kinds of research that might be conducted with stem cells: transplantation research, research into the basic developmental functions and their disorders, research into targeted genetic therapies, research into tumor growth, therapeutic drug research and development using human stem cell lines, and research into addiction (National Institutes of Health 2001, ES 4–5). While stem cell therapies start with the process of tissue culture and tissue transfer, because they give rise to what could theoretically be a perpetual system of exchange, as introduced cells lead to the growth of new cells in situ, their focus is not on replacement but on regeneration.

To clarify the distinction between these two medical paradigms, let's return to Breuer's 1927 short story "The Man with the Strange

Head." Breuer's story provides a rich if metaphoric dramatization of the replacement approach to medicine. The story's protagonist, Mr. Anstruther, has a head that is oddly wrinkled, but a strong and vigorous body. The contrast is sufficiently usual that it prompts a neighbor to spy on him. Peering into Anstruther's living room, the neighbor finds him pacing the floor in a frightening combination of mechanistic regularity and seeming neurological dysfunction: "His head hung forward on his chest with a ghastly limpness. He was a big, well-built man, with a vigorous stride. Always it was the same path. He avoided the small table in the middle each time with exactly the same sort of side step and swing. His head bumped limply as he turned near the window and started back across the room. For two hours we watched him in shivering fascination, during which he walked with the same hideous uniformity" (Breuer 1927, 942).

The mystery of the "uncanny, machine-like exactitude of his movements" is solved when the man finally totters, sways, and collapses. The reader learns that Anstruther, seeking "a mechanical relief for his infirmity," had approached a Frenchman who was the creator of an automatic chess player and some "famous animated show window models, and persuaded him to build an automaton to encase Anstruther's aged body" (Breuer 1927, 970). The resulting total prosthesis for Anstruther's aged and disabled body has such a lifelike form and function that it is detected only upon its possessor's death, when it is dismantled by its creator, who "put his hands into the chest cavity, and as the assistants pulled the feet away, he lifted out of the shell a small, wrinkled, emaciated body; the body of an old man, which now looked quite in keeping with the well-known Anstruther head. The undertaker's assistants carried it away while we crowded around to inspect the mechanism within which were the arms and legs of the pink and live-looking shell, headless, gaping at the chest and abdomen, but uncannily like a healthy, powerful man" (970).

In true modernist fashion, Anstruther has responded to his own increasing debility by commissioning the creation of an entirely prosthetic body: a "pink and live-looking shell, headless," within which can fit his "small, wrinkled, emaciated body" (970). What was Anstruther's infirmity? Breuer's adjectives give it away: his movements are characterized by the tremulous, hunched, forward stance of Parkinsonism, "a clinical syndrome comprising combinations of motor problems—

"The Man with the Strange Head." Reprinted from Dr. Miles J. Breuer, "The Man with the Strange Head," *Amazing Stories* 1, no. 10 (January 1927).

namely, bradykinesia, resting tremor, rigidity, flexed posture, 'freezing,' and loss of postural reflexes."[26] Diagnosis proceeds by identifying a specific way of walking: "Parkinsonian gait is the distinctive unsteady walk associated with Parkinson's disease. There is a tendency to lean unnaturally backward or forward, and to develop a stooped, head-down, shoulders-dropped stance. Arm swing is diminished or absent and people with Parkinson's tend to take small shuffling steps (called festination). Someone with Parkinson's may have trouble starting to walk, appear to be falling forward as they walk, freeze in mid-stride, and have difficulty making a turn."[27]

When Anstruther assumes his machine torso, he doesn't so much transcend the Parkinson's disease as accommodate it. The machinic adjectives in Breuer's story are apt not only because the movements characteristic of Parkinson's disease *seem* like clockwork but even more because Anstruther's solution to his Parkinson's (and to the general problem of aging) is as mechanistic as his model of the life process. He hopes to replace his worn-out physical body with a new mechanical one and thus to change the speed at which his biological clock runs. Replacing his worn-down biological clock with a vigorous new mechanical clock, he hopes to gain more time.

Parkinson's disease is one of the most frequently cited targets for stem cell therapy, not only in the popular press but in scientific and governmental circles.[28] When we move from the Breuer short story in 1927 to stem cell research in 2001, we find that the theme of resetting the human clock jumps from a metaphoric (and machinic) to a literal (and organic) level. Stem cell researchers are investigating the notion that "intracellular clocks" regulate the rate of cell division, so that "progressive shortening of telomeres could act as a mitotic clock, counting off divisions before senescence."[29] Stem cells are particularly interesting as a potentially transferable tissue culture because they may contain telomerase, the key to rolling back the aging process. As the NIH report *Stem Cells* explains:

> A telomere is a repeating sequence of double-stranded DNA located at the ends of chromosomes. Greater telomere length is associated with immortalized cell lines such as embryonic stem cells and cancer cells. As cells divide and differentiate throughout the lifespan of an organism or cell line, the telomeres become progressively shortened and lose the ability to maintain their length. Telomerase is an enzyme that lengthens telomeres by adding on repeating sequences of DNA. . . . High levels of telomerase activity are detected in embryonic stem cells and cancer cells, whereas little or no telomerase activity is present in most mature, differentiated cell types. The functions of telomeres and telomerase appear to be important in cell division, normal development, and aging.[30]

A vivid symbol of the paradigm shift from replacement to regeneration, or from organ transplantation to stem cell therapy, was the founding in July 2001 of the McGowan Institute for Regenerative Medicine (MIRM) at the University of Pittsburgh, in the institutional site that had previously been a premier organ transplant center. As the McGowan Institute's Web site explains: "regenerative medicine is an emerging field that approaches the repair or replacement of tissues and organs by incorporating the use of cells, genes, or other biological building blocks along with bioengineered materials and technologies."[31] Subsuming the McGowan Center for Artificial Organ Development, this new institute testifies to the profound reorganization going on in medicine: the move to what one science fiction writer has aptly described as the "medical-industrial complex, dependent on stable grant procedures."[32]

Ventures in regenerative medicine are big business, even more potentially profitable than were the rejuvenation strategies of sixty years ago, and the culture of biomedical commodification has spawned an interesting new hybrid institution: the medical center that combines basic research with commercial and even medical-industrial applications. How does this hybrid institution differ from Strangeways Laboratory, which evolved from a hospital for malaria patients to a laboratory exploring disease processes in vitro, including the culture of tissues and organs? Bartley Griffith, the surgeon who directed the McGowan Center for Artificial Organ Development since 1992, describes the motivation for this new larger institute as "a natural extension of the McGowan Center's vision and missions, to ease the suffering of patients" (UPMC 2001). Yet more than patient care is going on here: as the University of Pittsburgh Medical Center Health System Web site explains, "It is expected that the new institute will devise innovative clinical protocols as well as pursue rapid commercial transfer of its technologies related to regenerative medicine" (UPMC 2001). Indeed, it is the economic rather than the scientific innovations that seem to be the most impressive products of the MIRM.

Two stories that appeared cheek by jowl on the MIRM Web site in 2001—"Human Heart Kept Alive outside the Body for First Time in Study of Portable Organ Preservation System" and "Stem Cells from Skeletal Muscle Can Restore Bone Marrow Function"—demonstrate the MIRM's incorporation of both medical paradigms. The first story introduces the Portable Organ Preservation System (POPS™) developed by the MIRM physicians working with TransMedics, Inc: "The system . . . weighs approximately 70 pounds and is shaped like a box that can fit in an ambulance or private jet. A smaller unit, weighing about 55 pounds, could fit on the seat of a commercial airplane. The system has a four-hour battery, built-in handles and detachable wheels. It consists of a portable electro-perfusion device, bio-compatible organ-specific disposable components and proprietary chemical solutions that bathe the organ" (UPMC 2001).

Recalling the accomplishment that inspired the tissue culture research at Strangeways—Alexis Carrel's accomplishment in culturing a living chicken heart for an extended time—this MIRM-developed experimental system makes it possible to keep a human heart alive outside the body. The MIRM Web site story emphasizes POPS™'s potential applications for human heart transplant procedures: enabling physicians to extend the possible transit time for donor organs, to perfuse and thus resuscitate previously expired organs, and even "to use hearts and lungs from nonheart-beating donors. We'd essentially be able to resuscitate these organs, then transplant them," according to Kenneth R. McCurry, M.D., assistant professor of surgery, director of Lung and Heart-Lung Transplantation at UPMC (UPMC 2001). Still, the sentence that follows the POPS™ story emphasizes an equally newsworthy fact: the economic potential of this discovery. "POPS™ is a trademark of TransMedics, Inc. All other brands and product names may be trademarks of their respective owners and are used here for reference only" (UPMC 2001). Unlike the anonymous laboratory culture flasks of Carrel, and the lectures, papers, and films of the Strangeways researchers, POPS™ is a proprietary possession of TransMedics, Inc., a start-up company manufacturing medical devices in Malden, Massachusetts. The uncanny potential of POPS™ is its ability to maintain a heart (or liver, kidney, or pancreas) alive and functioning without external stimulus; more than that, it can bring that organ back to life. As the TransMedics Web site boasts, "The POPS™ is the ONLY technology that is capable of resuscitating non-heart beating and marginal organs (20–30% of organs wasted/year)."[33]

As with Carrel's chicken heart experiment years earlier, news coverage on POPS™ relied on a range of discourses to convey the machine's unsettling ability to normalize processes of life extension, even rebirth. A writer in the *Augusta Chronicle* described the machine as "similar in size and height to a baby carriage," and Steve Silverman reported on *Red Herring*, the financial Web zine, that the machine is "ironically, about as big as a medium Igloo cooler."[34] Anne Barnard, reporter for the *Boston Globe*, reported in a story carried on the Web by *Tx News* that the "doctors, and research assistant Dennis Sousa, laugh when someone mentions *Futurama*, the cartoon television show in which the heads of past leaders, such as Henry Kissinger, are preserved in jars to advise future generations." Appropriately enough, Barnard describes POPS™ as "a cross between a vacuum cleaner and a robot" and says it "looks like something from science fiction—a heart pumping on its own, fed by a greenish-yellow fluid, with 50 top-secret ingredients, that the company calls 'High Energy Maintenance Solution.' "[35] According to another report, Dr. Waleed Hassanein, the thirty-one-year-old Massachusetts physician-inventor, had put his surgical career on hold to start TransMedics. After having tested his machine on more than five hundred animal hearts, he demonstrated it to physicians at Boston Medical Center, showing them "a disembodied pig heart in the container, beating."[36] But despite its mixture of humdrum domesticity and sci-fi edge, POPS™ is a success story with both human and economic dimensions. Avoiding traditional research grants (too slow!) to fund the development of his brainchild, Dr. Hassanein instead found support in industry, first working for Medtronic, Inc. (a manufacturer of a range of biomedical devices such as pacemakers), and finally banding together with two friends from Egypt to form TransMedics. While Hassanein was criticized by his cardiologist sister from Cairo ("You left your family to go to the US to be a cardiac surgeon, and now you're not going to be one"), his company was praised by *Red Herring* for carrying out the "Deal of the Day" on 25 September 2001, when it acquired venture capital funding of $8 million to develop POPS™.[37] And TransMedics went on to win the 2002 Design and Engineering Award from *Popular Mechanics* for this system that, as *Popular Mechanics* explained, "duplicates conditions inside a healthy human body. During transport to a transplant center, the donor organ maintains a normal physiologic functioning state. Hearts continue to beat, kidneys produce urine, and livers produce bile. POPS™ represents an imaginative application of biomedical, electrical and mechanical engineering. More important, it may save countless lives."[38]

In addition to the POPS™ story, with its echoes of cut-and-exchange or replacement medicine, the MIRM Web site in January 2002 featured a second story celebrating the exciting medical and economic potential of regenerative medicine: "Stem Cells from Skeletal Muscle Can Restore Bone Marrow Function."[39] The synopsis of a presentation at the annual meeting of the American Society for Cell Biology, this report details the findings of a series of experiments carried out by MIRM director Johnny Huard and colleagues on the transplantation of adult stem cells in mice. Working with a genetically engineered population of "mdx" mice, bred to provide a model for the human disease Duchenne muscular dystrophy because they lack the essential muscle protein dystrophin, Huard and his colleagues demonstrated that transplantation of adult stem cells could lead to regeneration of bone marrow tissue. "Intriguingly," they report, they even found that "these transplanted cells also turned up in blood vessel and peripheral nerve tissue, suggesting that they might have the capacity to contribute in several ways to the regeneration of functional muscle tissue, potentially including restored neurons and blood vessels."[40]

Just as the MIRM subsumed the previous occupant of its site, the McGowan Center for Artificial Organ Development, so too the relations between regenerative medicine and replacement medicine are less oppositional than they are opportunistically positioned. For example, only two months before the ASCB meeting at which MIRM's Johnny Huard presented his stem cell research, Peter Schwartz, chairman of the Global Business Network and partner in the venture capital firm Alta Partners, one of the venture capitalists behind the capitalization of TransMedics' POPS™, wrote an article in *Red Herring* trumpeting the paradigm-shifting importance of stem cell therapy to the new field of regenerative medicine. Drawing on a range of expansionist tropes to persuade his financier readers to look favorably on such biomedical ventures, Schwartz explained, "Regenerative medicine is the new frontier, unlocking the secrets of how the body generates itself. Biological systems are self-assembling, based on genetic code: if we cut our finger, our body knows how to repair them [*sic*] (if it's not too deep). Imagine if we could repair a diseased kidney or a failing heart the same way. Today we often return from the hospital with a bit less of ourselves, as pieces are snipped away. Tomorrow we will come home with regrown livers and reconnected spinal cords. We will even grow new brain tissue to repair the damage of a stroke."[41] Stem cell research is so important, he explained, because it represents "a fundamental change in the paradigm of medicine. . . . For medicine to be able to regenerate damaged and aging bodies, scientists must understand how stem cells work." Schwartz was playing both ends against the middle here, arguing on behalf of the new paradigm while funding the old.

The move from patient care to product development that characterizes the realm of regenerative medicine draws on a discursive strategy used a decade earlier in the shift from the new reproductive technologies to what is now called "assisted reproduction." While the earlier term emphasized novelty and technology, the current term effaces both of those aspects of the field, to stress instead the comfortable themes of *caring*: human assistance in the process of reproduction. The new term seems to obliterate the forbidding nature of specialized science and medicine itself, replacing it with the soothing notion of *art*. The self-proclaimed genesis of the new field of regenerative medicine involves a similar linguistic act of reframing and consolidation. Rather than emphasizing the highly engineered and technologically sophisticated mechanisms through which the human body is repaired or rebuilt, the label "regenerative medicine" suggests that the body just naturally repairs, or regenerates, itself.

FICTIONS OF REGENERATIVE MEDICINE

We have seen that the biomedical imaginary operates not only on the medical front but in imaginative literature as well, where it often addresses not only the possibilities offered but the responsibilities required or evaded by any new medical practice (Waldby 2000, 136–37). Regenerative medicine is surely no different; we can assess its perils and promises most fully if we build into our assessment the way it appears in fiction. Yet genre, authorship, intended audience, and market can all shape and limit fictional representation of regenerative medicine, as becomes apparent when we compare two very different works that address medical-technological responses to aging. Both were written around the turn of the twenty-first century, and both address issues of age-related debility, but the first reached a large mainstream market while the other addressed the small, intellectually and socially more marginal community of fringe theater. In addition to raising questions about the different potential of the genres of science fiction and theater, as well as the different demands of mainstream and alternative markets, these two texts also raise the following questions about the construction of regenerative medicine in the biomedical imagination. What are the implications of the technologies we choose to respond to aging? How does regenerative medicine, as currently practiced, both profit from and suffer the constraints of the biotechnologies that are its foundation? Finally, how do our different ways of conceptualizing technology, regeneration, and aging itself shape the options available to contemporary medical practice?

Bruce Sterling's *Holy Fire* is a mass-market science fiction novel by a prolific, bestselling author of highly topical, technologically inflected cyberpunk fiction.[42] Published in hardcover in 1996 and in a Bantam softcover edition just one year later, Sterling's novel was blurbed in *Wired* as "a book made entirely of ideas . . . big, fat, juicy technological extrapolations, presented with flair and enthusiasm. An intellectual feat, it is also a treat for the spirit and the senses" (Sterling 1997, jacket). In contrast, Anne Davis Basting's *TimeSlips* is a play written by a scholar-playwright in collaboration with a group of "outsider artist" writers and performed in 2001 Off-Off-Broadway, at the Here Arts Center in New York City.

HOLY FIRE: REGENERATIVE MEDICINE AND THE TECHNOLOGICALLY ENHANCED INDIVIDUAL

Although regenerative medicine's self-presentation (on the MIRM Web site, for example) stresses its *natural* processes, in *Holy Fire* (1997) Sterling represents those treatments as antinatural, for they transport human beings entirely "out of nature," as Yeats put it in "Sailing to Byzantium," the poem that provides Sterling's title. The resonances are ironic: regenerative medicine transports its patients to a posthuman Byzantium where they become would-be "sages standing in a holy fire," perpetually frozen into unnaturally endless life spans. The world Sterling introduces in this novel provides a detailed and critical portrait of regenerative medicine, extrapolating from the most prominent aspects of late-twentieth-century medicine: the reliance on animals as research and treatment reservoirs for human beings, and the increasing research interest in the notion that the process of aging (at the cellular and organismic level) can be interrupted or reversed through the processes of stem cell culture and tissue transfer. Though they are Sterling's inventions, each of these forms of medicine is rooted in actual early-twentieth-century interventions into the life span. Xenomedicine, the interwoven field of human and animal antiaging interventions that Sterling's characters explore, dates back at least to Julian Huxley's experimental rejuvenation of the axolotl, while NTDCD (neo-telomeric dissipative cellular detoxification), the fictional life extension procedure that Sterling's protagonist Mia Ziemann undergoes, builds on the tissue culture experiments of Carrel and Strangeways.

Yet the novel's preoccupation is (our) future, not our past. Ninety-four-year-old medical economist Mia Ziemann exemplifies the therapeutic, stochastic, and economic thrust of regenerative medicine, not just in the work she does but in her very embodiment:

> She had professionally studied the demographics of the deaths of millions of lab animals and billions of human beings, and she had examined the variant outcomes of hundreds of life-extension techniques. She'd helped to rank their many hideous failures and their few but very real successes. She had meticulously judged advances in medical science as a ratio of capital investment. She had made policy recommendations to various specific organs of the global medical-industrial complex. She had never gotten over her primal dread of pain and death, but she no longer allowed mere dread to affect her behavior much. (Sterling 1997, 8)

In xenomedicine, pets contribute to the life extension strategies of their owners by being subjected (first) to the same medical procedures that their humans will later undergo: "arterial scrubbing, kidney work, liver and lung work" in the pursuit of longer life for their owners and, incidentally, for themselves (Sterling 1997, 7). Like Maureen Duffy's Gor, who has been transformed surgically from mute gorilla-human hybrid to articulate human being, Sterling's novel features a hybrid creation of xenomedicine: Plato, "one of the most heavily altered dogs in human ownership" (6). Plato's many life extension surgeries have given him not only a machine-generated voice but—as the "verbal prosthetic for a canine brain" has gradually taken root in the brain and transformed it—language skills as well. "It took ten years for the wiring to sink in, to fully integrate. But now speech is simply a part of him." What does he talk about? "Nothing too abstract. Modest things. Food. Warmth. Smells" (7).

At the other end of the continuum from the modest extension in life and capacities available to Plato the dog is the "serious life-extension upgrade" that Mia Ziemann undertakes and that forms the centerpiece of the novel's investigation of posthuman existence. Now that care of the self has become a civic responsibility whose effectiveness has direct fiscal implications for the economic health of the whole, the decision to partake in any of the different strategies for attaining a longer life is far from a simple one. For one thing, individual health now has collective, civic, and even geopolitical implications. People who "destroyed their health . . . because they lacked foresight, because they were careless, impatient, and irresponsible," don't survive. "Careless people had become a declining interest group with a shrinking demographic share" (60). In contrast, people who have taken care of their health are considered deserving of high-tech interventions. Demographically they have clout, and biologically they are more able to profit physically from any medical treatments. Sterling's description of the menu of choices available to Mia dramatizes the medical trend away from care or cure, and toward a futures market in enhancement technologies, where values are registered less in the mind and body of the patient than on the stock exchange: "There were a hundred clever ways to judge a life-extension upgrade. Stay with the blue chips and you were practically guaranteed a steady rate of survival. Volunteer early for some brilliant new start-up, however, and you'd probably outlive the rest of your generation. . . . Novelty and technical sweetness were no guarantees of genuine long-term success. Many lines of medical advancement folded in a spindling crash of medical vaporware, leaving their survivors internally scarred and psychically wrecked" (58).

Freedom of choice is available not only to those who are choosing between various life extension possibilities but also to those "who morally disagreed with the entire idea of technologized life extension. Their moral decision was respected and they were perfectly free to drop dead" (61). The result is a gerontocracy that recalls the Catch-22 of Robert Silverberg's "Caught in the Organ Draft": only those who have profited from life extension attain political and civic control, and only those with political and civic control have access to life extension.

Mia Ziemann's encounter with regeneration medicine leaves her a marginal member of a posthuman society organized around the technological management of individual lives and life spans. Having been expelled by the regenerative medical program, and having rejected her daughter's offer to care for her in her now inevitable decline, she instead ends up as a wandering photographer, obsessed with the Amish.

In such a critique of technologically based regenerative medicine, the concluding turn to the Amish people seems both inevitable and facile. Sterling makes the obvious comparison between their *natural* way of living, aging, and dying and the unnatural life strategies of Mia's world. "Amish people around seventy. . . . The natural human aging process. . . . It's amazing and terrifying! And yet there's this strange organic quality to it. . . . The Amish are wonderful. They can tell I'm some kind of impossible monster by their standards, but they're so sweet and good about it. They just put up with us posthumans. Like they are doing the rest of us a favor" (349). Yet Sterling's choice of the Amish, with their drastically antitechnological and communitarian approach to life, presents an opposition so complete as to offer no alternative as the novel ends. The leap into posthumanity seems inevitable, and Mia's yearning for the Amish only a nostalgic delusion.

Earlier I argued that it can be productive to consider the impact of generic conventions on literary representations and to ask whether a different kind of literary text might enable the articulation of different aspects of the biomedical imaginary. I want to do that in a moment, by shifting from this look at a science fiction novel to a very different sort of text. Having considered *Holy Fire*, the product (and I use that word deliberately) of Bruce Sterling, a prolific author of extrapolative cyberpunk fictions that typically explore the social impact of cutting-edge technologies and technological theories, I turn for contrast to another representation of medical-technological responses to aging, this one an Off-Broadway play written by the performance art scholar Anne Davis Basting.[43]

While different literary genres may enable the imagination of different worlds, the temporal context of any performance or text will also shape its vision. Performance artist and theorist Charles Garoian has argued that historical context will determine whether a performance enforces or interrupts a tendency toward isolation, individuation, and disempowerment.[44] Modernist performance art was exclusionary in its stress on isolation and disjunction, he holds, while postmodern performance art is inclusive, stressing civic belonging. "For the early modernists . . . performance art served as a liminal space, a virtual laboratory, where the body's preexisting modes of art production were challenged with the dynamic ideas, images, and processes of modern industrial culture" (Garoian 1999, 9). Modernist performance art focused on "the disjunctive character of the machine age—its new technologies, rapid modes of mass production, and consumer capitalism," while in contrast, postmodern performance artists focus in their work on "critical citizenship, civic responsibility, and radical democracy" (9).

In its focus on the tension between individual and community, *techne* and nature, Garoian's model illuminates the contribution that performance art and theater can make to our understanding of the ultimate liminal moment in human life: the experience of aging. Yet by restricting postmodern performance art to the theme of critical citizenship, Garoian's model neglects those forms of postmodern performance art that address experiences of aging outside or beyond that theme. Gender, race, and level of ability can be very restrictive if the individual is (albeit unconsciously) conceptualized as stable, autonomous, and possessed of easy access to public life and a meaningful vote. What of those who, in mind and/or in body, depart from the normate model of the fully functional body and fully present, functional, and conscious mind? If it is to address the real challenges of embodied aging, performance art must move out of such a potentially gender-, race-, and ability-restrictive, normative model of the civic public sphere. That performance art can do so—that it is doing so—is demonstrated by Anne Davis Basting's play *TimeSlips*. As I understand it, this play combines features of both modernist and postmodern performance art: the modernist notion of the liminal laboratory of performance, where the body met the challenge of technology, and the postmodern notion of performance as radical democracy forging a new notion of civic participation. Along the way, it articulates an alternative model for a regenerative medicine that can address age-related dementia, one that I want to read *in contrast* to the model provided by stem cell research.

TIMESLIPS: REGENERATIVE NARRATIVES AS LOW-TECH MEDICINE

TimeSlips was "inspired by stories told by people with Alzheimer's disease in New York City."[45] Rather than an individual performance of aging mediated by medical technologies, *TimeSlips* offers us a collaborative performance of aging, mediated by a reconceptualized notion of both medicine and technology. The play is drawn from a project of the same name: an interactive, intergenerational storytelling project undertaken with older people suffering from Alzheimer's and related dementia, the same age-related maladies addressed in fiction by Dr. Miles J. Breuer in 1927 ("The Man with the Strange Head") and by stem cell researchers today. In four adult day centers, two in Milwaukee, Wisconsin, and two in New York, Anne Basting and a group of trained volunteers (undergraduate students as well as adult artists) met in small groups for an hour to catalyze collaborative storytelling. Here is Basting's description of a typical meeting:

> Center staff helped us assemble the storytellers in a circle of chairs. After greeting the storytellers individually, the students presented an image to the group. One student sat in the middle of the circle with a large sketch pad and a box of brightly colored markers, ready to capture the group's responses. The other student and the day center staff sat in the circle and asked questions based on the image, like "What should we call her?" "Where should we say they are?" The questions began simply, but quickly built into vivid, complex, non-linear stories that give us a rare glimpse into the experience of living with dementia.[46]

Photographer Richard Blau was present to document the storytelling meetings, and his photographs capture the mixture of technological simplicity and interpersonal electricity that made the storytelling project so popular with the day center patients, staff, and families.

Like the dementia storytelling groups on which it is based, the play *TimeSlips* begins with a group of dementia patients sitting in a circle of chairs and being shown some simple photographs to prompt their storytelling. Yet rather than the simple large sketch pad and the box of bright markers wielded by the student helpers who assisted the storytelling workshops as transcribers, the play gives us Polly, the nursing assistant, who works alone to assemble the patients in a circle as they drift in, and to orient the patients: *Who can tell me what day it is?* When Polly shows the patients a prompting photograph, asking, "Do you remember anyone like this?" the conversation drifts from the disjointed and inane into the imaginative. Met with denials ("I wouldn't know someone like him. Don't insult me"), Polly admits, "You know what? I don't know him either," and instead urges them, "Let's make it up." "Oh boy, here we go!" says John. Over the course of four days, as the patients work together to respond to Polly's prompts, five new charac-

(*top photo*) The TimeSlips Storytelling Process.
(*bottom photo*) The collaborative storytelling group. Both photographs by Dick Blau. Reprinted by permission.

ters emerge from the patients' collaborative storytelling: Thomas Rex, "a cad of a singing cowboy"; Godfreya, the talking horse who loves Thomas Rex; Maryanne, a determined swimmer; a man who reads and carries around piles of books, overwhelmed nearly to distraction by their contents; and a cancan dancer who recalls her triumph on the Paris stage. The stories of these characters all intertwine as Polly notes the patients' contributions to the evolving story on her clipboard. Together they generate a story that spirals wildly (and wonderfully) into imaginative whimsy, moving them from their bewildered isolation into witty humor:

Sarah: Godfreya is a talking horse. She's been trained.
John: I say Thomas Rex is married.
Sarah: Yes, but he's very available.
Marie: His smile could make you swoon.
Sarah: He is attracted to beautiful women—like us.
Marie: I heard that too. From the talking horse.

Even the nurse's assistant joins in as the dementia patients animate their characters. Initially her collaboration is propelled by an encounter with her own aging. The stage directions are concise: "Polly enters. She looks for drugs she must dispense, comparing them to the list on her clipboard." Despite the list of drugs that can nudge her memory ("Dopamine? Acetaminaphin [*sic*], Aspirin, Cevimeline . . ."), Polly can't remember what she has come to fetch. She drops her clipboard, focuses "on her own aging hands and face, realizing her own proximity to this disease," and exclaims, "Lord deliver me. I forgot what I was looking for" (scene 6). But one morning, her contribution holds not fear but pleasure. When the unintelligible intercom announcement propels them into storytelling group, "Polly is excited to finally tell one of her own stories. This is new for her" (scene 8). "I remember one time when I was a girl, I crawled into my grandmother's closet. . . . Now my grandmother was dead mind you. She died before I was born. But my mother, she couldn't throw away her clothes. So there she hung, 30 of her at least, in pressed ensembles, legless, headless, suspended by a wire hook. I crawled in and sat among her shoes, looking up the hems of her skirts. They were like curtains on a dark stage. Well, I was there for what seemed like hours, telling stories about who the skirts had seen, where the shoes had been" (scene 8). As she recalls her childhood experience of animating the dresses and shoes of her grandmother until her mother opened the closet and screamed, thinking that Polly's grandmother "had come back to give her a good kick in the behind," it is the shared nature of the story, and not its accuracy, that is important (scene 8). As she confides to the others when she concludes the tale, "I think that's right. You know the last time I told that story, I forgot the ending entirely. I had to call my mother to get it right. Or maybe it was my sister" (scene 8).

The nursing assistant's story—the only one told by someone *not* suffering from dementia, yet generated by her empathy for their experiences of aging, and demonstrating the collaborative nature of all knowledge and memory—is emblematic of the whole process that Basting's play catalyzes. For the audience of *TimeSlips*, too, the animated characters generated in the collaborative storytelling give us a "kick in the behind," forcing us to abandon our notion that to suffer from dementia is to be "legless, headless, suspended." Instead we glimpse the joy, humor, and jealousy of which they are capable, and we acknowledge the extent to which human experience must be cocreated. Yet we do so without denying the central truth of the experience of aging. Despite its energetic and collaborative force, the process of collaborative storytelling can't hold back the slippage of time. When the play ends, the dementia patients have all but forgotten the wild story they told together. They deny it when Polly explains to them, "You told stories today / About a horse and a cowboy / About a can-can dancer / from England." "You must be thinking of someone else, honey," says Marie. But Polly answers, "No, it was you. It was all of you."

In its curve from vacant forgetfulness through energetic collaborative encounter to gradually increasing forgetfulness and lights-out, Basting's play enacts a response to aging that enriches it experientially without denying its pains. In this, the play captures the therapeutic storytelling project in which it originated: "The stories captured an array of personal memories, popular culture references, life experiences, and nonsensical answers born of the illness. All answers were validated and woven into the stories in order to build the storytellers' trust in the process and to develop a story that represents the full and often tragic arc of the disease."[47] In so doing, the play meets the criterion for a successful performance of aging, as Basting herself set it forth in her study *The Stages of Age*: "There is a fine and sensitive line between

Enjoying the storytelling process. Photograph by Dick Blau.
Reprinted by permission.

alleviating the physical and emotional pains of older Americans and further devaluing old age by recreating it as an imitation of young adulthood. If there is to be a real restructuring of how Americans value old age, we cannot rely on technology and the Baby Boom to provide us ways to dodge the label. We must create an intergenerational effort to build the value of old age anew."[48]

What can we learn if we consider *TimeSlips* as another in that chain of biotechnological interventions in aging that I have traced in this chapter? This is a chain that extends from the early-twentieth-century experiments in replacement medicine exemplified by the work of Carrel, Strangeways, and Voronoff to the early-twenty-first-century experiments in regenerative medicine, both those carried out by Pittsburgh's MIRM and those given imaginative extrapolation in Sterling's *Holy Fire*. First, we can acknowledge that in Basting's play we find a convergence of the worlds earlier kept separate. Something that begins as a therapeutic, medical intervention (the storytelling groups) produces a work of the imagination (the play) that ultimately exists as a performance in the Off-Broadway theater. Perhaps because of its hybrid trajectory through the realms of both medicine and literary creation, Basting's play provides an alternative model for our performances of aging, whether we think of them as the biomedical intervention in aging I have traced from the modern through the postmodern period, or if we focus on them as imaginative constructions in fiction and science fictions of the same eras. This is not to say that Basting's play departs from the features characterizing the modern performances of aging I have surveyed in this chapter: the reliance on technology and the use of prosthetics to extend the body, whether negatively (replacing a failing part) or positively (extrapolating from and thus extending human capacities).[49] Nor does her play steer clear of the postmodern strategies of technologically enabled renewal central to the examples I have drawn from twenty-first-century biomedicine and literature.

Yet while Basting's play shares these qualities, it does so with a difference. Technologies figure prominently in the modern rejuvenation performances (both fact and fiction) with which this chapter began: scalpel, syringe, and clamps; x-ray, photomicrograph, and petri dish; telegraph keys, typewriter, and film. And technologies also figure prominently in the factual and fictional enactments of regenerative medicine: from MIRM's new invention for organ culture, the POPS™, to fictional Mia Ziemann's "neo-telomeric dissipative cellular detoxification," carried out in a gigantic tissue culture involving a "gelatinous tank of support fluids," "magnetic resonance techniques," and cell splicing (Sterling 1997, 62–64). Basting's play incorporates technology, too, yet the technologies it employs are even more primitive, even more naturalized, than the ones used by early-twentieth-century rejuvenation. Basting's project uses folding chairs, a sketch pad, and colored magic markers. The play based on the project reduces the sketch pad and colored markers to a clipboard and a pen. If we recall how the modernist performances of aging were modeled on the cut (which removes the body part or body from its context and isolates it) and the exchange (which subjects the body to expert technical manipulation in the name of some invested future gain), we will see that the primitive technologies in Basting's model also perform the cut and the exchange, but notice how differently. The folding chairs cut the group from the ordinary flow of time and space of the adult center, and the process of working with markers and sketch pad to produce a collaborative story enacts the exchange of words, images, popular culture references, even nonsense, between and among the group of storytellers. In Basting's

(*top photo*) Technologies of the TimeSlips process. (*bottom photo*) Anne Davis Basting and a photo prompt for collaborative storytelling. Both photographs by Dick Blau. Reprinted by permission.

play, technologies function not to heighten the atomization, isolation, and investment in some deferred future that we found in the other performances, in laboratories, or in literary texts. Rather, they function to increase community in the present. Joining a specific kind of collaborative mental and social replacement to cognitive regeneration, the play dramatizes a prosthetic intersubjectivity. Rather than engaging in the prosthetic repair or enhancement of an individual, *TimeSlips* remedies lack and extends capacity through shared memory and shared experience.

Of course, Anne Basting's *TimeSlips* has a liminal relation to both the medical and theatrical communities. The centers for adult dementia sufferers in which it began (funded by Brookdale Foundation, a gerontology funding agency), like the Off-Broadway theater in which the play had its first performance, have the paradoxical flexibility and freedom of liminality itself. The enabling mixture of biomedicine and storytelling imagination that produced *TimeSlips* articulates a model of regenerative medicine based not on shoring up the individual but on enhancing community. Perhaps it is no accident that such storytelling falls into the category of "outsider art": like science fiction, the tales woven by Alzheimer's patients not only articulate the possibilities of life at the limen; they are themselves liminal.

Whether aging is being treated with replacement medicine (as it was throughout the twentieth century) or with regenerative medicine (as it has begun to be now at the beginning of the twenty-first century) seems to matter less than the goal that medicine addresses. Faced with the diminishment of aging, confronted with the boundaries of the human life span and the (until now inevitable) end written into the human plot, we seem to have but two choices. We can try to replace organs that no longer work, or we can try to catalyze renewal, the growth of newer, younger tissues in organs and bodies that are old. But in either case, our focus seems inevitably to be only on the individual: we can repair the aging body, or we can regenerate those aging tissues. The liminal moment of aging is conceptualized as a problem of the individual, and whether the biotechnologies with which we address the problem have replacement or regeneration as their goal, the context in which they are put in play is still the individual body.

Yet *TimeSlips* models a third alternative response to aging that occurs not on the level of the individual but on the level of the community.

Working together, the individuals in this project/play replace each other's flawed memories with new ones, generating in the process a remarkable renewal of imagination, life, and hope. As the characters are created, they come together, producing not only physiological and psychological renewal but new plots. Like the new voices and new memoirs of Alzheimer's disease enabled by earlier diagnosis, the new characters created in the stories and play of TimeSlips have an impact both on the individual people carrying the diagnosis of dementia and more broadly on our entire society, as it makes choices between different ways of responding to aging and death.

In its choice of technology, as in its prophylactic and therapeutic responses to aging, TimeSlips both resembles and differs from treatments getting much attention in contemporary medicine as well as contemporary fiction. For example, Bruce Sterling's *Holy Fire* (1997) imagined a complex system of antiaging medicine that harnessed the capacities of embryonic stem cells, so that his novel offers a fictional parallel to the contemporary hopes of curing Alzheimer's and Parkinson's diseases with stem cell therapy documented in the NIH stem cell report of 2001. In contrast, although pharmacotherapy is much in evidence both in the Alzheimer's residences and onstage, both Basting's geriatric project and her Off-Broadway play *TimeSlips* makes their major intervention in the process of cognitive degeneration using technologies so basic as to be invisible: paper, pen, camera, language itself. By foregrounding the therapeutic efficacy of collaborative storytelling, Basting's project reconfigures the central twentieth- and twenty-first-century medical strategies of replacement and regeneration, bringing medicine and literature together in a remarkable healing performance of our ultimate rite of passage: aging and dying.

Coda

THE PLURIPOTENT DISCOURSE OF STEM CELLS: LIMINALITY, REFLEXIVITY, AND LITERATURE

When it comes to the ability to control our genetic inheritance, there is little precedent to serve as guide—just science-fiction stories.—Ralph Brave, "Governing the Genome"

What if the study and crafting of fiction and fact happened *explicitly*, instead of covertly, in the same room, and in all rooms? —Donna Haraway, *Modest_Witness@Second_Millennium. FemaleMan_Meets_OncoMouse_: Feminism and Technoscience*

In September 1999, the Alpha congress "ART, Science, and Fiction" met in Copenhagen. "ART" stands for assisted reproduction. As we know, this used to be called "reproductive technology" until "technology" retreated into hiding as the unvoiced last letter of a newly normalized procedure whose new acronym is, significantly, ART. Sponsored by Alpha, the international society for scientists in reproductive medicine, this conference brought together more than 250 reproductive biologists to hear sessions addressing the techniques of ART with titles ranging from "Selection of Gametes and Embryos," "Cryobiology," and "Biological Risks and Limits," to "Embryonic Stem Cells" and "The Future of ART." Most interesting for our purposes was the talk by Dr. Robert Edwards, who with Dr. Patrick Steptoe pioneered the technique of in vitro fertilization and serves as honorary Alpha president. Edwards's talk "The Future of ART" put into circulation the curious relation between science and fiction in the construction of the newest of liminal lives: the embryonic stem cell. A description of that session was posted on the Web for interested reproductive biologists:

> On Friday morning Professor Edwards, the honorary president of Alpha, officially opened the meeting by stipulating [*sic*] on the future of ART and the role of the scientists in its further development. He stressed that embryologists will drive ART and that the input from cell research, animal studies and veterinary sciences is very important. It will lead to a better understanding of the human embryo. The homology between mammals and lower species should not be ignored. Different techniques like de- and re-differentiation of cells, transfer of animal cells to the human, reconstructive surgery, cloning, etc. will find realistic application in the human. The ethics and laws surrounding this type of treatment will have to be guided. Evolutionary consequences should not be ignored.[1]

As summarized, Edwards's talk makes three rhetorical moves worth comment. First, note the contradiction between his prediction that "embryologists will drive ART" and his recommendation that the ethics and laws "surrounding" a number of different techniques "will have to be guided"—though by what guiding force, he does not say. Second, note the attention paid to technology transfer between veterinary science and animal studies and human medicine. As is revealed by the intertwined involvement of twentieth-century biomedicine and literature with questions of hybridity, cross-species fertilization, and xenotransplantation, animals too have functioned as liminal lives. Because they have been perceived (until recently) as combining biophysiological proximity to human *corporeality* with ethical, social, and cognitive distance from full human *being*, animals have served as essential resources for the project of replotting human life. (Recently this instrumental position on the animal/human relationship has received serious scrutiny from scholars in disciplines as widespread as medical anthropology, cultural studies, and psychology. But the fundamental role of animals as research reservoirs in biomedicine remains essentially unchallenged.) Finally, note that among the hot-button issues such as xenotransplantation, reconstructive surgery, and cloning, the issue leading off his survey of techniques requiring ethical and legal oversight is "de- and re-differentiation of cells" (Nijs 1999). Edwards is talking here about research with stem cells: cells that have the capacity to divide for an unlimited number of times in culture, and to form almost any kind of specialized cell. These cells are described as *pluripotent*, meaning that they can develop into cells from the mesoderm, endoderm, and ectoderm, the three germ layers from which all the cells of the body arise. Their unmatched developmental flexibility accounts for the great research interest in these cells, since there is hope that they can be coaxed to develop into replacements for human cells damaged by a number of degenerative illnesses, from Alzheimer's and Parkinson's disease to diabetes and liver disease. Edwards's talk addressed some of the most important issues in biomedicine today: the tension between science and ethics, the role of animal and veterinary science in the biomedical reconstruction of the human, and the crucial role played in both of these by stem cells. And curiously, it did so at a conference with the intriguing title "ART, Science, and Fiction." While much has been made of the crucial role played by the flexibility of stem cells in future medical practice, the Alpha congress's title invokes another kind of flexibility that we would do well to examine. That is *discursive flexibility*, the flexibility of fiction.

This book has explored the ways that fiction has worked to negotiate between biomedical and social strategies, to articulate wishes and enable disavowals. In this coda, I want to consider how the discursive flexibility provided by fiction can serve as a guide to "the ethics and laws" we must generate in response to novel biomedical strategies, as Edwards himself suggests. To explore how this kind of discursive flexibility might function, let's reenter the debate over stem cell research in 2002: a year after the Nightlight Christian Adoptions agency sued the U.S. government to prevent federal funding for research on embryonic stem cells, and three years after Edwards spoke at the Alpha congress in Copenhagen, in order to see how the discourses of ethics and fiction are functioning now in relation to stem cells.

In the 4 February 2002 issue of the *New Yorker*, "Talk of the Town" columnist Dr. Jerome Groopman attacked Dr. Leon R. Kass, head of the new National Bioethics Commission, for opening the commission's first meeting "not with facts but with fiction." Kass talked about Nathaniel Hawthorne's tale "The Birthmark," in which a brilliant scientist married to a woman whose extraordinary beauty is marred only by a blemish on her cheek becomes obsessed with eliminating imperfection, with disastrous results. The story has all the gothic conventions: bubbling beakers, arcane tomes, elixirs of immortality, a stunted, apelike assistant. In the end, the scientist's treatment cures the blemish but kills the wife.[2]

Groopman, whose laboratory is active in the study of neurological degenerative diseases such as Alzheimer's and multiple sclerosis, and who thus has reasons to support stem cell research, clearly holds views at odds with the biomedical conservatism that made Kass the Bush appointee to head the commission.[3] In his devastatingly witty *New Yorker* column, Groopman uses Kass's literary turn as the occasion for a lesson in what can, and what can't, count as bioethics. First, what *can*: he surveys the scientific background on the issues of stem cell research and cloning: the crucial distinction between the two types of human cloning (reproductive and therapeutic); the medical uses that may ultimately be found for the stem cells produced by therapeutic cloning (to treat Parkinson's and Alzheimer's diseases, juvenile diabetes, and spinal cord injuries); the recommendation issued by the National Academy of Sciences that therapeutic cloning be supported and reproductive cloning be banned. Then what *can't*: after this summary of the issues at stake for Bush's new National Bioethics Commission, Groopman delineates Kass's particular approach to bioethics: " 'the wisdom of repugnance,' . . . which states, basically, that if you find something repugnant—if you just don't think it's right—then it must be wrong" (Groopman 2002, 24).

That Groopman differs from Kass is no news; what interests me is the ground they choose to contest. Groopman critiques Kass for his turn to fiction, and for the quality of the fiction he chooses: fiction with embarrassingly gothic conventions—those bubbling beakers, arcane tomes, elixirs of immortality, even the stunted, apelike assistant—fiction that seems closer to *science fiction* in its portrait of science. In this chapter's first epigraph, Ralph Brave joins Groopman in dismissing the bioethical guidance that can be found in literature, particularly in science fiction stories. I will suggest that this impasse over the status and role of fiction in the debate about stem cells has important social and medical implications.

In his *New Yorker* column, Groopman criticizes Kass's turn to literature both for its recourse to gothic clichés and for the cast of mind that he seems to think the literature breeds. As he explains, "Using literature to warn against the scientific search for perfection is a hallmark of Kass's approach to bioethics. (Hawthorne, Homer, and Huxley are among his touchstones.) So is a reflexive suspicion toward the enterprise of biotechnology" (Groopman 2002, 23). He frames the problem as a disciplinary or generic one: if Kass were more familiar with the enterprise of biotechnology and less infatuated with literature, Groopman suggests, he might be better informed. As it is, he offers a vision "dismally remote from what actually goes on in the nation's laboratories [where] there are no wild-eyed wizards with perfection potions" (24). "While Kass conjures a world of lab-bred James Bonds, two hundred thousand Americans live with spinal-cord injuries, a million and a half have Parkinson's, and four million have Alzheimer's" (24). When fiction faces off against fact, the outcome seems not only inevitable but right: in the battle for our sympathies, those lab-bred James Bonds in all their science-fictional unreality are no match for the literal, enumerated Americans suffering with spinal cord injuries, Parkinson's and Alzheimer's diseases. If we weren't already in Groopman's corner by then (and given Kass's tendency to inflammatory potshots at everything from autopsies to abortion, it is likely that we are), we would be when Groopman comes to his conclusion. Acknowledging that the Bioethics Commission is stacked with conservative presidential appointees, he closes with the hope that someone like Dr. Janet D. Rowley, one of the few "working scientists" in the group, will be able to "help shape a medical guideline that is based on fact, not on literature or aesthetics—one that distinguishes real science from science fiction" (24).

Note what has happened here. As Groopman articulates it, the purpose of the National Bioethics Commission has become not a consideration of ethics but the drafting of a medical guideline, while the purview of bioethics has been limited—narrowed—to "fact, not literature or aesthetics" (24). As the granddaughter of a woman who died of Parkinson's, and the daughter-in-law of a woman who died of Alzheimer's, I have a personal stake in the search for a cure for both of these devastating neurological diseases, which cast a genetic shadow on me and my children. But I want to make two points about this act of rhetorical narrowing.

First, the devaluation of fiction that is at its core stands in stark contrast to the *investment* in another kind of rampant fictionalizing: the act of imaginative extrapolation that drives contemporary discussion of stem cell research, whether scientific, governmental, or journalistic. It is a familiar trope: stem cells *may provide the cure for* . . . and then fill in the diseases—the list always includes Parkinson's, Alzheimer's, and spinal cord injuries, and it often includes juvenile diabetes. The ritual recitation of that list is worth investigating as a site of pluripotential rhetoric, performing networking in the Latourian sense. As the target of stem cell therapy shifts to correspond to the rhetorical situation, from ailments of the elderly to those of the young, for example, the procedure enrolls more allies. And this occurs not under the aegis of fiction (though its central move is the extrapolative projection of a present wish into a future narrative) but under that of medical fact. Second, the focus of a *bioethics* commission is not to find a cure; it's to open a discussion that can find an ethical position on the procedures used in the search for a cure. If we want to derive a principled bioethical position on the difficult case of stem cells, surely we need to move beyond the litany of wished-for results.

When we call, as Groopman does, for a purely factual basis for medical decision making, ruling that both literature and aesthetics are irrelevant to our inquiry, we lose this crucial additional sort of pluripotentiality: the rhetorical or fictional flexibility that I want to go on to discuss. The negative impact of Groopman's call for fact, not fiction (or aesthetics), is illuminated if we consider it in relation to three concepts: the notion of the biomedical imagination, the concept of an enabling negation, and the practice of generic policing.

THE BIOMEDICAL IMAGINATION

When Groopman enforces the boundaries between fact and fiction, or between "real" science and science fiction, he ensures that an important aspect of the bioethical context is omitted from the committee's deliberations: the biomedical imagination. To recap Catherine Waldby's formulation: this is the broader context in and from which biomedical creativity emerges: "the speculative, propositional fabric of medical thought, the generally disavowed dream work performed by biomedical theory and innovation. . . . [the] speculative thought which supplements the more strictly systematic, properly scientific, thought of medicine, its deductive strategies and empirical epistemologies."[4] Along with examining the biomedical imagination for its role in the management of anxiety caused by the process of disciplinary systematization, I want to suggest that we can also look there for expression of wishes and desires. As a zone of dream work and speculation, the biomedical imaginary is a rich space for the articulation of desire, since of course it is desire—not in the sexual but in the more broadly philosophical and psychoanalytic sense—that motivates all creativity, including scientific creativity. However, this zone of fantasy, dream work, and speculation is also subject to contradictions, tensions, and ambiguities that are foreclosed as a scientific project moves from the imagination into practice. Thus the articulation of desire in the biomedical imaginary is a complex process, operating not only through the direct wish or intention but also through the more indirect mode of negation and disavowal, as we can see if we turn to another work of fiction.

ENABLING NEGATION

"The Birthmark" is not the only Hawthorne story that Leon Kass could have considered.[5] Had he chosen the even more interesting "Dr. Heidegger's Experiment," he might have learned a bit more about how literature functions not only to dramatize our desires but also to articulate our defenses against exploring them. In this story, the curious Dr. Heidegger persuades three white-haired old men and a withered old woman to drink vials of water drawn from the Fountain of Youth, by showing them how the water rejuvenated a "withered and crumbling flower" until it looked "scarcely full blown."[6] The scientist watches, without himself partaking, as the old people drink the vials of water, become young again, grow giddy, and as the men finally fight among themselves for the young woman's attentions, spilling the precious vial of rejuvenating waters. In the meantime, Dr. Heidegger sits watching, pressing to his lips the cherished rose, which though rejuvenated has now withered once again during the course of the evening. His friends, too, are withering, and they bemoan the fact that "the changes of a lifetime [had] been crowded into so brief a space, and . . . they [were] now four aged people, sitting with their old friend, Dr. Heidegger." The story ends with a moral as Dr. Heidegger tells his friends what he has learned from the experiment: "Yes, friends, ye are old again . . . and lo! The Water of Youth is all lavished on the ground. Well—I bemoan it not; for if the fountain gushed at my very doorstep, I would not stoop to bathe my lips in it. . . . Such is the lesson ye have taught me!" As if to hammer the moral home, Dr. Heidegger presses the now-withered rose to his lips and asserts: "I love it as well thus as in its dewy freshness."

Sometimes the moral to a tale is less interesting than the conditions of its telling, for the way the tale is *framed* is crucial to our understanding of its outcome. As he sets the stage for his story, the narrator of "Dr. Heidegger's Experiment" anticipates the responses his tale might evoke: "if any passages of the present tale should startle the reader's faith, I must be content to bear the stigma of a fiction monger." *The stigma of a fiction monger*: both the form and the content of this act of disavowal are worth pausing over. Responding to the possibility of a reader's lack of faith in the tale he is telling, Dr. Heidegger protects himself from attack by claiming that his story is fiction. In making this claim, I would argue, he is engaging in an act of enabling negation. I mean this—to reiterate—in the sense that Sigmund Freud meant it in his essay "Negation," where he observed that "the subject-matter of a repressed image or thought can make its way into consciousness *on the condition that it is denied*" (Freud [1925] 1959, 182; italics mine). Dr. Heidegger can tell his story precisely because he is willing to back away from it, to disavow it as fiction (even though fiction is—as he acknowledges—a stigmatized genre).

There are two important aspects to this speech act. First, it has enabling properties. Dr. Heidegger wants to experiment with a youth serum, and to tell us about that experiment, and he makes it possible by being prepared to deny the truth of what he says. If we're interested in investigating desire—whether it is Dr. Heidegger's desire to experiment with a youth serum, or our contemporary medical desire to cure Alzheimer's and Parkinson's—it is worthwhile considering those aspects of the desire that embarrass us enough that we disavow them. We need to consider precisely those impulses that we disavow as "*just* fiction," for they are the ones that enable us to engage in a particular speech act (to articulate or perform something) without having to own (or own up to) the impulses that prompted us.

But now we need to turn to the question of the stigma attached to the genre of fiction. In a piece in the *American Prospect,* Chris Mooney traces the origins of the field of bioethics back to a response to notions of in vitro fertilization and cloning that not only relied heavily on fiction but was itself a "trumped-up" publicity stunt.[7] While the field of bioethics had its impetus from the horrible revelations of the Tuskegee syphilis study, the field really took off with the appearance of a 1972 *New York Times Magazine* cover story by Willard Gaylin, "The Frankenstein Myth Becomes a Reality—We Have the Awful Knowledge to Make Exact Copies of Human Beings." "The article," Mooney observes, "(published fully 25 years before the unprecedented birth of Dolly, the cloned sheep) was a publicity stunt for the newly launched Hastings Center," the inaugural bioethics center.[8] Mooney reports that "[Leon] Kass and Paul Ramsey both borrowed ideas from *Brave New World* to challenge in vitro fertilization in the early 1970s—Kass in the *New England Journal of Medicine* and Ramsey in the *Journal of the American Medical Association.* 'Both of these bioethicists, writing in the two leading American medical journals, drew on fictional associations of external fertilization to reinforce a slippery slope argument against the technique.' . . . But fiction is just that—fiction. It's questionable whether Kass's speculative futurism counts as a responsible, or even helpful, approach to bioethical dilemmas" (Mooney 2001, 7).[9]

While it is fascinating to consider that bioethics as a field may be engaged in consolidating its disciplinary position precisely through distancing itself from its genre of origin, I'm most interested now in Mooney's argument that fiction, precisely *as fiction,* is unlikely to be a helpful approach to bioethical dilemmas. This blanket approach ignores not only the fine gradations of difference between fiction and fact but more importantly the variety of different ways of using fiction. Clearly, fear-mongering recourse to specters from science fiction (Frankenstein *or* his monster) isn't helpful. And yet we can learn much from a careful look at popular culture, at science fiction, and at fiction itself. The essential nature of fiction is its open-endedness: the fact that the whole course of a story, from framing to conclusion, holds significance and can springboard discussion. When we ban fiction—as a genre—from a discussion, we not only do away with that open-ended approach but are making some important choices about what counts as, and can contribute to, knowledge.

GENERIC POLICING

Mooney's blanket statement is an overt instance of generic policing. What work is being done by invoking the genre of fiction in this bioethical discussion? To answer this question, we first need to move beyond thinking of genre only as a classificatory term used in literary and aesthetic discourse. Instead we need to think of genre as a category that has sociological and cultural meaning.

Genre regulates and it brings into being; it shapes how we can enter and engage in a preexisting social practice, and it constitutes that social practice, giving us ways of understanding it, as well as the conventions that make such practice possible. This process of regulation and constitution is complex, and is carried out when we label different genres according to the demands of different rhetorical situations, defining one set of things as *fictional* and another as *factual,* or one set of representations as *realistic,* in contrast to others that may be *gothic* or *science fiction.* In the process of living with them, those labels become not only descriptive but prescriptive: they specify not only how we understand those objects, but how we behave in relation to them. Recall that, as Anis Bawarshi has demonstrated, "as individuals' rhetorical responses to recurrent situations become typified as genres, the genres in turn help structure the way these individuals conceptualize and experience these situations, predicting their notions of what constitutes appropriate and possible responses and actions. This is why genres are both functional and epistemological—they help us function within particular situations at the same time they help shape the ways we come to know these situations."[10] Bawarshi's point is that genres govern the structures within which we interact, as well as our actions within those social structures.[11]

Groopman, Brave, and Mooney all engage in generic policing as part of their response to the bioethical dilemma posed by stem cell research.

Mooney argues that the very genre of fiction makes it unlikely to contribute to a bioethics discussion. Groopman and Brave go even further. For Groopman, it's not so much the fictionality of Kass's examples but their particular location within the genre of fiction, the fact that they represent the subgenre, paragenre, or debased genre of science fiction (with its embarrassing clichés and its lack of verisimilitude), that makes them unable to contribute to a successful bioethics assessment. And for Brave, the irrelevance of science fiction stories to our search for a principle to guide bioethical activity is simply self-evident.

How does this relate to the bioethics committee's attempt to come to grips with the stem cell debate? Because genre has a regulatory and constitutive function extending beyond the realm of aesthetic into the social, the dismissive labels of fiction and science fiction function to regulate and constitute not only the social relations and practices at play when Kass and his committee discuss stem cell research but also *the subject positions and social interactions available to all of us*—Kass, Groopman, Brave, Mooney, and ourselves—as we approach the stem cell question. Whether Kass turns to fiction or Groopman turns away from it, both are being produced and are answering to the law of genre. As the law of genre divides fiction (and especially science fiction) from fact, it is producing two different ecosystems, each with its own way of responding to the questions of hope for a cure, of individual and social choice, of hope for our future individually and collectively, that are so powerfully invoked in the stem cell debate. Though this generalization is ultimately too simple, we could say that the fiction ecosystem values ambiguity, complexity, and open-endedness, while the law of fact values clarity, simplicity, and closure. Thus the policing of genre, separating the realm of fact from the realm (and regulatory operations) of fiction, produces a narrowing of the conversation by its very utterance.

One could argue that this need not be the case if we invoke genre not as a law to be obeyed but as a site for playful investigation. If we are playing, rather than going to fiction for Truth, we will be less likely to dwell on the moral of one work of fiction than to examine a range of perspectives drawn from several different fictions. Then we could answer Kass's choice of "The Birthmark" with other fictions that represent the biomedical reconfiguration of human beings not as a doomed attempt to attain perfection but as a complex and ambiguously charged response to human frailty. I am thinking here of Octavia Butler's *Xeno-*

genesis trilogy, with its portrait of the gene-trading Oankali, who are attempting to modify the human species away from aggression to ensure our survival.[12] But perhaps even more relevant to the stem cell debate are Naomi Mitchison's *Solution Three* and *Memoirs of a Spacewoman*, which represent cloning and interspecies grafting as practices that oscillate between the potential for emancipation and control, and thus challenge human beings to remain continually alert to the unconscious fantasies embedded in seemingly neutral technologies.[13]

Just as fiction can give us access to a *thick description* that can provide a ground on which to debate moral questions in this era of cultural pluralism, so too it can also provide a richer, more complex approach to bioethics.[14] This approach is exemplified in Tod Chambers's *The Fiction of Bioethics*, which takes as its premise Nietzsche's question "Why couldn't the world that concerns us—be a fiction?" In a study that considers bioethics cases from a range of different aesthetic and critical perspectives, not attempting to achieve a unitary and exhaustive analysis but rather mingling discourse analysis, media studies, reader-response criticism, feminism, semiotics, and poststructuralist criticism, Chambers demonstrates that the *framing* of the bioethics case contributes in crucial ways to its outcome. "Reading cases with attention to their fictional qualities, that is, their constructedness, in turn reveals how dilemmas are framed in ways that conceal as well as reveal other ways of seeing."[15] To apply both of these models to the stem cell case, this suggests that taking fiction seriously (as Groopman, Brave, and Mooney fail to do), while approaching it not as a truth category (as does Kass) but as a realm of desire as well as disavowal, could indeed give us a more powerful instrument for bioethical knowledge.

FICTION, KNOWING, AND UNKNOWING

What is the broader impact of generic policing, not just in the stem cell debate but for society more broadly understood? As I am formulating it, this process of generic policing is articulated with the way social relations are produced and reproduced in this era of reflexive modernity. As sociologist Anthony Giddens formulates this concept, "Modernity's reflexivity refers to the susceptibility of most aspects of social activity, and material relations with nature, to chronic revision in the light of new information or knowledge."[16] Bawarshi's notion of genre as something that governs our structures of interaction intersects with Giddens's structuration theory. Giddens argues that social relations are shaped in a particular way, over time, by the rules that human beings generate to govern our social interactions. "Rules basically consist of the stock of knowledge we need to enter into social interaction, and to respond to various social circumstances" (Dodd 1999, 187). These rules are both enabling and restrictive, according to Giddens: we use rules both to reflexively monitor and to continuously order and regulate our patterns of social behavior. Because rules regulate our stock of knowledge, as well as the ways we interact together and respond to that stock of knowledge, they both produce us (as knowing beings) and constrain us (as beings attempting to know more). And here's how the rules of genre come into play: to the extent that genre organizes and categorizes knowledge most broadly along the fiction/fact divide, as well as within that (as novel/poem/play, or as documentary/memoir/scientific report), genre's regulatory function collaborates with a major strategic function of reflexive modernity: the construction of "expert" knowledge, as distinct from nonexpert, "lay," or popular knowledge.[17]

We may not be surprised that the genres on the "fiction" side of the fact/fiction generic divide are held to offer little to the production of knowledge, but what is significant (and more surprising) is that they are also held to have little to contribute to the exploration of doubt, the other half of the social agenda in reflexive modernity. This relative indifference to the exploration of what isn't known, especially on the part of institutions charged with making biomedical policy, may reflect a semantic confusion in the circulation of the notion of reflexive modernity itself. As Ulrich Beck explained it in 1999, the term "reflexivity" has circulated with two overlapping but distinctly different meanings: (1) knowledge of, or *reflection* on, the process of modernization (including its foundations, consequences, and the problems it generates); and (2) exploration of the *unintended* consequences of modernity, or of what we didn't anticipate, didn't know, or were unaware.[18] Beck points out that the first meaning leads us to what Giddens has called "institutional reflexivity": "the circulation of scientific and expert knowledge on the foundations of social action," an endlessly precise feedback loop on the material conditions of any action (Beck 1999, 111). In contrast, the second meaning will bring us to a reflexivity closer to that envisioned by Pierre Bourdieu: "systematic reflection of the unconscious preconditions (categories) of our knowledge" (111). In short, these two meanings of reflexivity skirt the materialist/constructivist debate: with the first, we focus on increasing the technical precision of what we know; with the second, we step back to contemplate the way that our unconscious categories construct (structure, delimit, constrain) our potential awareness.

While fiction may not increase the technical precision of our knowledge, it can indeed illuminate the unconscious categories shaping our awareness. And because of the feedback loop central to reflexive modernity, that process of assessing our unconscious knowledge categories in turn produces material effects that we will then attempt to know with greater and greater precision, *if* we are aware of them. For social authority to be granted to fiction, as a contributor to the reflexive assessment both of the material conditions of our social reality as well as of the forces helping to construct that reality, we need to achieve a convergence of the two poles of the realism/constructivism debate. According to Beck, we could get to that point if we became aware of the interpretive power of what he calls *reflexive* realism and *reflexive* constructivism, or "symbolically mediated social relations with nature" (Beck 1999, 27).

While recently a number of scholars have offered examples of such a reflexive convergence of realism and constructivism, it seems fitting to end this book, as I began it, with an example from embryology.[19] In "Science Matters, Culture Matters," Anne Fausto-Sterling describes a conflict she had with her department over an embryology course and traces in the process her own journey from naive realism to scientifically informed, reflexive constructivism.[20] When she first taught embryology, she tells us, the course seemed straightforward: "a lecture on the development of the central nervous system . . . included a description of the morphogenetic events leading to its formation and a discussion of embryonic induction" (Fausto-Sterling 2003, 111). However, after spending more than a decade working in feminist and science studies, the same topic seemed anything but simple. "The neural tube appeared to me embedded in a matrix of epidemiological, medical, historical and social questions" (111). Fausto-Sterling's solution to this pedagogical and epistemological challenge was to generate what she calls her "non-modern anthropology course" (117). Teaching embryology in relation to its social context, she provides her students with a set of what she calls "knowledge webs"; to that she also adds a new set of collaborative assignments, "web expansion units," for which "students chose topics such as *in vitro* fertilization, cloning, conjoined twins, stem cells, and anencephaly. They designed web sites that address the history, the science and the ethics and social context of their topics" (118, 120).

The habit of genre policing produced by our disciplinary boundaries is a resilient one, requiring repeated reflexive consideration to overcome it. This is illustrated by the fact that even Fausto-Sterling participates in it, in her otherwise inspired rethinking of the embryology course. Although she admiringly cites Donna Haraway's vision, "What if the study and crafting of fiction and fact happened *explicitly*, instead of covertly, in the same room and in all rooms," when she describes the issues the students would network back to embryology, her model for those web expansion units includes history, science, and ethics, but not literature, or fiction in any of its forms (Fausto-Sterling 2003, 19).

From ART and the relations between science and fiction, through the debate over the role of fiction in stem cell research, to Fausto-Sterling's omission of fiction in her otherwise quite sweeping rethinking of the relations between science and culture, these instances highlight the disputed position of fiction, and more precisely literature, in cultural practice. Of course *literature* has itself participated in a form of generic policing known as canon formation that grants to some kinds of literature cultural cachet while stigmatizing others. We can see this form of generic policing embedded not only in the discomfort Groopman expresses at "the gothic conventions: bubbling beakers, arcane tomes, elixirs of immortality, a stunted, apelike assistant" of the story Kass chooses but also in the second Hawthorne story, where Dr. Heidegger pronounces himself willing to assume the "stigma of a fiction monger" in order to tell his tale.

I have been arguing that there are distinct advantages in adopting a more flexible approach to the question of what counts as useful for the production of knowledge. Derrida has observed, "As soon as genre announces itself, one must respect a norm, one must not cross a line of demarcation, one must not risk impurity, anomaly or monstrosity."[21] But it's precisely the risk of impurity, anomaly, even monstrosity, that we must confront if we are to examine the social and scientific implications of stem cell research. As an example of the kinds of wide-ranging and usefully ambiguous issues that can be addressed in fiction, consider an example from perhaps the most highly stigmatized fictional genre: graphic fiction, or comic books. Ruben Bolling's episode of the graphic fiction *Tom the Dancing Bug* entitled "Bad Blastocyst: Was It Bad? Or Just Misunderstood?"—in which a cleaning woman is knocked on the head and killed by a petri dish inhabited by a blastocyst, thus unleashing a national debate about the nature of the punishment that should be meted out to the bad blastocyst—shows how graphic fiction can both demonstrate and mock the rhetorical pluripotentiality of stem cells.[22] Yoking the genres of science fiction to true crime and detective fiction, with a witty closing nod to questions of human cloning, this brief comic strip reveals how stem cells can be used to raise issues as diverse as the legal definition of a rights-bearing subject, the questions of psychological and social responsibility, and the rationale for, and goal of, scientific research.[23] As Bolling's graphic fiction demonstrates, stem cells can provide the anchor for debates over capital punishment, abortion, disability studies, and animal experimentation. Because it enables us to enter, imaginatively, all those complex, ambiguous debates, fiction—canonical and noncanonical, debased, even stigmatized—can indeed serve as a valuable guide as we address the question of stem cell research.

We have come full circle in our exploration of the liminal lives that are replotting the human. From Nightlight Christian Adoptions with its adoptive embryos to the fictions that circulate around—and illuminate—debates over stem cell research, we have explored a set of literary and biomedical interactions that have themselves proven to be fissured, unstable, in the process of change. The liminal lives we have been following in the course of this book all show the mark of the move from zoë (the fact of being alive) to bios (life given a specific form by biomedicine). As a brief recapitulation will demonstrate, even if they are of animal origin, the liminal lives have cooperated with a reconceptualization of the course and boundaries of human life. The tissue-cultured cells at Strangeways, producing avian bone or teeth, caused Honor Fell to acknowledge the essentially constructed character of death: the term "death" means one thing, she told her medical students, when used in a medical setting, and something quite different

"Bad Blastocyst." Cartoon by Ruben Bolling, *Salon*, 8 August 2001. Reproduced courtesy of Salon.com.

when used in the laboratory. As Strangeways demonstrated in his final experiment, when he cultured a sausage purchased in town and produced a living cell line, "death" can be undone, replaced by laboratory life. Similarly, the tissue culture experiments demonstrated the flexibility of species boundaries: though the researchers focused on avian and animal cells, the results of their work were crucial to the task of reshaping human life. The technique of in vitro fertilization is based in the technique of embryo culture with which the Strangeways researchers were preeminent. Rhetorical liminality was also characteristic of the research at Strangeways. If the Strangeways experiments demonstrated the flexibility and potential of cultured cells, they also demonstrated another sort of liminality: their liminal identification with the subjects of their scientific research. Although in lectures and scientific papers the researchers focused on the technical and therapeutic control that tissue culture enabled, in their poetry they tended rather to identify with the cultured cells, to explore parallels between the processes of cellular and human growth, and to stress the rhetorical force and showmanship made possible by the visualization technologies (microscopy, photomicrography) available to them.

The embryo culture central to Strangeways research in the 1920s and 1930s returned again to public awareness more than half a century later, when members of the British government's Warnock Committee (1984) and the United States NIH Human Embryo Research Panel (also known as the Muller panel) (1994) debated the ethics of embryo research. Here again, just as the poems of the Strangeways researchers articulated a perspective on tissue culture different from the one available either in the press or in scientific writing, so too fiction has given us a different vision of the hybrid embryo than the one invoked briefly, only to ban it, in the government document on human embryo research. The trope of negation, by enabling in the Muller panel report the articulation of a fantasmatic cross-species fertilization, under the pretext of banning it, performed a similar function in that government document to that performed by fiction in the cultural imaginary. From Shelley's *Frankenstein* through the twentieth-century visions of cross-species reproduction of H. G. Wells, Maureen Duffy, and Doris Lessing, fictions challenged the limits of human life, exploring what aspects of experience (emotional and morphological) we share with other species. The exploration of xenogenic desire as it has circulated through the twentieth century reveals that the practice of policing biomedical science relies on another, more primary, kind of policing—of the generic boundaries that differentiate official government documents from popular journalism, fact from fiction, and literature from science fiction. A reading of the postmodern feminist fiction of Clarice Lispector reveals that even those structures through which boundaries are policed in biomedicine and culture—as sexuality is constrained to practices taking place (whether "naturally" or biomedically) between human beings rather than across the human-animal boundary—are the product of a prior boundary policing, the distinction between reproductive and replicative life-forms. The way that hybridity is deployed (both to negate and to articulate a xenogenic desire) in the embryo research debate recapitulates the Strangeways researchers' investigations of the boundaries of the species and the life span, the experiments in transspecies organ transplantation in the mid-twentieth century, and the early-twenty-first-century debates over cloning, stem cell research, and regenerative medicine.

The biomedical interventions into the beginning and the end of life explored in the first three chapters are structurally reflexive, in that they frequently loop into each other in a Möbius-like fashion (so death returns to life) and repeat each other at temporal intervals (so that the tissue culture experiments in the 1920s return as the embryo culture debated in the 1990s). In contrast, the biomedical and literary interventions into human growth that I explore in chapter 4 are made possible by a cross-disciplinary shift in the conception of growth, from a function of the individual (or a problem of the individual) to a function of the aggregate (and a problem that should thus be addressed at the level of the group). The process of graphing, and thus of plotting normal growth curves, functions methodologically in biomedicine, enabling researchers and physicians to identify and treat abnormal growth patterns. While the extrapolative function of scientific graphing produces a morphological future by projecting from existing data, the extrapolation of fiction gives us a more far-reaching vision of the future, demonstrating the impact of growth not only on an individual, or on one group, but on a whole society, including the ethical and social structures that subtend it. Thus in fiction (whether canonical literature or science fiction), graphs are deployed as structural or thematic tropes, enabling writers to imagine and articulate new plots for human life. Moreover, as Wells's *Food of the Gods* demonstrates, the fictional exploration of growth raises questions about the limits of scientific intervention not found in the scientific assessment of abnormalities in growth, or of mechanisms for replotting a growth curve, such as the administration of growth hormone. The guinea pigs, giant babies, dwarfs, and supersized men who circulate in the mid-twentieth-century interventions into human growth testify to a disciplinary struggle: between our imaginative and technical interventions in the course of human life, or (to put it another way) between the ability to identify a statistical norm and the tendency to transform it into an ideal. As Lennard Davis has observed in a discussion of Galton's creation of the norm through the application of statistical theory, "The new ideal of ranked order is powered by the imperative of the norm, and then is supplemented by the notion of progress, human perfectibility, and the elimination of deviance, to create a dominating, hegemonic vision of what the human body should be."[24] Any neo-eugenic notion that we can find in biomedicine of improving the size and powers of the human being, we find countered in fiction by plots of growth gone wrong and shackled strength.

If the new normalization of growth led to a set of more or less problematic fictional experiments with intervention in growth in the first half of the twentieth century, the same era also saw an exhaustive exploration of strategies for reshaping the beginning and the end of life in both fiction and biomedicine. Two new liminal lives took center stage during this period, both of them given most vivid form in science fiction *before* being enacted in biomedical practice. The "incubaby" anticipates the much later appearance of the human embryo created by in vitro fertilization, whereas the "rejuvenate" would ironically *disappear* in culture as the extensive range of biomedical rejuvenation techniques, from hormone replacement therapy to plastic surgery, became an invisible part of normal medicine, enabling otherwise aging men or women to blend into an increasingly youthful society. Eugenics, embryology, sexology, and gerontology: these newly consolidating sciences or pseudosciences were all instrumental in moving reproductive technology into the mainstream, and in reconceptualizing rejuvenation therapy as normal medical practice. But equally important in these negotiations were the fictions by C. P. Snow, Gertrude Atherton, and a range of other science fiction writers, because they provided an arena to explore the anxieties raised by the disruptive reconfigurations of what had previously been understood as fixed and firm life stages: the liminal phase of conception and gestation, and the equally liminal phase of aging and death.

Changes in the understanding of organ transplantation in the twentieth and twenty-first centuries continued this trend of making the boundary between life and death both porous and difficult to establish. Here too, science fiction had a crucial role in producing cultural acceptance (in Europe, England, and the United States) of an ethically and socially charged procedure. As notions of the location of human identity changed, from being vested throughout the body (and thus potentially disrupted by the replacement of body parts and solid organs) to being vested in the notion of a potentially eternal self (supported by an endlessly renewable stock of body parts and organs), the personal and social disruptions caused by organ transplantation were articulated in literature much earlier than they surfaced in biomedical practice. Organ art—the art that uses human organs, cadavers, body parts, embryos, and fetuses, either gilt or embedded in acrylic, cast in plaster, or cremated and subjected to diamond-producing pressure—disrupts the increasing tendency to value organs not for themselves (in their essence, either as part of a valued identity or subject or as aesthetic objects) but for their exchange value, as commodities enabling the leveraging of a perpetually extended future. The legal debates surrounding the differential access to human body parts and organs that have sprung up in response to organ art illuminate the fact that professionalization, and its inherent process of disciplinary division, has been used to shape access to products not only of the mind but also of the body.

The final liminal stage encountered by human beings is our own aging and death, and that is the passage being negotiated by the final group of liminal lives this study has explored: men and women who are aged and nearing death. Whereas their mental and physical debilities were treated by replacement medicine in the twentieth century, now in the twenty-first century they have become the subject of what has come to be known as regenerative medicine. And yet as we have seen, the connotations of replacement and regeneration can be broadened if we reconceptualize the purview of medicine from the narrowly focused surgical or pharmacological intervention to the far broader sociocultural intervention performed by Anne Davis Basting in her *TimeSlips* project. While the early-twentieth-century science fiction stories portrayed the horrors awaiting those who turned to biomedicine to renegotiate the limits of life, the early-twenty-first-century stories crafted collaboratively by elderly dementia patients reveal not only horror but also beauty and bewilderment. Moreover, as they are given voice in the play that addresses the beauty and bewilderment of their final days, the concept of liminality itself as a performance comes full circle.

We saw at the outset that the anthropological definition of liminality is generally restricted to cultural modes of negotiating life passages. Victor Turner described liminality as "the movement of a man through his lifetime, from a fixed placental placement within his mother's womb to his death and ultimate fixed point of his tombstone and final containment in his grave as a dead organism—punctuated by a number of critical moments of transition which all societies ritualize and publicly mark with suitable observances to impress the significance of the individual and the group on living members of the community. These are the important times of birth, puberty, marriage, and death" (V. Turner 1967, 94). And yet, as I argued in the introduction, Turner's restriction of liminal performances to cultural arenas relies on a disciplinary division between culture and nature that, we now know, fails to represent the complex ways that culture intervenes in and produces nature, while nature undergirds practices that we have come to think of as cultural. The therapeutic effects of Basting's collaborative storytelling yoke the biomedical and the cultural, fact and fiction, in a liminal performance that both accepts the limit death imposes on human life and challenges the fixity of that limit. The Alzheimer's patients transcend the limitations of their solitary minds and bodies by producing a shared space of mutuality, a liminal performance quite different from the regenerative medicine to which Bruce Sterling's Mia Ziemann is subjected.

If this look at the liminal lives of the twentieth and twenty-first centuries has revealed biomedicine to be an unstable, porous, and culturally implicated practice, what has it taught us about literature? First, it is increasingly apparent that not only are the boundaries of the human life being reconfigured, but so too are the relations between biomedicine and literature. Indeed, a convergence appears to be occurring—and in some places to already have taken place—between biomedicine and the most stigmatized genre of fiction: science fiction. There are a number of ways to understand this convergence. To some, it suggests that science fiction has been superseded (or bested) by life, as in the old adage "Truth is stranger than fiction." Columnist William Safire assumes this when he observes, in a discussion of the new move to neuroethics, the new field that "deals with the benefits and dangers of treating and manipulating our minds," that "our generation has outlived science fiction."[25] A similar understanding shapes the enthusiastic forecast of Robert Lanza, vice president of Advanced Cell Technologies, the research group most actively working to put stem cell therapy on the medical agenda: "If this research is allowed to proceed, by the time we grow old, this will be a routine thing. You'll just go and get a skin cell removed at the doctor's office, and they'll give you back a new organ or some new tissue—a new liver; a new kidney—and you'll be fixed. And it's not science fiction. This is very, very real.[26]

Alert to the sense of generic stigma and disavowal underlying both of these comments, however, I have a different perspective on the convergence of science fiction with contemporary biomedical science. The need for generic policing *within* literature (to keep high literature from low), like the need for generic policing between literature and science, has been decreasing during the twentieth and twenty-first centuries, until it often seems to have attained the vanishing point. Hybrids surround us, and not even the realms of politics or literature are exempt. As we have traced the replotting of human life, enabled by the production of liminal lives from conception through growth to aging and death, we have seen the erosion of that old distinction between fiction and literature. Once enforced by the hierarchies of aesthetic value and by the invocation of generic taxonomies, the special category of literature has vanished as a result of the contestatory and constructive relations between the new biotechnologies (technologies of recombination) and the new narrative forms and strategies (technologies of signification).

Now that human life can be engineered to our specifications, with the assistance of animals, embryos, and organ donors, so that its beginning, trajectory, and ending are all something not *given* or the product of chance but instead subject to biomedical manipulation, how are our stories not only reflecting, but actually leading, that reconceptualization of the course of human life? How have the plot, characters, intended audience, and methods of production and consumption changed? Drawing on Victor Turner's analysis of the ritual process, we can enumerate some of the qualities and practices that are likely to characterize the emerging literatures of liminality. First, just as liminal personae "slip through the network of classifications that normally locate states and positions in cultural space," so too the literatures of liminality are less likely to fit comfortably in the customary generic categories (canonical fiction, gothic, science fiction, romance, etc.) and are more likely to address issues and wander in realms not previously thought to be specifically literary.[27] These are likely to be hybrid works (children's literature; graphic fiction; documentary fiction; disability or illness memoirs) that speak to, or liberate aspects of, social engagement not currently accessible in the normal structural realms of politics, law, medicine, and religion, and hitherto unencountered in novels or short stories targeted to the literary reader. They will test the boundaries not just of literature and science but of fact and fiction, language and visual representation, demonstrating that, as with other sorts of liminal phenomena, "the opposites, as it were, constitute one another and are mutually indispensable" (V. Turner [1969] 1995). As such, like people or societies in their liminal phases, these texts will function as a "kind of institutional capsule or pocket which contains the germ of future social developments, of societal change."[28]

To be specific about the kind of texts I have in mind—because these are an emergent phenomenon, and so examples are already abundant—I am thinking of works such as Ruth Ozeki's novels *My Year of Meats*, which adapts documentary film techniques to fiction to critique the methods of the American meat business, and *All Over Creation*, which juxtaposes large-scale genetic engineering in agriculture (the production of BT potatoes) to small-scale ventures to preserve heirloom seeds; Paul Karasik and Judy Karasik's *The Ride Together*, in which memoir alternates with graphic fiction to convey the unique experience of life and death of their autistic brother David; and Michael Rowe's *The Book of Jesse: A Story of Youth, Illness, and Medicine*, in which a father's true story of his son's terminal illness is anchored to the fantastical comic strips Jesse drew.[29]

Certainly most overlooked by cultural analysts, and perhaps most important as well, are works of children's fiction, for they arguably shape the future by introducing child and young-adult readers to a wide range of ethically charged interventions into human life. Among these are Nancy Farmer's *The Ear, the Eye, and the Arm* (environmentally induced genetic mutations); Margaret Peterson Haddix's *Among the Hidden* (population control) and *Turn-About* (reverse aging), and Lois Lowry's *The Giver* and *Gathering Blue* (aging and emotional control of populations).[30] In conclusion, I want to focus on one specific work of children's literature that explicitly addresses the biomedical replotting of the entire life span, in order to suggest some of the qualities that make this literature—as illustrative of all kinds of liminal literature—worth taking seriously, both for its more complex portrayals of these life span interventions and for its agency in shaping the way a population will both deploy and respond to them.

Nancy Farmer's *The House of the Scorpion* is a young-adult book, explicitly aimed at children between ages eleven and fourteen, "Grades 6 and up," and marketed as "middle-grade fiction" by Atheneum Books for Young Readers. Unlikely thus to draw the attention of adult readers, to make it into the front pages of the *New York Times Book Review* ("Children's Literature" is grouped together at the back of the magazine), or otherwise to contend for any of the significant literary prizes, still—or perhaps precisely because of its marginal status to all adult literature—Farmer's book dramatizes some of the powerful reimaginations of the life span of which this new liminal literature is capable.

Addressing the global commodification of replacement and regenerative medicine, Farmer's work tells the story of a young boy, Matteo Alacran, who discovers that he was cloned to provide spare organs for a powerful drug lord. As we follow Matt's dawning understanding of his socially marginal position as valuable livestock, property of the powerful Patron who created him, we also come to appreciate the economic and social context within which human organ harvesting flourishes. We learn about the country of Opium, the stretch of poppy-farming country between what once was Mexico and the United States, where workers who escape the border maquiladoras and follow *coyotes* (or smugglers) in the attempt to cross over to the United States as illegal immigrants are rounded up by border patrols and made to toil in the poppy fields that stretch "from the Pecos River to the Salton Sea" (Farmer 2002, 196). We gradually observe the existence of a culture of the very rich, whose money has been made in a complex drug cartel, in which Mexico and the United States have acceded to the "pact made in hell" offered by Matt's creator, the "Patron," to solve the problems of illegal immigration and the market for illegal drugs:

> Matteo Alacran formed an alliance with the other dealers and approached the leaders of the United States and Mexico. *"You have two problems," he said. "First, you cannot control your borders, and second, you cannot control us."* He advised them to combine the problems. If both countries set aside land along their common border, the dealers would establish Farms and stop the flow of Illegals. In return, the dealers would promise not to sell drugs to the citizens of the United States and Mexico. They would peddle their wares in Europe, Asia, and Africa instead. (Farmer 2002, 169)

Within this global culture of enormous wealth produced through the enslavement of the socially marginal, the commodification of life enhancement, rejuvenation, and life extension medicine flourishes. From fetal brain implants and hair follicle grafts to kidney and heart transplants, these new biomedical technologies are available to the highest bidder irrespective of their legality, precisely because the buyers and sellers coexist in a socially, politically, and ethically liminal zone. Moreover, we learn that the technologies designed to achieve control over the human life span are linked to another sort of biomedical intervention designed to subject the body itself to external control. In disturbingly provocative extrapolations from contemporary experiments in cochlear implants for deafness and electrode implants to treat epilepsy, Farmer introduces the practice of implanting computer chips, either in human beings to produce "eejits" (computer-chip-controlled "idiots") or in animals to produce "safe horses" and other docile creatures.

Through the eyes of the young clone Matt, we come to see how this culture of organ harvest and biological control is articulated into a broader culture of biomedical and technological exploitation: of the adult poor (the Illegals who have been surgically converted into eejits in order to work in the fields of the vast drug farms or work as housekeepers, groundspeople, cooks—virtual slaves on the large estates), of children (constantly subject to adult control, either reduced to eejits who sing in church choirs or act as ring bearers in weddings, or groomed to adulthood and forced to marry to secure economic alliances), of livestock (horses whose brain implants make them walk in straight lines and eat and drink only on command; cows used as surrogate wombs to gestate the human clones farmed for organs), of the water (diverted from the Gulf of California, purified of its dead fish and chemicals, to supply water for the Alacran estate), and of the land (now "wastelands" where the eejits live, surrounded by the toxic sludge from the purified water) (Farmer 2002, 171).

Farmer's novel insistently links current social and political practices into her extrapolation of these new biomedical reformulations of the human being, exploring not only what might be done by medicine in the future but where the impulse originated, why it might be put into practice, by whom and to whom. We learn that the Patron, who as the major drug lord of Opium exerts a chillingly total control over all the people around him, is himself the product of a life of terrible poverty in a country village, where "nothing grew . . . except weeds, and they were so bitter that they made the donkeys throw up. Even roaches hitchhiked to the next town. That's how bad it was" (140). When he grew to adulthood, he created his estate as a deliberate copy of the wealthy landholdings he had seen, from the outside, as a child: "As a boy, El Patron had observed the grand estate of the wealthy rancher who owned his village. . . . In every respect . . . he tried to duplicate that memory, only of course being vastly more wealthy, he could have dozens of statues, fountains, and gardens" (136). In the Patron's development from resistance to land-based inequity, to exploitation of cash-based inequity and finally of biological inequity, Farmer's fiction explores the way power harnesses biomedicine to its purposes. Moreover, she asks what the social implications are of these age extension, rejuvenation, and life enhancement strategies, as well as of these technologies of biological control: not only for those who may profit from them but for those who are subjected to them and for those to whom they will never be available.

The ambiguity and social complexity of Farmer's work—her refusal to animate the biotechnology without considering its human origins and effects—is clearly what Ursula K. Le Guin was responding to when she contributed a glowing notice to the back cover: "It is a pleasure to read science fiction that's full of warm, strong characters—people who are really fond of one another, children who are ignorant and vulnerable, powerful evildoers whom one can pity, good people who make awful mistakes. It's a pleasure to read science fiction that doesn't rely on violence as the solution to complex problems of right and wrong. It's a pleasure to read science fiction that gets the science right. It's a pleasure to read *The House of the Scorpion*."[31]

Le Guin's praise helps to clarify what makes this work of children's fiction remarkable: its insistent return from the novum to the familiar; its determination to address the ambiguous social origins and outcomes of even the most frightening biomedical reformulations of human life; its attention to the experiences of those subject to these innovations, rather than simply to those who wield them. Yet most interesting of all is the way such a complex exploration of the new biomedical choices available—for a fee, in a specific context—is marketed explicitly to young readers who are poised between childhood and adulthood. As Farmer explains the goal of her extrapolations, she explicitly relates her own past to their futures: "The events portrayed in *The House of the Scorpion* are not new. As a child I often saw illegals scurrying through alleys, and occasionally they stayed at my father's hotel. Ranches in the Southwest depended on them. Dope smuggling and drug empires aren't new either. But I hope I've created an environment and a story that makes readers think about these old problems in a new way. And then maybe even be inspired to change them, one day" (Farmer 2002, frontmatter).

By addressing a readership that is itself in transition, by portraying biomedical interventions as part of broader human practices and relations, these liminal literatures exemplify Turner's important notion of *communitas*, a symbolically encoded set of behaviors and experiences that are *in between* in the sense both of being in transition and of being held in common. As Turner explains, "Communitas is a fact of everyone's experience, yet it has almost never been regarded as a reputable or coherent object of study by social scientists. It is, however, central to religion, literature, drama, and art, and its traces may be found deeply engraven in law, ethics, kinship, and even economics."[32] Like other liminal phenomena in being viewed as beneath repute, juvenile, aesthetically undervalued, or even incoherent, this literature participates in what we have called the rubbish function: it can transform or unleash powerful energies by giving them space to be articulated (Thompson 1979). Such literature is likely to be "anti-structure," in the sense that it provides "not a structural reversal . . . but the liberation of human capacities of cognition, affect, volition, creativity, etc., from the normative constraints incumbent upon occupying a sequence of social statuses" (V. Turner 1982, 44).[33] In an era where J. K. Rowling, the author of the *Harry Potter* novels, is now wealthier than the queen of England, and where graphic fictions appear in the *New Yorker* and *Salon.com*, we do well to pay attention to all sorts of liminal literatures, for they are clearly having a huge social impact.[34] As they help us explore the meanings of the newly engineered human life span for our relations with each other, with other species, and with the earth, these literatures fulfill an increasingly important social function: reflecting the plots, possibilities, and perils of our liminal lives.

Notes

INTRODUCTION

1 "U.S. Sued over Stem Cell Research," *InfoBeat*, 9 March 2001, http://news@infobeat.com.

2 There is one standout in the list of othered entities gleaned from the metaphors in the *InfoBeat* report. In the dictionary, "cull" carries connotations of both positive and negative selection: thus its list of meanings includes "to choose from a number or quantity; to select, pick. Now most frequently used of making a literary selection," as well as "any refuse stuff; as, in bakeries, rolls not properly baked." *The Compact Edition of the Oxford English Dictionary*, vol. 1 (Oxford: Oxford University Press, 1976), s.v. "cull."

3 Indeed, Latour's deconstruction of the notion of modernity suggests that this nature/culture opposition may never have held true. The naturalization of technology simply enables us to posit a preexisting nature to our current culture. What has changed is less the *reality* of our constructedness as human beings than our ability to *perceive* that reality.

4 The hopes for successful therapies for Parkinson's disease based on injections of fetal cells into the brains of disease sufferers were dashed in March 2001, when the *New England Journal of Medicine* reported that a controlled study of the procedure not only failed to demonstrate any benefit but showed an alarming and steep increase in uncontrolled movements by patients receiving the treatment. "In about 15 percent of patients, the cells apparently grew too well, churning out so much of a chemical that controls movement that the patients writhed and jerked uncontrollably." Gina Kolata, "Parkinson's Research Is Set Back by Failure of Fetal Cell Implants," *New York Times*, 8 March 2001.

5 The liminal lives that are my focus share the in-betweenness of Latour's hybrids, as well as their origin in the habit of erecting epistemological and taxonomic categories (science/culture or culture/nature; but also human/animal, alive/dead, valuable/worthless).

6 In so doing, her work contributes to the study of agnatology, the study of "how ignorance has been understood, created, and ignored, [linked] also to allied creations of secrecy, uncertainty, confusion, silence, absence, impotence, etc.—especially as these pertain to scientific activities and outcomes." Robert Proctor and Londa Schiebinger, organizers, description for "Agnatology: A Cultural Politics of Ignorance," conference held at Pennsylvania State University, 25–26 April 2003. As one of the participants in that conference, and a member of the Science, Medicine, Technology, and Culture group at Penn State along with Schiebinger and Proctor, I understand *Liminal Lives* as a contribution to the emerging literature on agnatology.

7 Ann Curthoys and John Docker, "Is History Fiction?" manuscript courtesy of the authors.

8 They are citing Hayden White (1978).

9 In *Outside Literature*, Bennett refutes a view of "the role of literature in the constitution of social relation [as] . . . essentially epiphenomenal" (Bennett 1990, 108).

10 Bawarshi illustrates this point by considering the Patient Medical History Form, whose genre enables "the patient and doctor [to] reproduce the sociorhetorical conditions within which they interact. For instance, the genre reflects how our culture and science separate the mind from the body in treating disease, constructing the patient as an embodied object" (Bawarshi 2000, 354).

1. LITERATURE FOR FEMINIST SCIENCE STUDIES

1 Asimov and Asimov 1980, 113. As the authors explain, specialized fiction "falls into a particular category and usually has its own dedicated readers."

2 "The apparent drop in the sperm count is so sudden and steep that it has caused some scientists to wonder whether the human species is approaching a fertility crisis. . . . Some scientists are calling for bans on chemicals that may inhibit sperm production; others claim that there is not yet enough evidence even to know whether the sperm count is actually declining, much less that the human race is edging toward extinction" (Carlsen et al. 1992, 612–13).

3 The work of Michel Foucault has sensitized us to the three modes or practices by which human beings are turned into subjects: "classification practices, dividing practices, and self-subjectification practices" (Katz 1996, 17; borrowing from Dreyfus and Rabinow).

4 Although institutional consolidation always follows a period of more

diffused but sustained work, during which an area is coming into being, the field of literature and science in the United States can be traced back to the foundation of the Society for Literature and Science and the establishment of the Division of Literature and Science of the Modern Language Association, respectively. In Britain, it was formalized with the first lecture on literature and science sponsored by the Royal Society, the Royal Society of Literature, and the British Academy (Beer 1996; see also Collini 1993).

5 Of course, Haraway was not trained, nor does she teach, in a department of literature. Notably absent is the important work of literary scholars who do have that institutional training and affiliation: Gillian Beer and N. Katherine Hayles (Lederman and Bartsch 2001).

6 As series editors Philip Thurtle and Robert Mitchell describe the series: " 'In Vivo: the cultural mediations of biomedical science' will focus specifically on the relationship between cultural mediations and the content, practice, and applications of biomedical science while fostering an awareness of the importance of the interrelationship of biological, technological, and medical disciplines in western society. Specific subjects could include the uses of rhetoric in biomedical research and applications, the use of film in medical analyses, changes in conceptions of human embodiment resulting from changes in representational practices, the application of virtual reality technologies to medicine, the relationship of genomics to informational processing, or the institutions and rhetorical 'technologies' that enable organ donation in a consumer society" (In Vivo proposal, courtesy of the series editors).

7 Evelyn Fox Keller, *Reflections on Gender and Science* (New Haven: Yale University Press, 1985), and *Secrets of Life, Secrets of Death: Essays on Language, Gender, and Science* (New York: Routledge, 1992). The change in titles between these two works—the addition of the term "literature"—signals the increasing prestige of discourse analysis as a methodology within feminist science studies, a change that Keller herself has helped bring about.

8 Harding (1986) describes the origin myth of science as follows: "The origins myth for our scientific culture tells us that we came into existence in part through the kind of critical thought about the social relations between medieval inquiry and society that is subsequently forbidden in our scientific culture. This is a magical—perhaps even a religious or mystical—conception of ideal knowledge-seeking. It excludes itself from the categories and activities it prescribes for everything else. It recommends that we understand everything but science through causal analyses and critical scrutiny of inherited beliefs" (36).

9 Literature scholars Laura Otis and Richard Nash, and rhetoricians Richard Doyle and Celeste Condit, are increasingly contributing to the analysis of literature and science, but here again their work has generally not been taken up by scholars working in feminist science studies.

10 Several responses concern debates about the statistical models used by

Carlsen et al. to interpret their data. Three are concerns about the search method used to sample the historical data (the reliance on Medline and Index Medicus, which leave out "books and reports and other grey literature"), critiques of the statistical models used to analyze the aggregate data, and concerns about bias in the selection of patient populations (Bromwich et al. 1994, 19; Farrow 1994, 1). For an extensive recent review of the scientific and popular press response to the Skakkebaek article, see Krimsky 2000.

11 Continuing to trace the transformations to which Carlsen's article is subjected as it moves beyond the disciplinary boundary into popular press venues such as the *New York Times*, *Time*, the *New Scientist*, and even the *New Yorker*, I find a range of responses, from straightforward reporting of the debate generated by the article (discussions of statistical methods, sampling techniques, relationships between sperm counts and fertility, and the possibility of a link to environmental toxins) to analyses that cross the disciplinary divide, noting the convergence of James's fictional scenario with scientific findings of testicular anomalies and sperm motility decline. The convergence of literature and science is noted in Vines 1995 and Wright 1996.

2. THE CULTURED CELL

This chapter is in memory of Ian Crispin.

1 P. White 1954, vi–vii.

2 The decision would disappoint "thousands of people awaiting transplants. . . . It will also be a great disappointment to the Cambridge-based company, Imutran, the world leader in so-called xenotransplantation, which might now lose its pre-eminence to American, Japanese or Italian rivals." Jeremy Laurance, "Transplant of Pig Hearts to Be Banned," *Times* (London), 16 January 1997, 1.

3 Ibid.

4 There was a press outcry in August 1996, when nearly four thousand human embryos were destroyed in order to comply with the Authority's five-year storage deadline, as Sandra Goldbeck-Wood (1996) reported.

5 A good survey of the uses of tissue culture is Philip White's *Cultivation of Animal and Plant Cells* (1954). As a hypothetical technique, tissue culture dates back at least to 1812, when physiologist Le Gallois observed that "if one could substitute for the heart a kind of injection . . . of arterial blood, either natural or artificially made, . . . one would succeed easily in maintaining alive indefinitely any part of the body whatsoever" (Carrel and Lindbergh 1935, 621–22).

6 Advisory Group 1997, 163.

7 Harrison's full paper on the subject was published in 1910, and according to J. A. Witkowski, the tissue culture technique spread through Europe in the 1910s and 1920s (Witkowski 1979; Fell 1936; Carrel and Lindbergh 1935, 622).

8 Russell 1969. The cases of xenotransplantation performed to date cover a wide range of organ and tissue culture practices. Solid organs such as the heart, kidney, liver, and lungs have been transplanted from chimpanzees, pigs, baboons, and sheep into humans. Other animal-to-human tissue transplants have included the pancreatic islets of Langerhans from fetal sheep, neural tissue from fetal pigs, and baboon bone marrow. Skin, corneas, and blood from other species, although not currently being transplanted, are all being studied as potential xenotransplantation possibilities (Advisory Group 1997).

9 "Tissue cultivated in vitro shows two forms of growth—sometimes called *unorganised growth* and *organised growth* respectively" (Fell 1936, 3). This was one of a series of lectures on tissue culture delivered by Dr. Fell between 1936 and 1938.

10 See Catherine Waldby and Robert Mitchell, "Tissue Economies" prospectus, courtesy of the authors.

11 As the chairman's foreword to *Animal Tissue into Humans* explains, "Those who need tissue for transplant, whether it be a solid organ such as a heart [or a lung] . . . or cells such as pancreatic islets . . . may look to xenotransplantation as, at last, the answer to their needs" (Kennedy 1997, vii).

12 See Featherstone and Wernick 1995; Katz 1996. The social sciences are also involved in this reconceptualization of the life span. For example, "experimental gerontology . . . tries to manipulate natural aging to extend the life span of experimental animals and eventually that of man" (Medvedev 1991, 9–17, 10).

13 The cloning of Dolly has catalyzed a reevaluation of how age is measured: "Our 7-month-old lamb actually has a 6-year, 7-month-old nucleus in all her cells. It's going to be interesting to see what happens with the aging of this animal," notes Grahame Bulfield, director of the Roslin Institute. "Ewe Again? Cloning from Adult DNA," *Science News* 151 (1 March 1997): 132. See also Kaplan and Squier 1999.

14 By the term "paraliterature," I refer to those genres of fiction and nonfiction that are not authorized as part of the literary canon but nonetheless (and perhaps precisely because of their nonauthorized status) play a major role in expressing the biomedical unconscious of a culture. The relationship between literature and science is reciprocal rather than unidirectionally mimetic because literature functions not only to consolidate scientific fields and assimilate scientific discoveries but frequently to set the epistemological agenda for science. For an extended analysis of the two-way traffic between literature and science in the construction of the new field of reproductive technology, see Squier 1994.

15 Hunter 1991.

16 Holquist 1989, 21.

17 For analyses of the role of representation in shaping "facts" in medicine

and bioethics, see Donnelly 1996; Chambers 1996; Poirier, Rosenblum, and Ayres 1992.

18 Fell 1936, 1.

19 As E. D. Strangeways (the scientist's widow) later put it, he dedicated himself to "the study of living cells by means of tissue culture *in vitro* and *in vivo*, a field in which he was one of the chief pioneers." *History of the Strangeways Research Laboratory (Formerly Cambridge Research Hospital), 1912–1962*, in archives of the Strangeways, Contemporary Medical Archives Centre: SA/SRL/J.3., 12. (Hereafter Contemporary Medical Archives Centre is abbreviated CMAC.)

20 "T. S. P. Strangeways," obituary, *Lancet*, 1 January 1927, 56.

21 E. D. Strangeways, "1905–1926," in *History of the Strangeways Research Laboratory (Formerly Cambridge Research Hospital), 1912–1962*, 7–12; L. Hall 1996, 40.

22 "So far as I know, your lab is the only one in this country devoted entirely to tissue-culture." Letter from N. M. Hancox of the Department of Cytology and Histology at the University of Liverpool to Honor Bridget Fell, 25 November 1932.

23 Dusa Waddington McDuff, personal communication.

24 It is the unorganized nature of the growth, not whether the tissues are embryonic or mature, that provides the tissue with its potential immortality. Both embryonic tissue in unorganized growth, and cancerous tissue, result in the similar process of outgrowth, or as Waddington put it, "the induction of a cancer can be homologized with the induction of an embryonic tissue" (Waddington 1936b, 111). The continually growing tissue can be subdivided, with new cultures placed in fresh culture medium, and for certain tissues the process can continue, Fell observes, "indefinitely." Fell's definition is helpful here: "tissue undergoing unorganised growth *loses its normal histological structure*. For example, kidney tubules undergoing organised growth merely spread out over the glass of the culture vessel as an indifferent sheet of epithelium . . . and completely loses [*sic*] the anatomical features of the original tubules. This loss of histological structure is sometimes known as *de-differentiation*" (Fell 1936, 3).

25 The phenomenon of scaling was identified by Kenneth Wilson, Nobelist for physics in 1982, and later elaborated in nonlinear mathematics by Benoit Mandelbrot. "Scaling, as Mandelbrot uses the term, does not imply that the form is the *same* for scales of different lengths, only that the degree of 'irregularity and/or fragmentation is identical at all scales'" (Hayles 1990, 166).

26 As N. Katherine Hayles has described it in her study of nonlinear dynamics in literature and science, "Chaos theory looks for scaling factors and follows the behavior of the system as iterative formulae change incrementally" (Hayles 1990, 170).

27 Fell 1937, 10.

28 Wells 1926.

29 Honor B. Fell's chapter "Cell Biology" in the official history of the Strangeways Research Laboratory reveals that the following different kinds of cells were cultured there: chick embryonic cells, mammalian sternum cells, ear, mammary gland, ovary, salivary gland, pancreas, hair, and teeth cells (animal of origin unspecified), human thyroid cells, fetal rat skin cells, chicken skin cells, rabbit red blood cells, rat cancer cells (rhabdomyosarcomata), human fetal lung cells, rat salivary gland cells, human cancer cells, and many varieties of bacterial cells. Fell, "Cell Biology," in *History of the Strangeways Research Laboratory (Formerly Cambridge Research Hospital), 1912–1962*, in archives of the Strangeways, CMAC: SA/SRL/J.3., 19–33.

30 In the passage from his textbook on tissue culture that I have used at the beginning of the chapter, Philip White echoes this stress on the dramatic and performative aspects of the technique: "Every student can not only see these things but have the thrill of preparing them for himself" (P. White 1954, vii).

31 Her image recalls an illustration from Charles Kingsley's *The Water Babies* in which two scientists gaze at a water baby in a jar. This image, I argue in *Babies in Bottles*, shaped the scientific maturation of zoologist Julian Huxley, for whom it invoked the scientific drive to control of the processes of life and death (Squier 1994).

32 In particular, feminists have critiqued the tendency of scientists to argue from experimental findings to the macrosocial context without taking into account the processes of social construction that—as sociologists and anthropologists of science document—are integral to the construction of scientific facts. See Harding 1986; Longino 1990; Spanier 1995; Keller 1992; Latour and Woolgar 1979.

33 As *Nature* described it in 1937, "the Strangeways Laboratory of Cambridge . . . has become one of the most distinguished places in the scientific world for the prosecution of morphogenetic studies." The passage continued, "Here the technical methods, tissue culture *in vitro*, tissue grafting, the growth of large explants under a variety of conditions, embryological experimentation, the registration of results by the Canti-cinema method . . . have been used abundantly. . . . Dr. Fell and the Strangeways Laboratory have placed British anatomists deeply in their debt." "Morphogenetic Factors of Bone," *Nature*, 19 June 1937, 1036.

34 "Strangeways Research Laboratory," clipping service extract from the *British Medical Journal* (London), 1936, CMAC: SA/SRL. The primitive streak—"a linear thickening, visible with minimal magnification, that lies in the head-to-tail axis of the embryo-to-be," is the first marker of developmental singleness, according to embryologist Clifford Grobstein (Grobstein 1988, 26). Among the researchers involved in embryo culture, according to Honor Fell,

were Mr. Michael Abercrombie, Dr. Aron Moscona, Dr. J. Grover, Dr. B. McLoughlin, Dr. L. Weiss, Dr. P. D. F. Murray, Dr. A. Hughes, and Dr. Fitton Jackson (Fell, "Cell Biology," 21–22).

35 Waddington 1936a, 812.

36 Waddington 1935, 606.

37 As Honor Fell put it, "probably my main contribution to science has been the development and application to biomedical research of the organ culture technique" (Vaughan 1987, 239). The unpublished paper was titled "Evaluation of Research," and Vaughan speculates it was written around 1982.

38 Dr. Fell summarized the work in this area undertaken at Strangeways in her contribution to the laboratory's history: "Progressive development of many different organs was obtained: ear rudiments (H. B. Fell), mammary gland (M. Hardy), ovary (P. N. Martinovitch), salivary gland (E. Borghese), pancreas (J. Chen), hair (D. H. Strangeways and later M. Hardy), and teeth (S. Glasstone)" (Fell, "Cell Biology," 21). As the *British Medical Journal* summarized the Strangeways Research Laboratory report of 1936, "Miss S. Glasstone has shown that whole tooth-germs from rat and rabbit embryos explanted before cusp formation had begun [to] form cusps *in vitro.*" "Strangeways Research Laboratory," clipping service extract from the *British Medical Journal* (London), 1936, CMAC: SA/SRL.

39 "Rats' Teeth Grown in Laboratory in Cell-Growth Study by British," *New York Herald*, 13 March 1936, Strangeways Research Laboratory File, PP/FGS/C.17 Spear: Strangeways 1936, item 2816, CMAC, Wellcome Institute for the History of Medicine, London.

40 "She" appeared in their midst just as Dr. Fell had read a telegram praising her as "a remarkable person in many ways. Not only is she justly celebrated for her princely generosity and sparkling intelligence, but she is also reputed to be one of the loveliest women in England" (PP/FGS C.18 Spear: Strangeways 1937, CMAC, Wellcome Institute for the History of Medicine, London).

41 The gender transgression involved in Honor Fell's work surfaced in headlines such as "Woman Scientist Cultivates Life in Bottles," *Daily Express* (London), 16 March 1936, by the special correspondent, who was probably Charlotte Haldane, science writer, novelist, and wife of the eminent geneticist J. B. S. Haldane.

42 As Dame Janet Vaughan has observed, "the earliest and pioneering work on radiobiology originated in the Strangeways. . . . It has been said 'British Radiobiology stems from the Strangeways and there must be hundreds of people who owe their education on cells and radiation to the Strangeways School' " (Vaughan 1987, 249).

43 During World War II, Strangeways researchers worked extensively on the problem of wound healing. They also carried out a series of experiments focusing on "the metabolism of cell cultures grown under different ex-

perimental conditions," under the auspices of the British Empire Cancer Campaign. However, the most prominent cancer-related work in the laboratory was probably the work in radiobiology, which was made the focus of weeklong courses in radiobiology mounted by the laboratory between 1947 and 1949.

44 Harrison Hardy, "Any Tin in the Sun?" *Daily Mirror* (London), 20 March 1937.

45 Norah Burke, "Could You *Love* a Chemical Baby? For That's What Science Looks Like Producing Next," *Tit-Bits*, 16 April 1938.

46 Ibid.

47 Powell 1940.

48 Dudley Barker, "Cameraman of the Medical World: Filmed Cells through a Microscope: Lord Horder's Tribute to Dr. Canti," *Evening Standard* (London), 9 January 1936, 2.

49 Foxon 1976; *Evening Standard* (London), 9 January 1938.

50 The normally staid *Lancet* enthused about the photomicrographic apparatus, describing it as "a marvel of ingenuity," and praised Canti for "bringing the behavior of tissue cultures within the range of ordinary vision." "Cinematograph Work," R. G. Canti, CMAC: SA/SRL 23/H.2 (Strangeways Research Laboratory Scientific 1932 Cinematograph Work).

51 "Scientists May Now Watch Living Connective Tissue Reproduce Its Ultimate Cells," *World*, 12 June 1921, 2nd news section; italics mine.

52 We can trace a lineage (and it *is* patriarchal) back from these male images at the other end of the tissue culture micrograph to Anton van Leeuwenhoeck's discovery of tiny animals in sperm, revealing what Lisa Cartwright has called the "masculine fascination with the exaggerated image of minuscule bodily organisms" (Cartwright 1995, 84).

53 "The Seer," *World Radio* (London), 11 February 1938.

54 "What We Miss!" *Birmingham Mail*, 16 February 1938.

55 These poems are held in the file PP/FGS/C18–19, CMAC, Wellcome Institute for the History of Medicine, London.

56 Waddington 1936a, 811–12.

57 For a discussion of two literary examples of the genre, see Squier 1991. For visual examples of the same genre, see Henderson 1991; K. Newman 1996.

58 "I should like to argue that a scientist's metaphysical beliefs are not mere epiphenomena, but have a definite and ascertainable influence on the work he produces" (Waddington 1969, 123). Waddington's daughter also reminisces of her father, "He liked 'all over' paintings with no particular focus (Sam Francis, Jackson Pollock) and the idea of diversity, and always told me how much A. N. Whitehead influenced him with his idea of different nexuses of events, and fundamentally distinct (but overlapping) points of view with no one better than any other" (Dusa McDuff, personal communication).

59 Waddington's daughter observes: "My father did believe in progress—

in fact he wrote a book, *The Ethical Animal*, which was (as far as I remember) an attempt to formulate what progress is from a scientific (evolutionist) point of view, though he may not have remained with that belief" (Dusa McDuff, personal communication).

60 Honor Fell, "Cell Biology," 22, 19.

61 A parallel exists to the powerful role played by narrative in medicine. As visualization technologies increasingly dominate modern medical practice, we may find a conflict between two distinct and different narrative models, the earlier drawn from literature and culture, and the more recent influenced by film. See Hunter 1991.

62 The shift from original to replicant as privileged object of science has been inflected in fascinatingly different ways by contemporary feminist theorists. Donna Haraway has written movingly in celebration of a new life built not on origin stories but on replication with a difference, while Vandana Shiva sees the movement to patent intellectual and genetic property (in order to replicate it) as part of the centuries-old enclosure movement, which engages in "piracy from the mothers and grandmothers of the third world" (Haraway 1991; Shiva 1997).

63 My thanks to Dusa McDuff for this observation (personal communication).

64 See Lesley A. Hall's invaluable article "The Strangeways Research Laboratory: Archives in the Contemporary Medical Archives Centre" (L. Hall 1996, 235).

65 "The Clarks literally retooled their animals' bodies and habits to make them conform to experimental procedures. . . . In an attempt to render a part of the rabbit body more easily observable under the microscope, they physically implanted an instrument for observing growth within the living ear" (Cartwright 1995, 100).

66 This anticipates—though due to its inbuilt hierarchy does not finally attain—contemporary representations of the multiple, nested human subject. As Dorion Sagan (1992) has observed, "The body is not one self but a fiction of a self built from a mass of interacting selves" (370).

67 For a sweeping, if at times triumphalist, survey of the cultural response to twentieth-century innovations in medical imaging, see B. Kevles 1997. See also Cartwright 1995; Treichler, Cartwight, and Penley 1998; and Waldby 2000.

68 A troubling intersection exists here between my research into the Strangeways Laboratory's role in the development of the technique of tissue culture and the most celebrated line of cultured tissue cells, the HeLa cell line. Henrietta Lacks's carcinoma was misdiagnosed (epidermoid carcinoma of the cervix rather than the actual, far more malignant, adenocarcinoma), leading to an untimely death in her case, as in the case of my friend Ian Crispin. See Landecker 2000.

3. THE HYBRID EMBRYO AND XENOGENIC DESIRE

1 Warnock 1985, 70.

2 *Final Report of the Human Embryo Research Panel,* National Institutes of Health, 27 September 1994, 95–96. Because the panel is also known under the name of its chair, Stephen Muller, Ph.D., president emeritus of Johns Hopkins University, it will be cited hereafter in the text as "Muller panel."

3 Freud [1925] 1959, 182.

4 As Michelle Cliff (1990) has pointedly observed: "Did you know . . . that Thomas Jefferson held the popular view that the Black race was created when Black women mated with orangutans? (I do not know where the original Black women were supposed to have come from)" (273).

5 Schiebinger 1993, 95. Schiebinger is citing William Cohen, *The French Encounter with Africans: White Responses to Blacks, 1530–1800* (Bloomington: Indiana University Press, 1980), 242. Schiebinger notes, "As Cohen pointed out, this was a rumor started by the English" (239).

6 Young 1995, 6.

7 "Hybridity as a cultural description will always carry with it an implicit politics of heterosexuality. . . . The reason for this sexual identification is obvious: anxiety about hybridity reflected the desire to keep races separate, which meant that attention was immediately focussed on the mixed race offspring that resulted from inter-racial sexual intercourse, the proliferating, embodied, living legacies that abrupt, casual, often coerced, unions had left behind" (Young 1995, 25).

8 Ibid., xii.

9 The term "xenogenesis" first received mass public attention when it appeared in Octavia Butler's *Xenogenesis Trilogy,* the science fiction trilogy exploring interspecies communication and reproduction. The three volumes of the trilogy are *Dawn* (New York: Warner Books, 1987), *Adulthood Rites* (New York: Warner Books, 1988), and *Imago* (London: Victor Gollancz SF, 1989). As Donna Haraway (1989) observes in one of the earliest and best analyses of Butler's trilogy, "Butler's fiction is about miscegenation, not reproduction of the One" (378–79).

10 Wells [1896] 1988, 76.

11 Dahl 1962.

12 Newman 1991, 92.

13 While Latour (1993) never explicitly defines hybrids, he speaks of them as "entirely new types of beings, hybrids of nature and culture." Though the examples he gives are predominantly networked combinations of natural and technical objects, the engineered conjunction of human and animal species, or of two divergent life-forms, is clearly also part of his understanding of hybridity (10, 1–2).

14 Barbara Duden (1993) has explored the simplifications built into the increasing use of this term, as have Evelyn Fox Keller (1995) and Richard Doyle (1997).

15 Pateman (1988) does not discuss how the sexual contract is shaped by racialization, but the work of Robert Young (1995) suggests that a similar contract of sexual access to the bodies of disenfranchised racial others is integral to the institution of slavery.

16 For a fictional exploration of this scenario, see Cussins 1999. My thanks to Charis Cussins and Angela Lintz for their comments on an earlier version of this essay, "Interspecies Reproduction: The Feminist Implications of Hybrids," delivered at the conference "Women, Gender, and Science," at the University of Minnesota, Minneapolis, Minnesota, 12–14 May 1995.

17 One way of gauging the influence of the context on the meaning of xenogenesis is to consider another sort of hybridity very much in the news currently: the prospect of creating human-animal hybrids through cloning to provide a source of tissue and organs that carries less risk of xenotransplantation rejection.

4. GIANT BABIES

1 The passage continues: "But growth is not single. It is multiple, the combined expression of developmental and incremental factors. Increase in cell number is as much growth as is increase in cell size. Increase in cellular specialization is as much growth as is increase in cellular segregation. The laws which govern proliferation, differentiation, organization, mass increment, and all or any other growth expression of common possession may be assumed to be the same for all growing objects regardless of their structural, functional, or other distinguished attributes" ("Introduction to Growth" 1937, 1).

2 Kingsland 1997, 420; Porter 1997, 108.

3 "The Physiology of Growth," review of *The Fundamentals of School Health*, by Dr. James Kerr, *Nature*, 19 February 1927, 269–70.

4 Although the volume begins with a survey of heredity that includes those staples of eugenic discourse, the Jukes and Kallikak families, still the *Nature* reviewer taxed Dr. Kerr with overemphasizing the environmental influences on growth while underestimating the genetic. Nonetheless the reviewer praises the book for its "denunciation of the general practice of treating averages as standards" and labels it "to all time a classic" ("Physiology of Growth" 1927, 270–71).

5 Of course, this general rule exists to be broken, by avant-garde, experimental, and parodic fictions. However, the exceptions prove the rule. But in general, their two crucial ingredients are protagonists who are human beings

rather than ideas, as Roger Shattuck has observed, and "a sense of time, of the dimension in which people grow and change and strive" (Shattuck 2001, 11).

6 Wells's novel offers a satiric portrait of modern scientific interventions in human (and animal) nutrition. As one critic describes the novel, "[In] *The Food of the Gods* (1904), technological change, symbolized as 'bigness,' is represented as not merely strategic but wholly desirable. Here, as in the subsequent series of blueprints for a better society, thinly disguised as novels, Wells endows his scientist rulers with moral as well as technological supremacy" (Haynes 1994, 182). Other allegorical readings of this novel seem to me similarly off-base, as for example David Smith's argument that "Wells's attacks on science and on the routine scientific establishment mind . . . are enhanced by the discussion of the symbolic Food, which we can take to be socialism" (D. Smith 1986, 71).

7 By this, I mean an approach that traces both kinds of knowledge back to the communities and practices in and by which they are formulated. In this chapter, I draw particularly on the first rule for studying literature and science, formulated in chapter 1: We study literature and science in action and not ready-made literature or science. To do so, either we arrive before the disciplines, facts/interpretations, and machines/texts are black-boxed, or we follow the controversies that reopen them.

8 Latour 1987, 249.

9 I draw the term "life strategy" from Zygmunt Bauman's important book *Mortality, Immortality, and Other Life Strategies* (Stanford: Stanford University Press, 1990), where it connotes those culturally and socially mediated activities and institutions that enable human beings to avoid recognition of our mortality through focusing on projects of bodily control and improvement (9).

10 Wells [1904] 1965, 16.

11 *The Compact Edition of the Oxford English Dictionary*, vol. 2 (Oxford: Oxford University Press, 1971), 2959.

12 Poovey 1993, 263, 275, 256.

13 Tufte 1983, 32–34.

14 Bauman 1990, 58.

15 *The Compact Edition of the Oxford English Dictionary*, vol. 1 (Oxford: Oxford University Press, 1971), 1119, col. 280.

16 As Bauman (1990) observed, "The rulers' *biographies* become *history*. Unlike the lives of ordinary mortals, who enter history, if at all, as *statistics*—depersonalized and 'demographized,' they will be considered worthy to be carefully recorded, studied, written and taught about, interpreted and reinterpreted" (9).

17 "An ensemble formed by the institutions, procedures, analyses and reflections, the calculations and tactics, that allow the exercise of this very specific albeit complex form of power" (Foucault 1979, 20).

18 Caterham, "one of the most promising of English politicians . . . taking the risk of being thought a faddist, wrote a long article in *The Nineteenth Century and After* to suggest the total suppression of the food" (Wells [1904] 1965, 68).

19 Charles Sedgwick Minot, *The Problem of Age, Growth, and Death* (1908), reprinted from *Popular Science Monthly*, June–December 1907.

20 As TelDB, the World Wide Web telomere information center supported by the NIH, explains, "Telomeres are the physical ends of linear eukaryotic chromosomes. They are specialized nucleoprotein complexes that have important functions, primarily in the protection, replication, and stabilization of the chromosome ends. . . . Shortened telomeres appear to lead to cell senescence. Eventually telomeric sequences can shorten to the point where they are not long enough to support the telomere-protein complex protecting the ends and the chromosomes become unstable. These shortened ends become 'sticky' and promote chromosome rearrangements. Some rearrangements may contribute to the development of cancers." *http://www.genlink.wustl.edu/teldb/teldb.html* (accessed 21 July 2003).

21 Travis 2000, 279.

22 "Robert P. Lanza of Advanced Cell Technology in Worcester, Mass., and his colleagues report that they've cloned cows from aged cells. They find that cells from the clones have longer DNA tips, or telomeres, than the original cells and show other signs of youthfulness" (Travis 2000, 279).

23 *Dictionary of Scientific Biography*, ed. Charles Coulston Gillispie, vol. 9 (New York: Charles Scribner's Sons, 1974), 416.

24 Gilford's work stressed the socially constructed nature of the boundaries between different stages of life and, implicitly, between different species as well. He contended that the fetal/human boundary is not so much biological as social, a product of the social context: "The foetus is, indeed, not a human being at all, but is simply a foetus. . . . Prior to birth the life of the foetus is equivalent to that of a lower animal, and is held of very little account, especially during the earlier months. The father of a family seldom grieves when his wife miscarries, and, indeed, nowadays, as we all know, many a mother has very little compunction in taking the life of her unborn child, provided the killing can be done at a fairly early stage of its career. She is, too often, glad to escape from the incubus of child-bearing at the cost of a mere miscarriage. But immediately after birth the foetus becomes a human being, and its life is prized accordingly, so that the mother, who would without scruple cut short the career of her pre-natal child, tenderly cherishes that of her post-natal child. Should she deliberately destroy it, she is regarded with horror as a murderess" (Gilford 1911, viii). The passage begins with a metaphor that once again invokes the parallel between literature (this time, sacred literature) and science (this time, human evolution): "[We] know perfectly well that human development does not end at birth. On the contrary, it may

be said only to begin at that epoch. The pre-natal period is no more than a genealogical introduction to the forthcoming chapters of development. Like the opening verses of the New Testament it has no direct bearing on the narrative, but consists solely in a condensed summary of that which happened in earlier days" (viii).

25 Charles H. Rector, "Crystals of Growth," *Amazing Stories* 2, no. 9 (December 1927): 874–77.

26 Philip Wylie, *Gladiator* (New York: Alfred A. Knopf, 1930).

27 Although Wylie's novel was published before the end of the war, it achieved cult status afterward, going into twenty printings by 1955. http://www.library.csi.cuny.edu/dept/history/lavender/momism.html (accessed 17 July 2003).

28 Wylie 1930, 4, 5.

29 Notice that the transgression of morphological norms is accompanied by a transgression of generic norms as well, erasing the distinction between this tale of a cat's death throes and the death of heroes in the epics of Homer.

30 Ed Earl Repp, "The Gland Superman," *Amazing Stories* 12, no. 5 (October 1938): 8–29.

31 Frank Patton, "The Test Tube Girl," *Amazing Stories* 16, no. 1 (January 1942): 16.

32 In a striking return to the theme of this science fiction short story, the Land Institute, an environmentalist agricultural research center in Salina, Kansas, dedicated to "natural systems farming," or sustainable low-impact agriculture, is experimenting with transforming hybridizing food crops such as wheat, corn, and oats to convert them from annuals to perennials in order to reduce the energy costs of food production. See Cox et al. 2002; *http://www.landinstitute.org/vnews/display.v/ART/2002/06/01/3dcbf8a7874a8?8?in_archive* (accessed 21 July 2003).

33 Cover page statement, *Growth, Development, and Aging* 52, no. 1 (1988).

34 Elliott 1998.

35 In addition to the giants that are the focus of the science fiction works I survey in this chapter, twentieth-century fiction is peopled by numerous dwarfs and midgets. One work that demands a more careful exploration for its representation of the social and ontological problem of growth is Walter de La Mare's *Memoirs of a Midget* ([1921] 1982). As Angela Carter observes in her preface to the 1982 edition, "*Memoirs of a Midget* is a minor but authentic masterpiece, a novel that clearly set out with the intention of being unique and, in fact, is so; lucid, enigmatic, and violent with the terrible violence that leaves behind no physical trace" (viii–ix). Like the meaning of gigantism in the science fictions of Wells and Wylie, the implications of de La Mare's protagonist's size, Carter suggests, are sexual and social rather than medical: "I should say that Miss M. herself, in her tiny, bizarre perfection, irresistibly reminds me of a painting by Magritte of a nude man whose sex is symbolized

by a miniature naked woman standing upright at the top of his thigh. . . . Miss M.'s actual size, therefore, is not within the realm of physiological dimension; it is the physical manifestation of an enormous *difference*" (xiv–xv).

36 Kolata 1986, 23; bracketed insertion mine.

37 Schulman and Sweitzer 1993, 62–63.

38 "If classic criteria for diagnosing GH deficiency were applied, this would appear to limit GH distribution to those children with a subnormal growth rate (typically <4 to 5 cm/y) and peak GH levels of less than 7 or less than 10μg/L after provocative testing" (Lantos et al. 1989, 1021).

39 The very concept of a normal growth curve is called into question by this FDA decision, which is aimed at the 400,000 children between ages seven and fifteen who meet those conditions. Eli Lilly predicted "that only 10 percent will receive the growth hormone because of eligibility restrictions and because six shots a week are required for years." Unlike Wells's careful demonstration of the unequal distribution of the food of the gods, Eli Lilly's brief statement makes no mention of the impact of class, race, sex, and geographic origin on access to Humatrope, though clearly those factors will influence the distribution of the drug and thus the resulting distribution of the change in height across human populations. Associated Press, "A Hormone to Help Youths Grow Is Approved by FDA," *New York Times*, 7 July 2003. For a discussion of the research into the height-enhancing and age-retarding effects of growth hormone, see Melvin Konner, "One Pill Makes You Larger: The Ethics of Enhancement," *American Prospect Online* 10, no. 42 (January 1999).

40 Science studies scholars N. Katherine Hayles (1993) and Brian Rotman (2000) have weighted the trajectory of this shift toward the digital and informatic.

5. INCUBABIES AND REJUVENATES

I thank Paul Brodwin and Stephen Katz for their comments on an earlier version of this chapter.

1 "Ape-Child?" *Time*, 16 August 1926, 16.

2 Young 1995.

3 Voronoff 1943, 111.

4 This is apparent in Voronoff's "The Grafting of Women," which appears in his 1943 volume *The Sources of Life*: "The woman had been a widow for twenty years. He had a blind son of thirty, from whom she wished never to be separated. Having no private income, she had succeeded in earning her son's living and her own by literary work, translations, newspaper articles, and so on. As age advanced, however, she became more and more easily tired, and steady intellectual or physical work became difficult. Old age came, bringing the breakdown of the body and general enfeeblement. To climb the stairs to

her sixth floor flat became a torture; working till late hours left her exhausted. Her earnings were diminishing, poverty was facing her, and she came to me with a cry of distress, asking me to restore her strength and her energy for work, in order that she might be saved from definitely sinking and being thenceforth unable to support her son. . . . The grafting of an ovary from a female chimpanzee on June 10, 1924, which I performed with the assistance of my friend, Dr. Dartigues, the well-known gynecologist . . . literally transformed this poor woman. I saw her again three years after the operation. Her figure had again become erect, her movements alert; the face no longer wore the expression of pain that made it look so old. . . . But what rejoiced the worthy woman most was that she was again able to climb lightly her six flights of steps, work twelve hours a day, and feel in herself a renewed strength which restored her moral courage to face the struggle for life. . . . I am profoundly grateful to her for having been the first woman to give me the opportunity to observe the changes that may result from the grafting of a young ovary in an aged woman's body" (224–25).

5 The *New York Times* for 1923 has extensive coverage of the controversy around gland grafting, particularly the work of Voronoff and his colleague Heckel. For a sample of the *Times* coverage, see "Paris Doctors Lift Ban on Voronoff," *New York Times*, 11 January 1923, 44:5; "Surgeon Revivifies Col. E. H. R. Green," 30 March 1923, 1:5; "Gland Treatment Spreads in America," 8 April 1923, 2:7, 8; "American Surgeons Active at Conference," 21 July 1923, 6:8; "Graft Gland on Horse," 30 July 1923, 13:4; "Rejuvenation Is Filmed," 9 October 1923, 17:1; "Doctors Who Scoffed Now Hail Voronoff," 13 October 1923, 15:3; "Declares Glands Do Not Rejuvenate," 14 October 1923, 6:1; "New Blood for the Aged," 23 November 1923, 3:6; "Old Timers Race in Prison: Show Up Well, Following Gland Transplanting Operations," 30 November 1923, 2:5; and, in the *Times Sunday Magazine*, M. B. Levick, "Pursuit of the Elixir of Life: Dr. Steinach's Predecessors Had Recipes for Changing Old Men into Young in 1600 B.C.," 9 September 1923, 4:9:1. For a wide-ranging popular discussion of these rejuvenation and age extension therapies, see McGrady 1968; Langone 1978; see also Lambert 1959; Voronoff 1928; Steinach 1940; Benjamin 1930.

6 Julian Huxley's survey of rejuvenation therapies, which appeared in 1922 in *Century Magazine*, was entitled "Searching for the Elixir of Life." In her survey of endocrine treatments, "Ain't Nature Gland!", Ruth F. Wadsworth (1929) assessed the rejuvenating effects of gland transplantation, under the heading "The Fountain of Youth," with the dismissive conclusion that "in the present state of our knowledge, operations for rejuvenation are only pitiful. Ponce de Leon found more in Florida than we can find in the operating-room" (48). The tropes extend to the present day, too. A newspaper story on the discovery of the gene that causes Werner's syndrome, a disease of premature aging, quoted the principal investigator, Gerard Schellenberg of the Seattle Veterans Affairs Medical Center: "A kind of Holy Grail of aging research has

been to find this gene" (Lauran Neergaard, "Scientists Discover Gene Linked to Aging," *Centre Daily Times*, 12 April 1996, 1).

7 Achenbaum 1995.

8 D. Kevles 1985, ix.

9 D. Kevles 1985, 251–68; Nelkin and Lindee 1995. See also Spanier 1995, 102–3.

10 For example, Frances Seymour, M.D., medical director of the National Research Foundation for Eugenic Alleviation of Sterility, Inc., enumerated the "dysgenic" traits that led her to rule out what she called "cross-artificial insemination" (or donor insemination) for infertile couples: "We refuse to aide [*sic*] a couple where the wife has an I.Q. less than 120 . . . or has any physical inheritable stigmata, any family history of inheritable disease or a short life expectancy." When she sent this letter to C. P. Blacker, general secretary of the British Eugenics Society, he wrote Lord Horder, president of the Eugenics Society, recommending against Seymour's scheme for artificially inseminating European women with semen from American donors: "the central project [is] . . . impracticable, uncalled for, and intrinsically absurd." Note that Blacker does not quarrel with Seymour's definition of dysgenic traits, only with her scheme for postwar repopulation through international donor insemination. Frances Seymour, M.D., "Artificial Insemination, Gynecology, Eugenics, and Their Relation to the Post–World War II Rehabilitation Plan"; C. P. Blacker, letter to Lord Horder, 2 November 1943, Contemporary Medical Archives Center, Wellcome Institute for the History of Medicine, London, England.

11 Witkowski 1987, 258.

12 Oudshoorn 1994, 20–21.

13 Lisser 1925, 14.

14 Ibid.; italics mine.

15 Oppenheimer 1967, 9; Keller 1995; Oudshoorn 1994.

16 Wolff 1986. See also Tiefer 1988.

17 Hirschfeld 1930, xii; Squier 1994, 102–5.

18 Benjamin 1930; Schmidt 1930.

19 Metchnikoff, whose brother was the subject of Tolstoy's "The Death of Ivan Ilych," first used the term "gerontology" in *The Nature of Man: Studies in Optimistic Philosophy* ([1903] 1908) (Achenbaum 1995, 23–25).

20 Achenbaum 1995; Katz 1996.

21 Katz 1995.

22 These miraculous accounts include the stories of supposed centenarian Luigi Cornaro or Englishman Thomas Parr, who supposedly lived to 152 years (Katz 1995, 62, 65). The miraculous, I would argue, did not so much disappear as transmute itself. With the advent of modern-age extension and rejuvenation technologies, the impulse fueling pre-mid-nineteenth-century tales of the miraculous was redirected to a more acceptably modern, scientific venue:

science fiction. So what had once produced tales of miraculous births, phenomenal longevity, and the fantasy of a marvelous eradication of death now produced stories of scientific wonders such as the novels and stories of H. G. Wells and the flood of science fiction tales appearing in magazines such as *Amazing Stories*.

23 See Squier 1994, 160.

24 Thus "self-identity and the body become 'reflexively organized projects' which have to be sculpted from the complex plurality of choices offered by high modernity without moral guidance as to which should be selected" (Shilling 1993, 181). As Brian Turner (1995) elaborates it, "that is, the notion that the self has a history with trajectories, with self-conscious lifestyles, with modes of operation and development: in this context therefore, the self is not to be taken for granted, not a fact of the person as it were" (255).

25 Schilling 1993, 44. By the middle of the nineteenth century, this focus on women's bodies was less a sign of ideological unanimity than the reverse. Thus Mary Poovey (1988) argues that "the representation of woman was also a site of cultural contestation during the middle of the nineteenth century. These contests reveal . . . the extent to which any image that is important to a culture constitutes an arena of ideological construction rather than simple consolidation" (9). See also Jordanova 1988; Schiebinger 1989. As I will discuss at greater length later, Nelly Oudshoorn (1994) has documented how doctors found easier access to a female than a male population pool, owing largely to the increasing medicalization of childbirth and the social acceptability of viewing the incarcerated female body as research reservoir (80).

26 For Freud, there seems to have been something scientistic about the very workings of fantasy. Removed from the life course, it nonetheless was a distillate of the very essence of psychic life, not unlike Alexis Carrel's 1930s experiments with tissue culturing that were so prominent a part of both reproductive technological and rejuvenation therapy discourse. "Freud presents phantasy as a unique *focal point* where it is possible to observe the process of *transition* between the different psychical systems *in vitro*" (Laplanche and Pontalis 1973, 316). The association of old age with childhood, in visual and verbal imagery, is a prominent characteristic of Western culture, as Jenny Hockey and Allison James (1995) have demonstrated. They argue that we choose to link the aged person metaphorically with the acceptably dependent child in order to "shield [ourselves] from the approaching vision of illegitimate social dependency in old age" (143). A similar link between child and aged person at times even functions as an uninterrogated trope in aging studies itself, as in the following passage: "Today there is a social construction of human lives which consists of the mild morning mist of childhood, the stage of education or training, the parallel stages of work and leisure, the stage of active retirement, and the dusk of old age" (Dahrendorf 1979, 92; cited in Conrad 1992, 68). My particular interest, however, is in the way that control

of embryological and fetal development is constructed as parallel to control of the aging process, because beginning-of-life and end-of-life changes are understood in relation to each other. Rejuvenation therapies—as the attempt to exert medical scientific control over the aging process—thus represent the other, untold, half of the story that I told in *Babies in Bottles* (1994).

27 McClure 1979, 98; [Snow] 1933. As Atherton describes the Steinach rejuvenation technique in *Black Oxen*: "The concentration of powerful Roentgen—what you call X-Rays—on that portion of the body covering the ovaries" (Atherton 1923, 138).

28 I draw my methodological model here from Latour 1987.

29 The passage continues, "Whatever the validity of his claims, whatever the truth of criticism raised by the conservatives of the medical world against Steinach and against the gland transplantations of Dr. Serge Voronoff, the idea which they have brought forth in scientific terms has seeped into the popular mind like water into sand. For the mind of man has thirsted for this secret for untold centuries, has evolved from it legends, myths, heroes, whole religious systems, creating in manifold form the hope that is built on the desire to become as immortal as the sun" (*New York Times Magazine*, 9 September 1923, 4:9:1).

30 "Voronoff and Steinach," *Time*, 30 July 1923, 19–20. Although Cartwright doesn't explore it extensively, an additional ingredient in this filmic mix was modern nationalism. This impulse was apparent on 9 October 1923, when the *Times* reported the inaugural screening of "a motion picture film of the scientific work of Dr. Eugen Steinach of Vienna" made at the urging of none other than Dr. Hainisch, president of the Austrian Republic.

31 "Doctors Who Scorned Now Hail Voronoff," *New York Times*, 13 October 1923, 15:3.

32 "Declares Glands Do Not Rejuvenate," *New York Times*, 14 October 1923, 6:1.

33 The title comes from Yeats's "Countess Cathleen" (Bodeen 1982). As Alasdair D. F. Macrae (1995) reports: "In 1934 [Yeats] was sufficiently worried about a diminution of vigour, including sexual, that he arranged to undergo a surgical operation. The Steinach operation was widely rumoured to transplant glands from monkeys into the patient (hence the Dublin joke about Yeats as the Gland old Man!); in fact, what he had was a vasoligature and vasectomy of some sort. Whatever the sort, he immediately felt a surge of physical and psychological energy" (118).

34 "Censorship Up in Far Monroe," *New York Times*, 4 October 1923, 22:6.

35 The passage continues: " 'Why isn't your head turned?' Clavering asked her one day when the sensation was about a month old . . . 'You are the most famous woman in America and the pioneer of a revolution that may have lasting and momentous consequences on which we can only speculate vaguely today' " (215–16).

36 I disagree here with Lois Banner's reading of the novel, which unaccountably sees it as depicting "the emotional damage caused by . . . the inevitable failure of [an affair with a younger man]" (Banner 1989–1990, 15).

37 De la Mothe 1992, 132–36.

38 Snow's optimism on this last point anticipates his argument in *The Two Cultures* that there should be more traffic between science and literature, for it implicitly constructs both disciplines as working by accumulation; science by an accumulation of experimental data, and literature by accumulated insight. As Pilgrim and Callan agree in *New Lives for Old*, science "simply . . . must get better. It's the only organised human activity . . . and so it's bound to get more complete, whoever does it. So long as someone does it." And literature improves because "consciousness is growing every day. It's growing by leaps and bounds since the discovery. That is the real effect of rejuvenation" (319, 321).

39 For a balanced look at the new promotion of testosterone therapy, see Greider 2003. I want to stress the surprising nature of both aspects of this phenomenon: the predominant focus on women in the medical applications of hormone therapy, and the reframing of this not as *rejuvenation* but as health maintenance. Not only does it contradict Snow's stress on male rejuvenation, but it diverges from what social historian Lois Banner (1989–1990) has characterized as "the constant leitmotif that aging men, as well as women, experience a climacteric." As she observes, "the notion of a 'grand climacteric' as a dangerous stage of life for men (occurring around the age of sixty-three) had been current for centuries, dating to Renaissance theorizations about life cycle development" (7).

40 As Oudshoorn (1994) observes: "In the 1920s and 1930s, the female body became the major object for hormone therapy. Female sex hormones became applied as universal drugs for a wide array of diseases in women. In this manner, sex endocrinologists constructed the image of the hormonal woman: it was the female body that became increasingly subjected to hormonal treatment. Compared to women, the introduction of sex hormones had rather minor consequences for men. Although endocrinologists created a market for male sex hormones, male sex hormone therapy was introduced for only a relatively small number of medical indications" (110).

41 Altman 1996. Francis Ford Coppola's *Jack*, appearing in summer 1996, capitalized on the surge of interest in the hormonal bases of aging that accompanied the discovery of this gene, giving the public the story of "a 10-year-old with a disease that makes his body age four times as fast as his mind" (Chris Hewitt, "*Jack*: Coppola a movie mismatch," *Centre Daily Times*, 9 August 1996, 15c).

42 Arguably, the popular press's romanticization of the plight of infertile women has made it possible for a wide range of research activities, from embryo experimentation to fetal tissue culturing, to garner funding and support under the aegis of "reproductive" technology, while the press hoopla

around unsubstantiated claims of rejuvenation and the resistance to notions of extending a woman's nonreproductive sexuality has made it necessary for endocrinology and gerontology to drastically de-emphasize the rejuvenation aspects of their project in order to maintain scientific status. Here I am drawing on Bruno Latour's 1987 analysis of the process of accumulating allies for a scientific project, especially chapter 4.

43 For a discussion of the debates between feminists about reproductive technology, see my introduction in Squier 1994.

44 As Oudshoorn (1994) points out, "It was only after 1929 that scientists could assess the identity of sex hormones with chemical methods, thanks to developments in organic chemistry in the area of steroid and lipoid compounds. Sex hormones—classified as steroids—could now be chemically identified and isolated (Long Hall 1975). Female sex hormones were first chemically isolated from the urine of horses and pregnant women in 1929. In 1932, English and German chemists classified female sex hormones as steroid substances, and a calorimetric test was developed to detect the presence of female sex hormone in organisms (Walsh 1985). Male sex hormones were first isolated from men's urine in 1931 and were classified two years later in the same group of chemical substances as female sex hormones: the steroids" (29).

45 Powell 1940.

46 William E. Schmidt, "Birth to a 59-Year-Old Raises British Ethical Storm," *New York Times*, 29 December 1993, 1, A6; "Why She's 61 and Pregnant," *Newsday*, 29 December 1993, 6; "Then: *Vanity Fair*; Now: *Modern Maturity*," *New York Times*, 3 January 1994, A22.

47 Those of us interested in the medical treatment of aging may be moved to repeat a question first posed in 1993, after the rash of postmenopausal pregnancies. We could even dedicate this question to Nora the chimpanzee, gland graft recipient and potential postmenopausal mother: "if women can have ovarian transplants, will it be even less acceptable for them to grow old without fighting it through medical intervention?" This question was first raised by Dr. Susan Sherwin, professor of philosophy and women's studies at Dalhousie University in Nova Scotia. As Dr. Sherwin has observed, "An enormous industry has grown up in recent years to postpone or prevent menopause through hormone replacement therapy; now reproductive life can also be prolonged. There are questions of what we value in women" (Gina Kolata, "Reproductive Revolution Is Jolting Old Views," *New York Times*, 11 January 1994, C12).

48 July 9, 2002, saw the early discontinuation of the Women's Health Initiative Study of combined hormone therapy in 16,608 postmenopausal women, when it became clear that the risks of hormone replacement therapy outweighed its benefits. This well-publicized event produced a full-scale medical retreat from the wisdom of hormone replacement therapy for all but the

most intractable symptoms of menopause (Humphries and Gill 2002, 1001). See also Laurie Barclay, M.D., "More Evidence Linking Estrogen Plus Progestin Therapy to Breast Cancer," *Medscape Medical News 2003*, *http://www.medscape.com/viewarticle/457781* (accessed 22 July 2003).

49 Oudshoorn 1996, 129; See also S. Niemi 1987.

50 Jane E. Brody, "Restoring Ebbing Hormones May Slow Aging," *New York Times*, 18 July 1995, C1, C3; "Growth Hormone Fails to Reverse Effects of Aging, Researchers Say," *New York Times*, 15 April 1996, A13.

51 Squier 1996, 530.

52 Jane E. Brody, "Hormone Replacement for Men: When Does It Help?" *New York Times*, 30 August 1995, C8.

6. TRANSPLANT MEDICINE AND NARRATIVE

1 Jane Wildgoose, "Who Really Owns Our Bodies?" *Guardian Unlimited*, 30 January 2001.

2 "The Von Hagens Interview," *Science Interviews*, http://www.scienceinterviews.com (accessed 16 November 2001).

3 Celia Hall, "Organs Scandal: Children's Heads Were Found in Hospital Archive," *Daily Telegraph*, 31 January 2001.

4 "Frequently Asked Questions," LifeGem Web site, http://www.lifegem.com/secondary/faq.htm. My thanks to Melissa Littlefield for alerting me to this remarkable new manipulation of cadavers, which interestingly breaches the human/animal barrier to memorialize pets as well as relatives and loved ones.

5 Scheper-Hughes et al. 2000.

6 Trull 2003, 1. For an extended discussion of the notion of human body parts and organs as transferable, see Andrews and Nelkin 2001.

7 The defense also argued that the Royal College of Surgeons was itself guilty of retaining anatomical specimens longer than the three years permitted by the Anatomy Act of 1984, and disposing of them improperly. The Anatomy Act of 1984 "allows bodies and body parts to be preserved for three years and used for medical purposes by people and premises licensed by the Department of Health. (Any other use is a crime.) After that period the remains are supposed to be given a proper burial. Also, a body can only be used if the person has given permission in their will" (Walker 1999, 225).

8 See von Hagens (2003), "Preservation by Plastination," on his scientific Web site, *http://www.kfunigraz.ac.at/anawww/plast/pre.html*.

9 The translation is courtesy of Google.com. Von Hagens (2001), "Current exhibition: 'Body Worlds. The fascination of the genuine one,'" *http://www.koerperwelten.com/berlin.hor* (accessed 30 April 2001).

10 "Science Interviews.com—the Von Hagens Interview," *http://www.scienceinterviews.com/generic.jhtml?pid=6*.

11 Ibid.

12 The phrase was used by Jane Wildgoose.

13 Celia Hall, "Organs Scandal: Children's Heads Were Found in Hospital Archive," *Daily Telegraph*, 31 January 2001. Note the intrusion of juvenile literature, as well as fantasy, into the labeling, which was felt to be particularly reprehensible. Textualization of the deceased also operates in the remediation suggested by the Donaldson report: "Where tissues or organs have been donated for teaching purposes, families will be invited to prepare a 'life book' on the person who has died. This would take the form of a scrap book of the child's life 'so students would be reminded that [the organ was] part of a real child, and so afforded appropriate respect." "Science and Medicine: 105,000 Body Parts Retained in the U.K., Census Says," *Lancet* 357, no. 9253 (3 February 2001).

14 "Science and Medicine," *Lancet*, 3 February 2001.

15 "British Organ Scandal Physicians Face Investigation," *Reuters Medical News*, 15 March 2001.

16 Barwick 2001b; see also 2001a.

17 Laurence 2001; Batty 2001.

18 Bunyan 2001; Laurence 2001, 2.

19 As Walker (1999) observes, "One issue raised by the Kelly case was the legitimacy of casting as an artistic technique. Casting is an ancient reproductive method and a historic method of manufacturing editions of sculptures. It is also an ancient method of producing three-dimensional records of the appearances of the dead. Yet in the nineteenth century it was considered suspect as a means of achieving first-order art. . . . It is also worth remembering that casts of body parts are made by medical practitioners and preserved in medical museums" (224).

20 "BMA Response to Alder Hey Inquiry Report," BMA Online Press Centre, 30 January 2001, issued by BMA London Office, *http://web.bma.org.uk/pressrel.nsf*.

21 "Each specimen container will bear a unique identifier linking it to the catalogue, but not normally to the deceased's identity (separate records may provide this linkage)" (RCP 2001).

22 Gawande 2001, 94.

23 That this is not the case in Japan has been studied by Margaret Lock (2002), Trevor Corson (2000), and Emily Ohnuki-Tierney (1994).

24 "Renal Transplant News," Saint Barnabas Health Care System, *www.sbhcs.com/services/renal/newsletter/transplantation.html* (accessed 17 July 2001). As Swazey and Fox observed in 1992: "The number, variety, and combination of solid organs and other body parts transplanted during the 1980s, along with the array of extracorporeal and implanted devices in regular use, being tested, or being designed has brought our society closer to the world of

'rebuilt people' *classically portrayed in science fiction*, in which humans are more and more composed of transplanted parts of one another, and of 'man-machine unions' that 'prosthetize' humans and humanize man-made organs" (xv; italics mine).

25 Terasaki 1990; Starzl 1992.

26 Susan L. Smith, R.N., Ph.D., "Progress in Clinical Organ Transplantation," *Medscape*, 1 January 2000, 2.

27 Ibid., 10; Swazey and Fox 1992, 24.

28 June 2003 saw some dramatic improvement in organ toleration with the development of a new, "tolerogenic" protocol for transplantation developed by Thomas Starzl. Following this protocol, the organ recipients have their T cells depleted with thymoglobulin intravenously before their lung transplant surgery, and thus postoperatively they can reduce the level of immunosuppressant drugs they must take: taking "tacrolimus—in some cases, on a less-than-daily basis—and low-dose prednisone . . . without an increase in organ rejection rates." Martha Kerr, "Tolerogenic Protocol Reduces Immunosuppressive Drugs after Lung Transplant," *Medscape*, 9 June 2003.

29 As David Edelstein observes, "The screenwriter, Steven Knight, is an English TV veteran who helped to create the original *Who Wants to Be a Millionaire?*, and the grasping/yearning quality of that infamous title is in every scene: The film is, *Who Wants to Be a Citizen—and What Part of Your Anatomy Will You Give Up?" Slate*, 18 July 2003, *http://slate.msn.com/id/2085813/* (accessed 24 July 2003).

30 Scheper-Hughes et al. 2000.

31 Mooney 2001.

32 Harrison 1999.

33 Awaya 1999.

34 Ross 1991, 108–9.

35 Alexander 1927, 1073.

36 "Consensus Statement on the Live Organ Donor" 2000.

37 Bowers 1931, 923.

38 "They are hardly 'corpses' in the traditional sense. Although they are 'dead patients' they do not resemble our other dead patients. The expression 'brain dead' is accurate but seems to avoid the crucial issues. Most would agree that these donors are no longer 'persons.' When the patient is admitted to the operating room, the recorded diagnosis is 'beating-heart cadaver'—a term that is offensive to many people. Gaylin coined the term 'neomort' ten years ago, but it has not become popular. *Perhaps we will only be able to give these artificially maintained organ donors an appropriate name when we ourselves have made the necessary emotional and cultural adjustments*" (Youngner et al. 1985, cited in Swazey and Fox 1992, 62–63; italics mine).

39 An alternative interpretation for this story would emphasize the hand's threatening liminality, understanding Van Puyster's death as a voluntary

murder of the Negro in himself, or to be more precise of his own racial indeterminacy.

40 Cohen 1999.

41 Altman 1998.

42 Steve Bailey, "Proposed Hand Transplant Ignites Medical and Ethical Controversy," *Los Angeles Times*, 18 October 1998, 29.

43 "Hand Transplant Surgery Prompts Debate," *American Medical News* 41, no. 48 (28 December 1998), 31.

44 Thomas H. Maugh II, "Science File," *Los Angeles Times*, home edition, 15 October 1998, 2.

45 "Hand Transplant Surgery Prompts Debate," 2.

46 Ibid. Note the ironic reversal of Bowers's story: now it is not the *donor* of the hand, but the hand recipient, who is the criminal. Thanks to Gretchen Helmreich for this observation.

47 Lemonick 1998.

48 Ibid.

49 Associated Press, "Giving These Doctors a Hand: Rare Surgery Reacquaints New Zealander with Fingers," *Boston Globe*, city edition, 16 October 1998.

50 Altman 2000.

51 One of the team of surgeons, Dr. Nadley S. Hakim, disclosed the transplant recipient's name and current prognosis in an interview in July 2003: "Denis Chatelier, who is the first ever person in the world to receive a double hand transplant, is doing very well since his transplant in January 2001. He is leading a normal life, back to his usual job in a factory, which involves using his hands. His function is back to over 65% of normal, which is a very reasonable result" (Barclay 2003).

52 "Transplanted Hand Amputated," *New York Times*, 4 February 2001.

53 A previous hand transplant was carried out in Ecuador in the 1960s, but the recipient lived "less than two weeks," leading the press overwhelmingly to refer to this as the first successful hand transplant. Thomas H. Maugh II, "Science File," *Los Angeles Times*, 15 October 1998, 2.

54 U. Voronoy, "Sobre bloqueo del aparato reticuloendotelial del hombre en algunas formas de intoxicacion por el sublimado y sobre la transplantacion del rinon cadaverico como metodo de tratamiento de la anuria conseutiva a aquella intoxicacion" [Blocking the reticuloendothelial system in man in some forms of mercuric chloride intoxication and the transplantation of the cadaver kidney as a method of treatment for the anuria resulting from the intoxication], *Siglo Medico* (1937): 97, 296–97, cited in Smith, "Progress in Clinical Organ Transplantation," 5.

55 Silverberg [1972] 1983.

56 "The Back Page by Tom Tomorrow," *New Yorker*, 2 July 2001, 88.

57 Thompson 1979, 9.

58 "Med Help International," *http://www.medhelp.org/forums/maternal/archive/3105.html* (accessed 25 March 2002).

59 "World's First Womb Transplant," *iafrica.com: only what you want*, http://www.iafrica.com (accessed 25 March 2002).

60 The Saudi surgical team, headed by Dr. Wafah Fageeh, issued a press release that downplayed the importance of immunosuppressant drugs. "The young woman was given drugs to prevent the womb being rejected by her body. The drugs used were those given to kidney transplant patients, hundreds of whom have had successful pregnancies afterwards." Sarah Boseley, "Surgeons Hail World's First Womb Transplant," *Guardian*, 7 March 2002. Yet Canadian researchers reported in the journal *Teratology* that "kidney transplant patients who became pregnant while receiving immune-suppressing therapy had a much higher than normal rate of stillbirths and preemie deliveries. Their babies were also more likely to be underweight at birth" (Marcus 2002, 2).

61 Jeremy Laurence, "Saudi Surgeons Announce First Womb Transplant," *Independent*, 7 March 2002.

62 Marcus 2002.

63 Richard Smith, consultant gynecologist at the Chelsea and Westminster hospital in London, speaking to Sarah Boseley, "Surgeons Hail World's First Womb Transplant," *Guardian*, 7 March 2002.

64 BBC News, "Womb Transplant Breakthrough Hope," *BBC News Health*, 7 March 2002, *http://news.bbc.co.uk* (accessed 7 March 2002).

65 "The reason our experiment worked is because we connected the vascular system of the implanted uterus directly to the existing blood supplies, rather than using stents which have caused other transplants to fail, Brannstrom told *New Scientist*" (Vince 2002; Hutchinson 2003). See also Bhattacharya 2003; Sample 2003.

66 The remarkable thing about the popular press representation of this tongue transplant is the way it elides the cadaver origin of the tongue while describing the immunological difficulties of tongue transplants in general: "Doctors investigating the possibility of tongue transplants in the past have encountered problems because the mouth is continually filled with foreign and potentially infectious bacteria" (Demetriou 2003). See also Associated Press, "Doctors in Austria Report First Human Tongue Transplant," *Atlanta Journal Constitution*, 22 July 2003; and Reuters Health Information 2003, "Tongue Transplant Man Doing Well," *Medscape*, 25 July 2003.

67 "Although it sounds like a plot from a science-fiction movie, surgeons in Britain say they will be able to carry out the first full facial transplant within a year. . . . But the main roadblock to the procedure may be the ethical questions surrounding the issue." "Face Transplants Not Just Science Fiction: Doctor: Procedure Technically Feasible but Ethically Ambiguous," CNN.com, 28 November 2002.

68 Thompson 1979, 8. My thanks to Carol Christ for suggesting that I read Thompson's work.

69 Suvin 1997; see also "A Colloquium with Darko Suvin," *Science Fiction: A Review of Speculative Literature* 16, no. 43 (2001): 3–33. I have enjoyed my e-mail conversations with you, Darko, and only regret that 9/11 made it impossible for us to meet in Thessaloniki.

70 Gibson-Graham 1996.

71 As this book was going to press, Catherine Waldby and Robert Mitchell shared with me the prospectus for their book in progress, "Tissue Economies: Gifts, Commodities, and Bio-value in Late Stage Capitalism." I want to thank them for their generosity, and I regret that it was too late to incorporate a response to it.

72 The article reported that "a favorable outcome of human embryonic dopamine cell transplantation in patients with Parkinson's disease" can be predicted by a good response preoperatively to therapy with L-dopa, the drug used to restore fluid movement to Parkinson's sufferers by replacing dopamine in the brain. Dr. Curt R. Freed, of the University of Colorado at Denver, reported on research showing that patients under sixty years of age who responded favorably to L-dopa therapy also responded well to the transplantation of embryonic mesenphalic tissue. For the transplantation of embryonic tissue, not even cyclosporine or other immunosuppressant drugs were needed, Freed reported: the patients received symptom relief even without immunosuppression therapy, with 85 percent of the transplant patients surviving. "Response to Dopamine Cell Transplantation Predicted by L-Dopa Response," *Reuters Medical News for the Professional*, 18 November 2001.

73 Charlie LeDuff, "Hauling the Debris, and Darker Burdens," *New York Times*, 17 September 2001, A1, A12.

74 Beck 1992.

75 Maureen Dowd, "From Botox to Botulism," *New York Times*, 26 September 2001.

76 Lane 2001, 80.

77 Chris Hewitt, "*Fears* Too Real to Enjoy," *Centre Daily Times Weekend*, 31 May 2002, C15.

78 Jay Boyar, "Freeman and Affleck Propel Exciting but Unsettling Movie," *Orlando Sentinel*, 31 May 2002, 17.

7. LIMINAL PERFORMANCES OF AGING

1 Gina Kolata, "An Early Sign of Alzheimer's Brings Fear, and New Insight," *New York Times*, 2 June 2002, 20.

2 Schechner and Appel 1990, 11.

3 Butler 1990, 24–25.

4 Victor Turner, "Are There Universals of Performance in Myth, Ritual, and Drama?" in Schechner and Appel 1990, 11.

5 Armandola 2003.

6 "The tissue to be cultivated should be taken from the animal immediately after death. This, however, is not essential and tissue has been known to grow quite well in vitro when taken from the body as much as a week after death or considerably longer if the body has been kept in cold storage. From this we can see that when a doctor pronounces a patient 'dead' he is only using the word 'death' in a restricted sense" (Fell 1936, 3). It is the unorganized nature of the growth, not whether the tissues are embryonic or mature, that provides the tissue with its potential immortality. Both embryonic tissue in unorganized growth and cancerous tissue result in the similar process of outgrowth, or as Waddington put it, "the induction of a cancer can be homologized with the induction of an embryonic tissue" (Waddington 1936b, 111). The continually growing tissue can be subdivided, with new cultures placed in fresh culture medium, and for certain tissues the process can continue, Fell observes, "indefinitely." Fell's definition is helpful here: "Tissue undergoing unorganised growth loses its normal histological structure. For example, kidney tubules undergoing organised growth merely spread out over the glass of the culture vessel as an indifferent sheet of epithelium . . . and completely loses [*sic*] the anatomical features of the original tubules. This loss of histological structure is sometimes known as de-differentiation" (Fell 1936).

7 Rolleston (1927) had presented a lecture on this topic at the Royal Institution on 13 May.

8 Crary 1999, 3.

9 While its main stress is on the effect of aging on the individual, the essay already anticipates the turn to population-based medicine that would come later in the century when it discusses the statistical basis for legends of remarkable longevity.

10 As T. S. P. Strangeways (1924) explained, "The earliest attempt to grow animal tissue in vitro appears to have been made by Leo Loeb, [who] . . . succeeded in growing epithelium on the surface of a blood clot contained in a glass vessel at the bottom of which had been placed a fragment of kidney. . . . Burrows (1910) and Carrel (1911) improved upon this method by using clotted plasma as a medium, and finally succeeded in maintaining growth by the addition to the plasma of extract of embryonic tissues" (ix–x).

11 Breuer 1927, 970.

12 The passage, which I quote in the body of chapter 6, reads: "Several years ago the German Professor, Dr. Walter Finkler, amputated the heads of various insects and transplanted them on others. Strange to say, the insects with the transplanted heads, after the new ones had grown, managed to get along the same as with their original heads. So the operation of exchanging

your old stomach for a new one may, after all, not be an impossibility, but you may get the surprise of your life if you ever make such an exchange. At least one millionaire who bought himself a new stomach found this out rapidly in totally unexpected results."

13 Huekels 1927.

14 Verrill 1927.

15 Pease 1938.

16 Gade 1940, 145. When Lawrence W. Smith, M.D., and Temple Fay, M.D., (1939) reported that their experiment confirmed the existence of a relationship between embryonic growth and the uncontrolled growth of cancer cells, they were repeating the modernist tendency to construct the beginning and the end of life in relation to each other.

17 Wade 2000, D1.

18 Department of Bioartificial Organs, Kyoto University, http://www.frontier.kyoto-u.ac.jp/ca04/text/eng/oindex_e.html (accessed 6 December 2001).

19 Lévy 1998, 40–41; see also Thacker 2000.

20 *Reuters Medical News for the Professional*, 3 January 2002.

21 Kolata 2002; F. Quaini, K. Urbanek, A. P. Beltrami, N. Finato, C. A. Beltrami, B. Nadal-Ginard, J. Kajstura, A. Leri, and P. Anversa, "Chimerism of the Transplanted Heart," *New England Journal of Medicine* 346, no. 1 (2002): 5–15.

22 "Levothyroxine Use Could Increase Number of Transplantable Donor Organs," *Reuters Health*, 3 January 2002. See also Salim et al. 2001.

23 *Tit-Bits*, 16 April 1938; Paul Recer, "Stem Cell Guidelines Issued," *Yahoo! News*, http://dailynews.yahoo.com (accessed 23 August 2000).

24 Sandra Blakeslee, "Lack of Oversight in Tissue Donation Raising Concerns," *New York Times*, 20 January 2002, 1, 20.

25 National Institutes of Health 2001.

26 S. Fahn, "Description of Parkinson's Disease as a Clinical Syndrome," PMID:12846969, *http://www.ncbi.nlm.nih.gov* (accessed 27 July 2003).

27 *http://www.neurologychannel.com/parkinsonsdisease/symptoms.shtml* (accessed 27 July 2003).

28 "Supporters of [stem cell] research say it holds the promise of curing ailments such as Alzheimer's, Parkinson's and juvenile diabetes." Kathy Kiely, "Broader Stem-Cell Research Sought," *USA Today*, 10 September 2001.

29 Watt and Hogan 2000, 1429.

30 National Institutes of Health 2001, C3.

31 UPMC Health System News Bureau 2001.

32 Sterling 1997, 43.

33 http://www.medequity.com/html/body_transmedics.htm (accessed 24 January 2002).

34 *http://www.augustachronicle.com/cgi-bin/printeme.pl* (accessed 24 January 2002).

35 Anne Barnard, "Organ Transport Device Unveiled," *Boston Globe*, 14 August 2001.

36 "Doctor Develops Device to Transport Functioning Organs to Recipients," *Boston Globe*, 14 August 2002.

37 Silverman 2003.

38 http://popularmechanics.com/science/research/2001/11/2002_design_and_engineering_awards (accessed 24 January 2002).

39 Huard 2001.

40 Ibid., 16.

41 Schwartz 2001.

42 Sterling 1997.

43 See Basting 1998.

44 Garoian 1999.

45 *TimeSlips*, by Anne Davis Basting, ran from 30 October to 18 November 2001 at the Here Arts Center, New York City; produced by Paul Lucas and Gail Winar; written by Anne Basting; directed by Christopher Bayes; "inspired by fictional stories told by people with Alzheimer's." Working script courtesy of the author.

46 Anne Davis Basting, *TimeSlips*, 25 September 2000, *http://timeslips.org/historyframe/text.html*. Another example of work with people suffering from dementia is DiPerna 2002. As John Langone reported in the *New York Times*, "A common perception of people with Alzheimer's disease or dementia is that they have limited ability to communicate. This is often the case, but as this moving little book of poetry demonstrates, thoughts and feelings remain and can emerge when properly and painstakingly nurtured. . . . As Lila Weisberger, president of the National Association for Poetry Therapy, says: 'There is a magic, even a clarity, in the unique words of someone who is not limited by conventional speech. While poets strive for a unique way to express a thought or a feeling, the person with dementia does this naturally' " ("Books on Health," *New York Times*, 22 January 2002, D8).

47 Anne Davis Basting, *TimeSlips*, 25 September 2000, *http://timeslips.org/historyframe/text.html*.

48 Basting 1998, 2.

49 " 'Prosthetics' is thus a useful heading under which to consider the general field of bodily interventions, technology, and writing in Modernism. But two senses of 'prosthesis' need to be distinguished. What I would label a 'negative' prosthesis involves the replacing of a body part, covering a lack. The negative prosthesis operates under the sign of compensation (Freud's 'suffering'). A 'positive' involves a more utopian version of technology, in which human capacities are extrapolated. From the nineteenth century, the

prosthesis in both these senses is bound up with the dynamics of modernity. Technology offers a re-formed body, more powerful and capable, producing in a range of modernist writers a fascination with organ-extension, organ-replacement, sensory-extension; with the interface between the body and the machine which Gerald Heard, in 1939, labeled *mechanomorphism*" (Armstrong 1998, 78).

CODA

1 Nijs 1999.

2 Groopman 2002, 23.

3 Though they are both well-known bioethicists, still Groopman and Kass differ substantially in the positions from which they approach stem cell research. Groopman, an oncologist and AIDS specialist who holds a chaired professorship at the Harvard Medical School, is a staff writer in medicine and biology for the *New Yorker.* He is the author (among other works) of *The Measure of Our Days*, a study of "the spiritual lives of patients with serious illness, and the opportunities for fulfillment they sometimes find." Kass, a chaired professor in the Committee on Social Thought at the University of Chicago and a fellow of the American Enterprise Institute, has an M.D. and a Ph.D. in biochemistry and is the author, with Amy Kass, of *Wing to Wing, Oar to Oar: Readings on Courting and Marrying* (Notre Dame, Ind.: University of Notre Dame Press, 2000). A practicing physician and researcher, Groopman has been active in working for AIDS awareness, while Kass, who according to Arthur Caplan "hasn't gone near a lab in 20 years," no longer practices medicine. http://www.jeromegroopman.com/bio.html (accessed 9 March 2002); http://olincenter.uchicago.edu/kass_cv.html (accessed 9 March 2002).

4 Waldby 2000, 136.

5 Nick Gillespie, "Birthmarks and Bioethics," *Reason Online*, 18 January 2002, http://reason.com/hod/ng011802.shtml (accessed 9 March 2002). Gillespie explores the relation between these two Hawthorne short stories and suggests that "Dr. Heidegger's Experiment" might be more appropriate a subject for the Bioethics Committee to consider. However, Gillespie's view of fiction is still too limited. He still understands the short story as articulating an unambiguous moral rather than providing a space in which to explore the ambiguities of what might be called technoscientific desire.

6 Nathaniel Hawthorne, "Dr. Heidegger's Experiment," http://www.classicreader.coom/read.php/sid.6/bookid.196 (accessed 22 February 2002).

7 Chris Mooney, "Irrationalist in Chief," *American Prospect*, http://www.pros pect.org/print/V12/17/mooney-c.html (accessed 9 March 2002).

8 Ibid., 3.

9 Mooney is referring to Jon Turney, *Frankenstein's Footsteps: Science, Genetics, and Popular Culture* (New Haven: Yale University Press, 1998).

10 Bawarshi 2000, 340.

11 Bawarshi illustrates this point by considering the Patient Medical History Form, whose genre enables "the patient and doctor [to] reproduce the sociorhetorical conditions within which they interact. For instance, the genre reflects how our culture and science separate the mind from the body in treating disease, constructing the patient as an embodied object" (Bawarshi 2000, 354).

12 Octavia Butler, *Dawn* (1987), *Adulthood Rites* (1988), and *Imago* (1989).

13 Mitchison 1995, [1962] 1985.

14 In this argument, I am drawing on Nigel Dodd's 1999 discussion of the postmodern social theories of Zygmunt Bauman and Richard Rorty, especially Rorty's reliance on "the thick descriptions of other cultures provided by literature" to "sensitize us to alternative perspectives" (175).

15 Chambers 1999, 178.

16 Giddens 1991, 20.

17 "This division of the world between experts and non-experts also contains an image of the public sphere" (Beck 1992, 57).

18 Beck 1999, 108.

19 For another example of reflexive materialism-constructivism, see Wilson 1998.

20 Fausto-Sterling 2003.

21 Derrida 1980, 203–4.

22 Ruben Bolling, "When Blastocysts Go Bad," 9 September 2001, http://archive.Salon.com/comics/boll/2001/08/09/boll/ (accessed 28 July 2003).

23 *Tom the Dancing Bug* is drawn by Ruben Bolling (pseudonym of Ken Fisher), who started the comic strip while at Harvard University Law School, from which he graduated in 1987. It is telling indeed that Fisher should turn to the comics as a more hospitable zone for ethical questioning and social critique than the law. http://www.amuniversal.com/ups/features/tom_dancing_bug/bolling.htm (accessed 28 July 2003).

24 Davis 1995, 35.

25 William Safire, "The But-What-If Factor," *New York Times*, 16 May 2002, A25.

26 Dunn 2002, 46.

27 Victor Turner, "Liminality and Communitas: Form and Attributes of Rites of Passage," in V. Turner [1969] 1995, 95.

28 "Excerpts from the Writings of Victor Turner," http://www.creativeresistance.ca/communitas/defining-liminality-and-communitas (accessed 14 July 2003).

29 Ozecki 1998, 2003; Karasik and Karasik 2003; Rowe 2002.

30 Farmer 1994; Haddix 2000, 2002; Lowry 2002, 2000.

31 Nancy Farmer, *The House of the Scorpion* (2002), jacket commentary by Ursula K. Le Guin.

32 Victor Turner, "Passages, Margins, and Poverty: Religious Symbols of Communitas," part 1, *Worship* 46, nos. 7–8 (1985): 393, cited in Nichols 1985, 402.

33 In "Passages, Margins, and Poverty," Turner points out that "major liminal situations are occasions on which . . . a society takes cognizance of itself, or rather where . . . members of that society may obtain an approximation, however limited, to a global view of man's [*sic*] place in the cosmos and his relations with other classes of visible entities" (cited in Nichols 1985, 400).

34 J. K. Rowling, *Harry Potter and the Sorcerer's Stone* (New York: Scholastic, 1988). This is not to argue that children's literature is being used this way for the first time. We have only to recall the way that Charles Kingsley's *The Water Babies* explored the relations between human beings and other species and catalyzed both Julian Huxley's work with the axolotl and Aldous Huxley's *Brave New World*, to see that the traffic between children's literature, science, and so-called canonical fiction has a long history (see Squier 1994). What does seem new, and with socially noxious effects, is the cordoning off of children's literature into its own genre, with its own standards, age grading, marketing and reviewing practices, and restriction to specific (and subordinate) awards categories. However, a recent review of *Harry Potter and the Order of the Phoenix* (New York: Scholastic Press, 2003) by mainstream reviewer John Leonard in the *New York Times Book Review* suggests that this trend may be reversing itself. Coming out as an enthusiastic fan of Rowling's fiction, Leonard slams "the nitpickers who disdain children's literature to begin with, which just means that they are tin-eared, tone deaf and born dumb. (Where do they think we begin to care about stories?). . . . And finally the world-weary and wart-afflicted who complain about the mediocre movies, the media hype, the marketing blitz, the embargo and maybe even the notion of a single mom becoming richer than the queen." John Leonard, "Nobody Expects the Inquisition," *New York Times Book Review*, 13 July 2003, late edition, 7:13.

Works Cited

Achenbaum, W. Andrew. 1995. *Crossing Frontiers: Gerontology Emerges as a Science.* Cambridge: Cambridge University Press.

Advisory Group on the Ethics of Xenotransplantation. 1997. *Animal Tissue into Humans: A Report by the Advisory Group on the Ethics of Xenotransplantation.* London: HMSO.

Alexander, W. 1927. "New Stomachs for Old." *Amazing Stories* 1, no. 22 (February): 1039–41.

Altman, Lawrence K. 1996. "Studying Rare Disorder, Scientists Find Gene Affecting Aging." *New York Times*, 12 April, A27.

——. 1998. "Surgeons in France Try Hand Transplant." *New York Times*, late edition (East Coast), 2 August, A18.

——. 2000. "Both Hands and Forearms Transplanted for First Time." *New York Times*, 15 January, A5.

Andrews, Lori, and Dorothy Nelkin. 2001. *Body Bazaar: The Market for Human Tissue in the Biotechnology Age.* New York: Crown Publishers.

Annan, Noel. 1990. *Our Age: English Intellectuals between the World Wars—A Group Portrait.* New York: Random House.

Armandola, Elena A. 2003. "Tissue Regeneration and Organ Repair: Science or Science Fiction?" *Medscape General Medicine* 5, no. 3. http://www.medscape.com.

Armstrong, Nancy. 1987. *Desire and Domestic Fiction: A Political History of the Novel.* Oxford: Oxford University Press.

Armstrong, Tim. 1998. *Modernism, Technology, and the Body: A Cultural Study.* Cambridge: Cambridge University Press.

Asimov, Janet, and Isaac Asimov. 1980. "In Praise of Specialized Fiction." In *How to Enjoy Writing: A Book of Aid and Comfort*, comp. Janet Asimov and Isaac Asimov. New York: Walker.

Atherton, Gertrude. 1923. *Black Oxen.* New York: Boni and Liverwright.

Awaya, T. 1999. "Organ Transplantation and the Human Revolution." In *Transplant Proceedings* 31: 1317–19.

Banner, Lois. 1989–1990. "The Meaning of Menopause: Aging and Its Historical Contexts in the Twentieth Century." *Working Paper* 3 (Fall–Winter): 7–15.

Barad, Karen. 1998. "Getting Real: Technoscientific Practices and the Materialization of Reality." *differences: A journal of feminist cultural studies* 10, no. 2.

Barclay, Laurie, M.D. 2003. "First Double Hand Transplant: A Newsmaker Interview with Nadley S. Hakim, M.D., Ph.D." *Medscape Medical News*, 7 July.

Barwick, Sandra. 2001a. "I Will Never Get over It, Says Distraught Mother." *Daily Telegraph*, 31 January.

——. 2001b. "Organs Scandal: They Laid 36 Parts of My Baby on a Table, Put Them into a Bag, and Ran." *Daily Telegraph*, 31 January.

Basting, Anne Davis. 1998. *The Stages of Age: Performing Age in Contemporary American Culture.* Ann Arbor: University of Michigan Press.

Basting, Anne Davis, Christopher Bayes, and Gail Winar, producers. 2001. *TimeSlips.* Here Arts Center, 10 November. http://timeslips.org/history frame/text.html.

Batty, David. 2001. "Alder Hey Organs Scandal: The Issue Explained." *Guardian Unlimited*, 12 March.

Bauman, Zygmunt. 1992. *Mortality, Immortality, and Other Life Strategies.* Stanford: Stanford University Press.

Bawarshi, Anis. 2000. "The Genre Function." *College English* 62, no. 3: 335–59.

Beck, Ulrich. [1986] 1992. *Risk Society: Towards a New Modernity.* London: Sage.

——. 1999. *World Risk Society.* Cambridge: Polity Press.

Beer, Gillian. 1983. *Darwin's Plots: Evolutionary Narrative in Darwin, George Eliot, and Nineteenth-Century Fiction.* London: Ark.

——. 1996. *Open Fields: Science in Cultural Encounter.* New York: Oxford University Press.

Benjamin, Harry, M.D. 1930. "The Reactivation of Women." In *Sexual Reform Congress*, ed. Norman Haire. London: Kegan Paul, Trench, Trubner.

Bennett, Tony. 1990. *Outside Literature.* London: Routledge.

Bhattacharya, Shaoni. 2003. "Womb Transplants for Women 'Three Years Away.'" *New Scientist.com* 18, no. 17 (1 July).

Blacker, C. P. 1943. Letter to Thomas Lord Horder, 2 November. Eugenics Society, Contemporary Medical Archives Centre, Wellcome Institute for the History of Medicine, London.

Bloom, Clive, ed. 1993. *Literature and Culture in Modern Britain.* Vol. 1: 1900–1929. New York: Longman.

Bodeen, DeWitt. 1982. "*Black Oxen.*" In *Magill's Survey of Cinema Silent Films*, ed. Frank N. Magill. Vol. 1. Englewood Cliffs, N.J.: Salem.

Bolling, Ruben. 2001. "When Blastocysts Go Bad." Cartoon. *Salon.com*, 8 August.

Bowers, Charles Gardner. 1931. "The Black Hand." *Amazing Stories* 5, no. 10 (January): 909–11.

Boyar, Jay. 2002. "Freeman and Affleck Propel Exciting but Unsettling Movie." *Orlando Sentinel*, 31 May, 17.

Braidotti, Rosi. 1994. "Of Bugs and Women: Irigaray and Deleuze on the Becoming-Woman." In *Engaging with Irigaray: Feminist Philosophy and Modern European Thought*, ed. Carolyn Burke et al. New York: Columbia University Press.

Brave, Ralph. 2001. "Governing the Genome." *Nation*, 10 December, 18–24.

Breuer, Miles J. 1927. "The Man with the Strange Head." *Amazing Stories* 1, no. 10 (January): 940–43, 970.

Broadhurst, Sue. 2000. "Liminal Aesthetics." *Body, Space and Technology*, 25 February 2004. *http://www.brunel.ac.uk/depts/pfa/bstjournal/1no1/journal.htm.*

Bromwich, Peter, et al. 1994. "Decline in Sperm Counts: An Artefact of Changed Reference Range of 'Normal'?" *British Medical Journal* 309: 1392–95.

Bunyan, Nigel. 2001. "Alder Hey Sold Tissue from Live Children." *News.telegraph*, 27 January. http://www.telegraph.co.uk/news/main.jhtml?xml=/news/2001/01/27/nald27.xml (accessed 2 August 2003).

Butler, Judith. 1990. *Gender Trouble: Feminism and the Subversion of Identity.* New York: Routledge.

Butler, Octavia. 1987. *Dawn.* New York: Warner Books.

——. 1988. *Adulthood Rites.* New York: Warner Books.

——. *Imago.* 1989. London: Victor Gollancz SF.

Carlsen, E., A. Giwercman, N. Keilding, and N. Skakkebaek. 1992. "Evidence for Decreasing Quality of Semen during Past 50 Years." *British Medical Journal* 305:609–13.

Carrel, Alexis, and Charles A. Lindbergh. 1935. "The Culture of Whole Organs." *Science* 81 (2002): 621–23.

Carter, Angela. 1982. Preface to *Memoirs of a Midget*, by Walter de La Mare. Oxford: Oxford University Press.

Cartwright, Lisa. 1995. *Screening the Body: Tracing Medicine's Visual Culture.* Minneapolis: University of Minnesota Press.

Chambers, Tod. 1996. "Dax Redacted: The Economies of Truth in Bioethics." *Journal of Medicine and Philosophy* 23:287–302.

——. 1999. *The Fiction of Bioethics: Cases as Literary Texts.* New York: Routledge.

Clarke, Bruce. [1969] 1996. *Dora Marsden and Early Modernism: Gender, Individualism, Science.* Ann Arbor: University of Michigan Press.

Clarke, Bruce, and Linda Dalrymple Henderson, eds. 2002. *From Energy to Information: Representation in Science and Technology, Art, and Literature.* Stanford: Stanford University Press.

Cliff, Michelle. 1990. "Object into Subject: Some Thoughts on the Work of Black Women Artists." In *Making Face, Making Soul: Haciendo Caras: Creative and Critical Perspectives by Women of Color*, ed. Gloria Anzaldúa. San Francisco: Aunt Lute Foundation Books.

Cohen, Lawrence. 1999. "Where It Hurts: Indian Material for an Ethics of Organ Transplantation." *Daedalus* 128, no. 4 (Fall): 135–65.

Cohen, William. 1980. *The French Encounter with Africans: White Responses to Blacks, 1530–1800.* Bloomington: Indiana University Press.

Collini, Stefan. 1993. Introduction to *The Two Cultures*, comp. C. P. Snow. Cambridge: Cambridge University Press.

"A Colloquium with Darko Suvin." 2001. *Science Fiction: A Review of Speculative Literature* 16, no. 1 (43): 3–33.

Conrad, Christoph. 1992. "Old Age in the Modern and Postmodern Western World." In *Handbook of the Humanities and Aging*, ed. Thomas R. Cole, David D. Van Tassel, and Robert Kastenbaum. New York: Springer.

"Consensus Statement on the Live Organ Donor." 2000. *Journal of the American Medical Association* 284, no. 22: 2919–26.

Corson, Trevor. 2000. "The Telltale Heart: Death and Democracy in Japan." *Transition* 9, no. 4: 78–96.

Cox, T. S., et al. 2002. "Breeding Perennial Grain Crops." *Critical Reviews in Plant Sciences* 21, no. 2: 59–91.

Crary, Jonathan. 1999. *Suspensions of Perception: Attention, Spectacle, and Modern Culture.* Cambridge: MIT Press.

Crawford, T. Hugh. 1993. "An Interview with Bruno Latour." *Configurations* 1: 247-69.

Curthoys, Ann, and John Docker. 2005. *Is History Fiction?* Sydney: University of New South Wales Press.

Cussins, Charis Thompson. 1999. "Confessions of a Bioterrorist: Subject Position and Reproductive Technologies." In *Playing Dolly: Technocultural Formations, Fantasies, and Fictions of Assisted Reproduction*, ed. E. Ann Kaplan and Susan Squier. New Brunswick, N.J.: Rutgers University Press.

Dahl, Roald. 1962. "Royal Jelly." In *Kiss, Kiss*, by Roald Dahl. London: Penguin.

Dahrendorf, Ralf. 1979. *Life Chances: Approaches to Social and Political Theory.* Chicago: University of Chicago Press.

Daston, Lorraine, and Galison, Peter. 1992. "The Image of Objectivity." *Representations* 40: 81–128.

Davis, Lennard J. 1995. *Enforcing Normalcy: Disability, Deafness, and the Body.* London: Verso.

DeKruif, Paul. [1926] 1954. *Microbe Hunters.* New York: Harcourt Brace Jovanovich.

de la Mare, Walter. [1921] 1982. *Memoirs of a Midget.* Oxford: Oxford University Press.

De la Mothe, John. 1992. *C. P. Snow and the Struggle of Modernity.* Austin: University of Texas Press.

de Lauretis, Teresa. 1987. *Technologies of Gender: Essays on Theory, Film, and Fiction.* Bloomington: Indiana University Press.

Du Plessis, Rachel Blau. 1985. *Writing beyond the Ending: Narrative Strategies of Twentieth-Century Women Writers.* Bloomington: Indiana University Press.

Demetriou, Danielle. 2003. "First Human Tongue Transplant a Success." *Independent Digital (U.K.),* 22 July.

Deleuze, Gilles, and Félix Guattari. 1987. *A Thousand Plateaus: Capitalism and Schizophrenia.* Minneapolis: University of Minnesota Press.

Derrida, Jacques. 1980. "The Law of Genre." In *Glyph* 7, ed. Samuel Weber. Baltimore: Johns Hopkins University Press.

Dictionary of Scientific Biography. 1974. Vol. 9, ed. Charles Coulston Gillispie. New York: Charles Scribner's Sons.

DiPerna, Marilyn Riso, ed. 2002. *Dispelling the Myths of Dementia through Poetry.* Kings Park, N.Y.: St. Johnland Nursing Center.

Dodd, Nigel. 1999. *Social Theory and Modernity.* Cambridge: Polity Press.

Donnelly, William J. 1996. "Taking Suffering Seriously: A New Role for the Medical Case History." *Academic Medicine* 71:730–37.

Doyle, Richard. 1997. *On Beyond Living: Rhetorical Transformations of the Life Sciences.* Stanford: Stanford University Press.

"Dr. R. G. Canti: Obituary." 1936. *Nature,* 15 February.

"Dr. Ronald George Canti, Obituary." 1936. *British Medical Journal,* 18 January, 137.

Duden, Barbara. 1993. *Disembodying Women: Perspectives on Pregnancy and the Unborn.* Trans. Lee Hoinacki. Cambridge: Harvard University Press.

Duffy, Maureen. 1981. *Gor Saga.* London: Eyre Methuen.

Dunn, Kyla. 2002. "Cloning Trevor." *Atlantic* 289, no. 6 (June): 31–52.

Eliot, Thomas Stearns. [1919] 1994. "Tradition and the Individual Talent." In *Contemporary Literary Criticism: Literary and Cultural Studies,* ed. Robert Con Davis and Ronald Schleifer, 27–33. New York: Longman.

Elliott, Carl. 1998. *A Philosophical Disease: Bioethics, Culture, and Identity.* New York: Routledge.

"Ewe Again? Cloning from Adult DNA." 1997. *Science News,* 1 March, 132.

Farmer, Nancy. 1994. *The Ear, the Eye, and the Arm.* New York: Orchard Books.

——. 2002. *The House of the Scorpion.* New York: Atheneum.

Farrow, Stephen. 1994. "Falling Sperm Quality: Fact or Fiction?" *British Medical Journal* 309:1–2.

Fausto-Sterling, Anne. 2003. "Science Matters, Culture Matters." *Perspectives in Biology and Medicine* 46, no. 1: 109–24.

Featherstone, Mike, and Mike Hepworth. 1991. "The Mask of Ageing and the Postmodern Life Course." In *The Body: Social Process and Cultural Theory,* ed. Mike Featherstone, Mike Hepworth, and Bryan S. Turner, 371–89. London: Sage.

Featherstone, Mike, and Andrew Wernick, eds. 1995. *Images of Aging: Cultural Representations of Later Life.* New York: Routledge.

Fell, Honor Bridget. 1936. "Bd. Ed. Voc. Course." 8 January. Strangeways Research Laboratory File, Contemporary Medical Archives Centre, Wellcome Institute for the History of Medicine, London.

——. 1962. "Cell Biology." In R. A. Peters, *History of the Strangeways Research Library (formerly Cambridge Research Hospital), 1912–1962*. Cambridge: Heffer.

——. 1936. "Lecture: Tissue Culture." Unpublished essay, 8 January. Strangeways Research Laboratory, File PP/HBF/E12, Contemporary Medical Archives Centre, Wellcome Institute for the History of Medicine, London.

——. 1937. "Lecture I: The Technique of Tissue Culture and Its Value in Research." Postgraduate School of Medicine lecture, 3 March. Strangeways Research Laboratory File, PP/HBF/E12, Wellcome Institute for the History of Medicine, London.

Foucault, Michel. [1963] 1994. *The Birth of the Clinic: An Archaeology of Medical Perception*. New York: Vintage Books.

——. [1966] 1973. *The Order of Things: An Archaeology of the Human Sciences*. New York: Vintage Books.

——. 1979. "Governmentality." *m/f: a feminist journal,* no. 3 (July): 5–21.

——. 1980. *The History of Sexuality*. Vol. 1: *An Introduction.* Trans. Robert Hurley. New York: Vintage Books.

Foxon, G. E. H. 1976. "Early Biological Film—the Work of R. G. Canti." *University Vision: The Journal of the British Universities Film Council* 15 (December): 5–13.

Frank, Pat. 1946. *Mr. Adam.* Philadelphia: Lippincott.

Freud, Sigmund. [1925] 1959. "Negation." In *Collected Papers,* ed. James Strachey, vol. 5, *Miscellaneous Papers, 1888–1938*. New York: Basic Books.

Frye, Northrop. [1949] 1994. "The Function of Criticism at the Present Time." In *Contemporary Literary Criticism: Literary and Cultural Studies,* ed. Robert Con Davis and Ronald Schleifer, 34–45. White Plains, N.Y.: Longman.

Gade, Henry. 1940. "Suspended Animation." *Amazing Stories* 14, no. 1 (January): 145.

Garoian, Charles. 1999. *Performing Pedagogy: Toward an Art of Politics.* Albany: SUNY Press.

Gawande, Atul. 2001. "Final Cut: Medical Arrogance and the Decline of the Autopsy." *New Yorker*, 19 March, 94–99.

Gennep, Arnold van. 1909. *The Rites of Passage*. Chicago: University of Chicago Press.

Gibson-Graham, J. K. 1996. *The End of Capitalism (As We Knew It): A Feminist Critique of Political Economy.* London: Blackwell.

Giddens, Anthony. 1991. *Modernity and Self-Identity: Self and Society in the Late Modern Age.* Stanford: Stanford University Press.

Gilford, Hastings. 1911. *The Disorders of Post-natal Growth and Development.* London: Adlard and Son.

Goldbeck-Wood, Sandra. 1996. "Europe Is Divided on Embryo Regulations." *British Medical Journal* 313, no. 31 (August): 512.

Greider, Linda. 2003. "Shazaam: Testosterone Therapy Can Revitalize Men, Say Proponents, but Critics See Dangers." *AARP Bulletin* 44, no. 7 (July–August): 16–18.

Grobstein, Clifford. 1988. *Science and the Unborn: Choosing Human Futures.* New York: Basic Books.

Groopman, Jerome. 2002. "Comment: Science Fiction." *New Yorker*, 4 February, 23.

Gross, Allan G. 1990. *The Rhetoric of Science.* Cambridge: Harvard University Press.

Grosz, Elizabeth. 2001. *Architecture from the Outside: Essays on Virtual and Real Space*. Cambridge: MIT Press.

Haddix, Margaret. 2000. *Among the Hidden.* New York: Aladdin.

———. 2002. *Among the Impostors.* New York: Aladdin.

Haldane, J. B. S. 1923. *Daedalus, or Science and the Future.* London: Kegan Paul, Trench, Trubner.

Hall, Lesley A. 1996. "The Strangeways Research Laboratory: Archives in the Contemporary Medical Archives Centre." *Medical History* 40:231–38.

Hancox, N. M. 1932. Letter to Honor Bridget Fell, 25 November. Contemporary Medical Archives Centre, Wellcome Institute for the History of Medicine, London.

Haraway, Donna. 1997. *Modest_Witness@Second_Millennium.Female Man_Meets_OncoMouse_:Feminism and Technoscience.* New York: Routledge.

———. 1989. *Primate Visions: Gender, Race, and Nature in the World of Modern Science.* New York: Routledge.

———. 1991. *Simians, Cyborgs, and Women: The Reinvention of Nature.* New York: Routledge.

Harding, Sandra. 1986. *The Science Question in Feminism.* Milton Keynes: Open University Press.

Harrison, Trevor. 1999. "Globalization and the Trade in Human Body Parts." *Canadian Review of Sociology and Anthropology* 36 (February): 21–35.

Hasta, M. M. 1926. "The Talking Brain." *Amazing Stories* 1, no. 5 (August): 440–45, 478–79.

Hawthorne, Nathaniel. 2002. "Dr. Heidegger's Experiment." *Classic Reader.com.* http://www.classicreader.coom/read/php/sid.6/bookid.196 (accessed 22 February).

Hayles, N. Katherine. 1990. *Chaos Bound: Orderly Disorder in Contemporary Literature and Science.* Ithaca: Cornell University Press.

——, ed. 1991. *Chaos and Order: Complex Dynamics in Literature and Science.* Chicago: University of Chicago Press.

——. 1993. "Virtual Bodies and Flickering Signifiers." *October* 66 (Fall): 69–91.

——. 1999. *How We Became Posthuman: Virtual Bodies in Cybernetics, Literature, and Informatics.* Chicago: University of Chicago Press.

Haynes, Roslynn D. 1994. *From Faust to Strangelove: Representations of the Scientist in Western Literature.* Baltimore: Johns Hopkins University Press.

Henderson, Andrea. 1991. "Doll-Machines and Butcher-Shop Meat: Models of Childbirth in the Early Stages of Industrial Capitalism." *Genders* 12 (Winter): 100–119.

Henry, Holly. 2003. *Virginia Woolf and the Discourse of Science: The Aesthetics of Astronomy.* Cambridge: Cambridge University Press.

Hertsgaard, Mark. 1996. "A World Awash in Chemicals." *New York Times Book Review*, 7 April: 25.

Hesse, Mary. 1966. *Models and Analogies in Science.* Notre Dame: University of Notre Dame Press.

Hirschfeld, Magnus. 1930. "Presidential Address: The Development and Scope of Sexology." In *Sexual Reform Congress.* London: Kegan Paul, Trench, Trubner.

Hockey, Jenny, and Allison James. 1995. "Back to Our Futures: Imaging a Second Childhood." In *Images of Aging: Cultural Representations of Later Life*, ed. Mike Featherstone and Andrew Wernick. New York: Routledge.

Holquist, Michael. 1989. "From Body-Talk to Biography: The Chronobiological Bases of Narrative." *Yale Journal of Criticism* 3, no. 1: 1–35.

Huard, Johnny. 2001. "Stem Cells from Skeletal Muscle Can Restore Bone Marrow Function." *Abstract from ASCB Meeting.* Proceedings of the American Society for Cell Biology 41st Annual Meeting, 8–12 December, Washington, D.C.

Huekels, Jack G. 1927. "Advanced Chemistry." *Amazing Stories* 1, no. 12 (March): 1127–29.

Hughes, Arthur Frederick William. 1938. Poem: ASWH to PM. January.

PP/FGS C.19. CMAC. Wellcome Institute for the History of Medicine, London.

Humphries, Karin H., and Sabrina Gill. 2002. "Risks and Benefits of Hormone Replacement Therapy: The Evidence Speaks." *Canadian Medical Association Journal* 168, no. 8 (April): 1001.

Hunter, Kathryn Montgomery. 1991. *Doctors' Stories: The Narrative Structure of Medical Knowledge.* Princeton: Princeton University Press.

Hutchinson, Martin. 2003. "Womb Transplant Baby 'within Three Years.'" *BBC News Online*, 1 July.

Huxley, Aldous. [1932] 1969. *Brave New World.* New York: Harper and Row.

Huxley, Julian. 1922. "Searching for the Elixir of Life." *Century Magazine*, February, 621–29.

"Introduction to Growth." 1937. *Growth: A Journal for Studies of Development and Increase* 1 (1937): 1–37.

James, P. D. 1992. *The Children of Men.* New York: Alfred A. Knopf.

Jordanova, Ludmilla. 1986. *Languages of Nature: Critical Essays on Science and Literature.* London: Free Association Books.

——. 1990. *Sexual Visions: Images of Gender in Science And Medicine between the Eighteenth and Twentieth Centuries.* Brighton: Harvester Wheatsheaf.

Kaplan, Caren. 1987. "Deterritorializations: The Rewriting of Home and Exile in Western Feminist Discourse." *Cultural Critique* 6: 187–98.

Kaplan, E. Ann, and Susan M. Squier, eds. 1999. *Playing Dolly: Technocultural Formations, Fantasies, and Fictions of Assisted Reproduction.* New Brunswick, N.J.: Rutgers University Press.

Karasik, Paul, and Judy Karasik. 2003. *The Ride Together: A Brother and Sister's Memoir of Autism in the Family.* New York: Washington Square Press.

Katz, Stephen. 1995. "Imagining the Life-Span: From Premodern Miracles to Postmodern Fantasies." In *Images of Aging: Cultural Representations of Later Life*, ed. Mike Featherstone and Andrew Wernick. New York: Routledge.

——. 1996. *Disciplining Old Age: The Formation of Gerontological Knowledge.* Charlottesville: University Press of Virginia.

Keller, Evelyn Fox. [1983] 1993. *A Feeling for the Organism.* New York: Freeman.

——. 1985. *Reflections on Gender and Science.* New Haven: Yale University Press.

——. 1986. "Making Gender Visible in the Pursuit of Nature's Secrets." In *Feminist Studies/Critical Studies*, ed. Teresa de Lauretis. Bloomington: Indiana University Press.

——. 1992. *Secrets of Life, Secrets of Death: Essays on Language, Gender, and Science.* New York: Routledge.

——. 1995. *Refiguring Life: Metaphors of Twentieth-Century Biology.* New York: Columbia University Press.

Keller, Evelyn Fox, Mary Jacobus, and Sally Shuttleworth, eds. 1990. *Body/Politics: Women and the Discourses of Science*. New York: Routledge.

Keller, Evelyn Fox, and Helen Longino. 1996. *Feminism and Science.* Oxford: Oxford University Press.

Kennedy, Ian. 1997. "Chairman's Foreword." In *Animal Tissue into Humans.* London: HMSO.

Kevles, Betty Holtzmann. 1997. *Naked to the Bone: Medical Imaging in the Twentieth Century.* New Brunswick, N.J: Rutgers University Press.

Kevles, Daniel. 1985. *In the Name of Eugenics: Genetics and the Uses of Human Heredity.* Berkeley: University of California Press.

Kingsland, Sharon. 1997. "Neo-Darwinism and Natural History." In *Science in the Twentieth Century*, ed. John Krige and Dominique Pestre. The Netherlands: Harwood Academic Publishers.

Kingsley, Charles. [1863] 1957. *The Water Babies: A Fairy Tale for a Land Baby.* London: J. M. Dent and Sons.

"Knockout Pigs Cloned." 2002. *Reuters Medical News*, 6 January.

Kolata, Gina. 1986. "New Growth Industry in Human Growth Hormone?" *Science* 234 (October): 22–24.

——. 2002. "Doctors Advance in Helping Body to Repair Itself." *New York Times*, 15 January, D1, D2.

Krimsky, Sheldon. 2000. *Hormonal Chaos: The Scientific and Social Origins of the Environmental Endocrine Hypothesis.* Baltimore: Johns Hopkins University Press.

Lambert, Gilles. 1959. *Conquest of Age: The Extraordinary Story of Dr. Paul Niehans.* New York: Rinehart.

Landecker, Hannah. 2000. "Immortality, in Vitro: A History of the HeLa Cell Line." In *Biotechnology and Culture: Bodies, Anxieties, Ethics*, ed. Paul E. Brodwin. Bloomington: Indiana University Press.

Lane, Anthony. 2001. "This Is Not a Movie: Same Scenes, Different Story." *New Yorker*, 24 September, 79–80.

Langone, John. 1978. *Long Life: What We Know and Are Learning about the Aging Process.* Boston: Little, Brown.

Lantos, John, M.D., Mark Siegler, M.D., and Leona Cuttler, M.D. 1989. "Ethical Issues in Growth Hormone Therapy." *JAMA* 261, no. 7 (February): 1020–24.

Laplanche, Jean, and J.-B. Pontalis. 1973. *The Language of Psychoanalysis.* Trans. Donald Nicholson-Smith. New York: W. W. Norton.

Latour, Bruno. 1987. *Science in Action: How to Follow Scientists and Engineers through Society.* Cambridge: Harvard University Press.

——. 1993. *We Have Never Been Modern.* Cambridge: Harvard University Press.

Latour, Bruno, and Steve Woolgar. 1979. *Laboratory Life: The Construction of Scientific Facts.* Princeton: Princeton University Press.

Laurance, Jeremy. 2001. "Alder Hey 'Sold' Body Parts to Drugs Firm for £10." *Independent*, 27 January.
Lederman, Muriel, and Ingrid Bartsch, eds. 2001. *The Gender and Science Reader.* New York: Routledge.
Lemonick, Michael D. 1998. "Sleight of Hand." *Time*, 18 October.
Lessing, Doris. 1988. *The Fifth Child.* New York: Alfred A. Knopf.
Levine, George. 1987. *One Culture: Essays in Science and Literature.* Madison: University of Wisconsin Press.
Lévy, Pierre. 1998. *Becoming Virtual: Reality in the Digital Age.* Trans. Robert Bononno. New York: Plenum Trade.
Lisser, H. 1925. "Organotherapy, Present Achievements, and Future Prospects." *Endocrinology: The Bulletin of the Association for the Study of Internal Secretions* 9 (January–February): 1–20.
Lispector, Clarice. 1988. *The Passion according to G.H.* Trans. Ronald W. Sousa. Minneapolis: University of Minnesota Press.
Lock, Margaret. 1995. "Transcending Mortality: Organ Transplants and the Practice of Contradictions." *Medical Anthropology Quarterly* 9:390–99.
——. 2002. *Twice Dead: Organ Transplants and the Reinvention of Death.* Berkeley: University of California Press.
Longino, Helen. 1990. *Science as Social Knowledge: Values and Objectivity in Scientific Inquiry.* Princeton: Princeton University Press.
Lowry, Lois. 2000. *Gathering Blue.* New York: Houghton Mifflin.
——. *The Giver.* 2002. New York: LaurelLeaf.
Ludovici, Anthony. 1929. *Lysistrata, or Woman's Future and Future Woman.* London: Kegan Paul, Trench, Trubner.
Macklin, Ruth. 1999. "Growth Hormones for Short Normal Children: An Ethical Analysis." ASBH *Exchange* 2, no. 4 (Fall): 1, 9–10.
Macrae, Alasdair D. F. 1995. *W. B. Yeats: A Literary Life.* New York: St. Martin's Press.
Marcus, Adam. 2002. "Womb Transplant's Success, Ethics Questioned." *HealthScout News, Yahoo!*, 25 March.
Martinovitch, Petar. Poem, "A Reply: Same Day." PP/FGS C.19. CMAC. Wellcome Institute for the History of Medicine, London.
McClure, Charlotte S. 1979. *Gertrude Atherton.* Boston: Twayne.
McGrady, Patrick M. 1968. *The Youth Doctors.* New York: Coward-McCann.
MedEquity. 2002. "TransMedics Secures $8 Million in Financing Round." http://www.medequity.com/html/body_transmedics.htm (accessed 24 January).
Medical Research Council. 1936. "The Medical Research Council's Report." *British Medical Journal*, 14 March, 534.
Medvedev, Zhores Alexandrowitsch. 1991. "The Structural Basis of Aging." In *Life Span Extension: Consequences and Open Questions*, ed. Frederic C. Ludwig. New York: Springer.

Metchnikoff, Eli. [1903] 1908. *The Nature of Man: Studies in Optimistic Philosophy.* New York: G. P. Putnam's Sons.

Minot, Charles Sedgwick. 1907. *The Problem of Age, Growth, and Death. Popular Science Monthly*, June–December.

Mitchison, Naomi. [1962] 1985. *Memoirs of a Spacewoman.* London: Women's Press.

——. *Solution Three.* 1995. New York: Feminist Press.

Moers, Ellen. 1976. *Literary Women: The Great Writers.* Garden City, N.Y.: Doubleday.

Monaghan, Peter. 2000. "Scholarly Watchdogs for an Ethical Netherworld." *Chronicle of Higher Education* 47, no. 6 (October): A23–A24.

Mooney, Brian C. 2001. "Tale of Two Livers: A Surgical Success Story." *Boston Globe*, 24 April.

"Morphogenetic Factors of Bone." 1937. *Nature*, 19 June, 1036–37.

National Institutes of Health. 1994. *Final Report of the Human Embryo Research Panel.* Comp. Stephen Muller. N.p.

——. 2001. *Stem Cells: Scientific Progress and Future Research Directions.* Comp. Department of Health and Human Services. N.p.

Nelkin, Dorothy, and M. Susan Lindee. 1995. *The DNA Mystique: The Gene as a Cultural Icon.* New York: W. H. Freeman.

Newman, Jenny. 1991. "Mary and the Monster: Mary Shelley's *Frankenstein* and Maureen Duffy's *Gor Saga*." In *Where No Man Has Gone Before: Women and Science Fiction*, ed. Lucie Armitt. London: Routledge.

Newman, Karin. 1996. *Fetal Positions: Individualism, Science, Visuality.* Stanford: Stanford University Press.

Nichols, J. Randall. 1985. "Worship as Anti-structure: The Contribution of Victor Turner." *Theology Today* 41, no. 4 (January): 401–9.

Niemi, S. 1987. "Andrology as a Speciality: Its Origin." *Journal of Andrology* 8:201–3.

NIH. *See* National Institutes of Health.

Nijs, Martine. 1999. "ART, Science, and Fiction." In *Congress Report from the Alpha Congress.* Proceedings of Alpha Congress, 16–19 September, Copenhagen. http://www.ferti.net/fertimagazine/congress/1999_11_01.asp (accessed 12 April 2002).

"Obituary: R. G. Canti." 1936. *Lancet*, 18 January. PP/FGS/C.17 Spear: Strangeways, Contemporary Medical Archives Centre, Wellcome Institute for the History of Medicine, London.

"Obituary: Ronald George Canti." 1936. *Times*, 9 January, n.p.

Ohmann, Richard. 1976. *English in America: A Radical View of the Profession.* New York: Oxford University Press.

Ohnuki-Tierney, Emily, et al. 1994. "Brain Death and Organ Transplantation: Cultural Bases of Medical Technology." *Current Anthropology* 35: 233–54.

Olsen, Gary W., et al. 1995. "Have Sperm Counts Been Reduced 50 Percent in 50 Years? A Statistical Model Revised." *Fertility and Sterility* 63: 887–93.

Oppenheimer, Jane M. 1967. *Essays in the History of Embryology.* Cambridge: MIT Press.

Otis, Laura. 2000. *Membranes: Metaphors of Invasion in Nineteenth-Century Literature, Science, and Politics*. Baltimore, Md.: Johns Hopkins University Press.

——. 2001. *Networking: Communicating with Bodies and Machines in the Nineteenth Century*. Ann Arbor: University of Michigan Press.

Oudshoorn, Nelly. 1994. *Beyond the Natural Body: An Archaeology of the Sex Hormones.* New York: Routledge.

——. 1996. "A Natural Order of Things? Reproductive Sciences and the Politics of Othering." In *FutureNatural: Nature, Science, Culture*, ed. George Robertson et al. London: Routledge.

Ozeki, Ruth. 1998. *My Year of Meats.* New York: Penguin Putnam.

——. 2003. *All Over Creation.* New York: Viking Penguin.

Pateman, Carole. 1988. *The Sexual Contract.* Stanford: Stanford University Press.

Patton, Frank. 1942. "The Test Tube Girl." *Amazing Stories* 16, no. 1 (January): 9–43, 53.

Pease, John. 1938. "Horror's Head." *Amazing Stories* 12, no. 5 (October): 42–57.

Peters, R. A. 1962. *History of the Strangeways Research Laboratory (formerly Cambridge Research Hospital) 1912–1962*. Cambridge: Heffer.

Poirier, Suzanne, Lorie Rosenblum, and Lioness Ayres. 1992. "Charting the Chart: An Exercise in Interpretation(s)." *Literature and Medicine* 11 (1992): 1–22.

"The Physiology of Growth." 1927. Review of *The Fundamentals of School Health*, by Dr. James Kerr. *Nature*, 19 February, 269–71.

Poovey, Mary. 1993. "Figures of Arithmetic, Figures of Speech: The Discourse of Statistics in the 1830s." *Critical Inquiry* 19 (Winter): 256–76.

——. 1998. *A History of the Modern Fact: Problems of Knowledge in the Sciences of Wealth and Society*. Chicago: University of Chicago Press.

Porter, Theodore M. 1997. "The Management of Society by Numbers." In *Science in the Twentieth Century*, ed. John Krige and Dominique Pestre. The Netherlands: Harwood Academic Publishers.

Powell, Jep. 1940. "The Synthetic Woman." *Amazing Stories* 14, no. 9 (September): 100–24.

Proctor, Robert, and Londa Schiebinger. 2003. "Agnatology: A Cultural Politics of Ignorance." In *Agnatology: A Cultural Politics of Ignorance.* Proceedings of Agnatology Conference, 25–26 April, Pennsylvania State University.

Rabinow, Paul. 1999. *French DNA: Trouble in Purgatory*. Chicago: University of Chicago Press.

Radford, Benjamin. 2001. "Urban Legend Makes International News." *Sceptical Inquirer* (Buffalo), May–June, 7–8.

Raloff, J. 1996. "Estrogenic Agents Leach from Dental Sealant." *Science News*, 6 April, 214.

RCP. *See* Royal College of Pathologists.

Rector, Charles G. 1927. "Crystals of Growth." *Amazing Stories* 2, no. 9 (December): 874–77.

Report of a Census of Organs and Tissues Retained by Pathology Services in England (conducted in 2000 by the Chief Medical Officer). 2001. London: Stationery Office.

Repp, Ed Earl. 1938. "The Gland Superman." *Amazing Stories* 12, no. 5 (October): 8–29.

Rocklynne, Ross. 1941. "Big Man." *Amazing Stories* (April), 70–87.

Rolleston, Sir Humphrey. 1927. "Concerning Old Age." In "Life and Death." Supplement. *Nature* 3009 (July): 2–12.

Rose, Gillian. 1993. *Feminism and Geography: The Limits of Geographical Knowledge*. Minneapolis: University of Minnesota Press.

Ross, Andrew. 1991. *Strange Weather: Culture, Science, and Technology in the Age of Limits*. London: Verso.

Rotman, Brian. 2000. "Going Parallel." *SubStance* 91:56–79.

Rowe, Michael. 2002. *The Book of Jesse: A Story of Youth, Illness, and Medicine*. Washington, D.C.: Francis Press.

Rowling, J. K. 1988. *Harry Potter and the Sorcerer's Stone*. New York: Scholastic.

Royal College of Pathologists. 2001. "National Summit on the Retention of Organs and Tissues Following Post-mortem Examination." *Royal College of Pathologists—News*, 17 March. http://www.rcpath.org/news/newsart2.html.

Russell, Bertrand. 1931. *The Scientific Outlook*. London: George Allen and Unwin.

Russell, K. 1969. "Tissue Culture: A Brief Historical Review." *Clio Medica* 4:109–19.

Ryman, Geoff. 1989. *The Child Garden, or A Low Comedy*. London: Unwin Hyman.

Sagan, Dorion. 1992. "Metametazoa: Biology and Multiplicity." In *Zone 6: Incorporations*. Cambridge: MIT Press.

Salim, A., P. Vassiliu, and G. C. Velmahos. 2001. "The Role of Thyroid Hormone Administration in Potential Organ Donors." *Archives of Surgery* 136, no. 12 (December): 1377–80.

Sample, Ian. 2003. "Womb Transplant Babies 'within Three Years': Scientists in Sweden Offer Alternatives to Surrogacy." *Guardian Online*, 2 July.

Schechner, Richard, and Willa Appel, eds. 1990. *By Means of Performance: Intercultural Studies of Theatre and Ritual.* Cambridge: Cambridge University Press.

Scheper-Hughes, Nancy, et al. 2000. "The Global Traffic in Human Organs." *Current Anthropology* 41 (2): 191–224.

Schiebinger, Londa. 1989. *The Mind Has No Sex? Women in the Origins of Modern Science.* Cambridge: Harvard University Press.

——. 1993. *Nature's Body: Gender in the Making of Modern Science.* Boston: Beacon Press.

——. 1999. *Has Feminism Changed Science?* Cambridge: Harvard University Press.

Schmidt, Peter. 1930. "Six Hundred Rejuvenation Experiments." In *Sexual Reform Congress*, ed. Norman Haire. London: Kegan Paul, Trench, Trubner.

Schulman, Neil, and Letitia Sweitzer. 1993. *Understanding Growth Hormone.* New York: Hippocrene Books.

Schwartz, Peter. 2001. "Scenarios: Regenerative Medicine Is the Future." *Red Herring*, 17 October.

"Science and Medicine: 105,000 Body Parts Retained in the U.K., Census Says." 2001. *Lancet*, February.

Seymour, Frances, M.D. 1943. "Artificial Insemination, Gynecology, Eugenics, and Their Relation to the Post–World War II Rehabilitation Plan." 23 August. D6/Eugenics Society Papers, Contemporary Medical Archives Centre, Wellcome Institute for the History of Medicine, London.

——. 1943. Letter to Thomas Lord Horder. 23 August. Eugenics Society Archive, Contemporary Medical Archives Centre, Wellcome Institute for the History of Medicine, London.

Shattuck, Roger. 2001. "A Brief History of Stories." In "Science Fiction and the Future of Medicine," special issue, *Literature and Medicine* 20 (1): 7–12.

Shelley, Mary. [1818] 1981. *Frankenstein*. New York: Bantam.

——. [1831] 1994. *Frankenstein*. New York: Dover.

Showalter, Elaine. 1977. *A Literature of Their Own: British Women Novelists from Bronte to Lessing*. Princeton, N.J.: Princeton University Press.

Shumway, David. 1994. *Creating American Civilization: A Geneology of American Literature as an Academic Discipline.* Minneapolis: University of Minnesota Press.

Shuttleworth, Sally, and John Christie, eds. 1989. *Nature Transfigured: Science and Literature, 1700–1900*. Manchester: Manchester University Press.

Shilling, Chris. 1993. *The Body and Social Theory.* London: Sage.

Shiva, Vandana. 1997. Lecture, Pennsylvania State University, 21 April.

Silverberg, Robert. [1972] 1983. "Caught in the Organ Draft." In *Caught in the Organ Draft: Biology in Science Fiction*, ed. Isaac Asimov, Martin H. Greenberg, and Charles G. Waugh. New York: Farrar, Straus and Giroux.

Silverman, Steve. 2003. "Dealflow: VCs Donate $8 Million to TransMedics." *Red Herring*, 25 September.

Smith, David C. 1986. *H. G. Wells: Desperately Mortal.* New Haven: Yale University Press.

Smith, Lawrence W., M.D., and Temple Fay, M.D. 1939. "Temperature Factors in Cancer and Embryonal Cell Growth." *JAMA* 113, no. 8 (August): 653–60.

Smith, Susan L. 2001. "Progress in Clinical Organ Transplantation." *MedScape*, 3 March.

Snow, C. P. 1933. *New Lives for Old.* London: Camelot.

——. [1959] 1993. *The Two Cultures and the Scientific Revolution.* New York: Cambridge University Press.

Spallone, Patricia. 1987. "Reproductive Technology and the State: The Warnock Report and Its Clones." In *Made to Order: The Myth of Reproductive and Genetic Progress*, ed. Patricia Spallone and Deborah Lynn Steinberg. Oxford: Pergamon Press.

Spanier, Bonnie B. 1995. *Im/partial Science: Gender Ideology in Molecular Biology.* Bloomington: Indiana University Press.

Spice, Byron. "Machine Keeps Heart Beating after Removal." *AugustaChronicle.com*, 10 October.

Squier, Susan M. 1991. "Fetal Voices: Speaking for the Margins Within." *Tulsa Studies in Women's Literature* 10, no. 1 (Spring): 17–30.

——. 1994. *Babies in Bottles: Twentieth-Century Visions of Reproductive Technology.* New Brunswick, N.J.: Rutgers University Press.

——. 1996. "Fetal Subjects and Maternal Objects: Reproductive Technology and the New Fetal/Maternal Relation." *Journal of Medicine and Philosophy* 21:515–35.

Starzl, Thomas E. 1992. *The Puzzle People: Memoirs of a Transplant Surgeon.* Pittsburgh: University of Pittsburgh Press.

Steinach, Eugen. 1940. *Sex and Life: Forty Years of Biological and Medical Experiments.* New York: Viking.

Steinmuller, Karlheinz. [1997] 2003. "Science Fiction and Science in the Twentieth Century." In *Science in the Twentieth Century*, ed. John Krige and Dominique Pestre, 339–60. New York: Routledge.

Sterling, Bruce. 1997. *Holy Fire.* New York: Bantam Books.

Strangeways, E. D. N.d. "1905–1926." In *History of the Strangeways Research Laboratory (Formerly Cambridge Research Hospital), 1912–1962.*

"Strangeways Research Library." Ms. CMAC:SA/SRL. Contemporary Medical Archives Centre, Wellcome Institute for the History of Medicine, London.

Strangeways, T. S. P. 1924. *Tissue Culture in Relation to Growth and Differentiation.* Cambridge: W. Heffer and Sons.

"Summary of Work Going on at Strangeways." 1937. *Lancet*, 18 September. Strangeways Research Library File, PP/FGS C.18 Spear: Strangeways, Contemporary Medical Archives Centre, Wellcome Institute for the History of Medicine, London.

Suvin, Darko. 1997. "Novum Is as Novum Does." *Foundation: The Review of Science Fiction* 69 (Spring): 26–42.

Swazey, Renee, and Judith P. Fox. 1992. *Spare Parts: Organ Replacement in American Society.* New York: Oxford University Press.

"Television Notes and Programmes: Viewers to Watch 'Artificial Immortality.'" 1938. *World Radio* (London), 11 February. Strangeways Research Library File, PP/FGS/C.19 Spear: Strangeways, Contemporary Medical Archives Centre, Wellcome Institute for the History of Medicine, London.

Terasaki, Paul I. 1990. *History of HLA: Ten Recollections.* Los Angeles: UCLA Tissue Typing Laboratory.

Thacker, Eugene. 2000. "The Post-genomic Era Has Already Happened." *Biopolicy* 3 (2): n.p.

Thompson, Michael. 1979. *Rubbish Theory: The Creation and Destruction of Value.* Oxford: Oxford University Press.

Thurtle, Philip, and Robert Mitchell. 2003. "In Vivo: The Cultural Mediations of Biomedical Science." MS.

Tiefer, Leonore. 1988. "A Feminist Perspective on Sexology and Sexuality." In *Feminist Thought and the Structure of Knowledge*, ed. Mary McCanney Gergen. New York: New York University Press.

Tomorrow, Tom. 2001. "The Back Page." *New Yorker*, 2 July, 88.

Travis, J. 2000. "'Cloning Extends Life of Cells—and Cows?'" *Science News*, 29 April, 279.

Treichler, Paula, Lisa Cartwright, and Constance Penley, eds. 1998. *The Visible Woman: Imaging Technologies, Gender, and Science.* New York: New York University Press.

Trull, D. 2003. "Body Part Sculptor." *Fortean Slips*, 2 August. http://www.parascope.com/articles/slips/fs_163.htm.

"T. S. P. Strangeways: Obituary." 1927. *Lancet*, 1 January, 56.

Tufte, Edward R. 1983. *The Visual Display of Quantitative Information.* Cheshire, Conn.: Graphics Press.

Turner, Brian. 1995. "Aging and Identity: Some Reflections on the Somatization of the Self." In *Images of Aging: Cultural Representations of Later Life*, ed. Mike Featherstone and Andrew Wernick. New York: Routledge.

Turner, Victor. 1967. *The Forest of Symbols: Aspects of Ndembu Ritual*. Ithaca: Cornell University Press.

——. 1977a. "Frame, Flow and Reflection: Ritual and Drama as Public Liminality." In *Performance in Postmodern Culture*, ed. Michel Benamou and Charles Caramello, 33–55. Madison: Coda Press.

——. 1977b. *The Ritual Process: Structure and Anti-Structure.* Ithaca: Cornell University Press.

——. 1982. *From Ritual to Theatre: The Human Seriousness of Play.* New York: Performing Arts Journal Publications.

——. 1985. "Passages, Margins, and Poverty: Religious Symbols of Communitas." Part 1. *Worship* 46, nos. 7–8: 393.

——. 1990. "Are There Universals of Performance in Myth, Ritual, and Drama?" In *By Means of Performance: Intercultural Studies of Theatre and Ritual*, ed. Richard Schechner and Willa Appel. Cambridge: Cambridge University Press.

——. 2003. "Excerpts from the Writings of Victor Turner." http://www.creativeresistance.ca/communitas/defining-liminality-and-communitas (accessed 14 July 2003).

Turney, Jon. 1998. *Frankenstein's Footsteps: Science, Genetics, and Popular Culture.* New Haven: Yale University Press.

"2002 Design and Engineering Awards." 2002. *Science Research*, 11 February.

UPMC Health System News Bureau. 2001. "Regenerative Medicine to be Focus of New Institute at University." 5 July. http://www.upmc.edu/newsbureau/magee/mirm.htm.

Vaughan, Janet. 1987. "Honor Bridget Fell." In *Biographical Memoirs of Fellows of the Royal Society*, 33: 237–59. London: Royal Society.

Verrill, A. Hyatt. 1927. "The Plague of the Living Dead." *Amazing Stories* 2, no. 1 (April): 6–20, 98.

Vince, Gaia. 2002. "Uterus Transplant Results in Live Births." *New Scientist.com*, 21 August.

Vines, Gail. 1995. "Some of Our Sperm Are Missing." *New Scientist*, 26 August, 22–25.

von Hagens, Gunther. 2001. "Current Exhibition: 'Body Worlds. The Fascination of the Genuine One.' " http://www.koerperwelten.com/berlin.hor (accessed 30 April).

——. 2003. "Preservation by Plastination." http://www.kfunigraz.ac.at/anawww/plast/pre.html (accessed 2 August).

Voronoff, Serge. 1928. *La conquete de la vie.* Paris: Bibliotheque-Charpentier.

——. 1943. *The Sources of Life.* Boston: Bruce Humphreys.

Waddington, C. H. 1935. "Letter." *Nature* 135:606.

——. 1936a. "Scientist at Work: What Controls the Development of Animals?" *Listener*, 28 October, 811–12.

——. 1936b. "Substances Promoting Cell Growth." *British Medical Journal* 28 (1936): 111.

——. 1937. Poem, "Happy the Egg, the Bubble Blastula." January. MS. Strangeways Research Laboratory, File PP/FGS C.18, Contemporary

Medical Archives Centre, Wellcome Institute for the History of Medicine, London.

——. 1969. "The Practical Consequences of Metaphysical Beliefs on a Biologist's Work: An Autobiographical Note." In *Towards a Theoretical Biology 2: Sketches*, comp. C. H. Waddington. Edinburgh: Edinburgh University Press.

Wade, Nicholas. 2000. "Teaching the Body to Heal Itself." *Science Times*, 7 November, D1.

Wadsworth, Ruth F. 1929. "Ain't Nature Gland!" *Colliers*, 7 December, 17, 46, 48.

Waldby, Catherine. 1996. *AIDS and the Body Politic: Biomedicine and Sexual Difference*. London: Routledge.

——. 2000. *The Visible Human Project: Informatic Bodies and Posthuman Medicine.* London: Routledge.

Waldby, Catherine, and Robert Mitchell. N.d. "Tissue Economies: Gifts, Commodities, and Bio-value in Late Stage Capitalism." MS.

Walker, John Allan. 1999. *Art and Outrage.* Sterling, Va.: Pluto Press.

Warnock, Mary. 1985. *A Question of Life: The Warnock Report on Human Fertilisation and Embryology.* London: Basil Blackwell.

Watt, Fiona M., and Brigid L. M. Hogan. 2000. "Out of Eden: Stem Cells and Their Niches." *Science* 287, no. 25 (February): 1427–30.

Wells, H. G. [1896] 1988. *The Island of Dr. Moreau.* New York: Signet.

——. [1904] 1965. *The Food of the Gods.* New York: Airmont.

——. 1926. "The Island of Dr. Moreau, Parts I and II." *Amazing Stories* 1, nos. 7–8 (October–November 1926): 636–55, 671–72, 702–23.

White, Hayden. 1978. "The Historical Text as Literary Artifact." In *The Writing of History: Literary Form and Historical Understanding*, ed. Robert H. Canary and Henry Kozicki. Madison: University of Wisconsin Press.

White, Philip R. 1954. *The Cultivation of Animal and Plant Cells.* New York: Ronald Press.

Williams, Raymond. 1977. *Marxism and Literature*. Oxford: Oxford University Press.

——. 1986. Foreword. *Languages of Nature: Critical Essays on Science and Literature*, ed. Ludmilla Jordanova. London: Free Association Books.

Wilson, Elizabeth. 1998. *Neural Geographies: Feminism and the Microstructure of Cognition.* New York: Routledge.

Witkowski, J. A. 1979. "Alexis Carrel and the Mysticism of Tissue Culture." *Medical History* 23:279–96.

——. 1987. "Optimistic Analysis—Chemical Embryology in Cambridge, 1920–1942." *Medical History* 31:247–68.

Wolff, Charlotte, M.D. 1986. *Magnus Hirschfeld: A Portrait of a Pioneer in Sexology.* London: Quartet.

Woodward, Kathleen, ed. 1999. *Figuring Age: Women, Bodies, Generations.* Bloomington: Indiana University Press.

Woolf, Virginia. [1929] 1991. *A Room of One's Own.* New York: Harcourt, Brace and World.

"World's First Womb Transplant." 2002. *Iafrica.com:only what you want,* 25 March. http://www.iafrica.com.

Wright, Lawrence. 1996. "Silent Sperm." *New Yorker,* 15 January, 42–55.

Wylie, Philip. 1930. *Gladiator.* New York: Alfred A. Knopf.

Yeats, W. B. 2003. "Countess Cathleen." *Project Gutenberg E-Book.* http://ibiblio.org/gutenberg/etext04/cntsc10.txt (accessed 2 August).

Young, Robert. 1995. *Colonial Desire: Hybridity in Theory, Culture, and Race.* New York and London: Routledge.

Youngner, S. J., M. Allen, E. T. Bartlett, et al. 1985. "Psychosocial and Ethical Implications of Organ Retrieval." *New England Journal of Medicine,* 1 August, 321–24.

The following portions of this book have already appeared in print, and I thank their editors and publishers for the permission to reprint revised versions of them:

"Transplant Medicine and Transformative Narrative, or Is Science Fiction 'Rubbish'?" in *Biotechnological and Medical Themes in Science Fiction,* ed. Domna Pastourmatzi (Thessaloniki: University Studio Press, 2002), 87–110.

"Aus der Sicht der Gewebekulturen: Neue Lebensspannen fuer den Menschen," in *Genealogie und Genetik: Schnittstellen zwischen Biologie und Kulturgeschichte,* ed. Sigrid Weigel (Berlin: Akademie-Verlag [Einstein Bücher], 2002), 101–39.

"Life and Death at Strangeways: The Tissue-Culture Point of View," in *Biotechnology and Culture: Bodies, Anxieties, Ethics,* ed. Paul E. Brodwin (Bloomington: Indiana University Press, 2000), 27–52.

"From Omega to Mr. Adam: The Importance of Literature for Feminist Science Studies," *Science, Technology, and Human Values* 24, no. 1 (Winter 1999): 131–57.

"Incubabies and Rejuvenates: The Traffic between Technologies of Reproduction and Age-Extension," in *Figuring Age: Women, Bodies, Generations,* ed. Kathleen Woodward (Bloomington: Indiana University Press, 1999), 88–111.

"Interspecies Reproduction: Xenogenic Desire and the Feminist Implications for Hybrids," *Cultural Studies* 12, no. 1 (1998): 360–81.

Index

SUSAN MERRILL SQUIER

is Brill Professor of Women's Studies and English

at Pennsylvania State University. She is author of *Babies in Bottles: Twentieth-Century Visions of Reproductive Technology;* editor of *Communities of the Air: Radio Century, Radio Culture* (Duke University Press, 2003); and coeditor of *Playing Dolly: Technocultural Formations, Fantasies, and Fictions of Assisted Reproduction* and *Arms and the Woman: War, Gender, and Literary Representation.*

She is past president and executive board member of the Society for Literature and Science.